# BASIC STATISTICS
# AND PROBABILITY

# BASIC STATISTICS AND PROBABILITY

## FOR BUSINESS AND ECONOMIC DECISIONS

Second Edition

**Milad A. Tawadros, Ph.D.**
Indiana University at South Bend

**KENDALL/HUNT PUBLISHING COMPANY**
2460 Kerper Boulevard,
Dubuque, Iowa 52001

B    402094    01

# CONTENTS

**v**

# PREFACE

The book is designed to satisfy different groups of students with various backgrounds. Quantitatively and nonquantitatively oriented individuals having business, economics, social sciences, or behavioral sciences backgrounds can benefit from the variety of topics presented in this book. A quantitatively oriented individual will find chapters on probability and probability functions very stimulating — and students with little background in mathematics can omit chapters on continuous probability without loss of continuity and without affecting the comprehension of the rest of the book.

Decision theory and econometrics require the use of matrix algebra. A brief review of matrices including addition, subtraction, and multiplication, is needed before the presentation of decision theory.

The current trend indicates that statistical applications in the decision making field depend heavily on probability, probability functions, expected value, and Bayesian theorem. All these topics are presented in Chapters 3 through 6 and are considered vital tools in management science, operations research, quantitative marketing, and quantitative economics. Chapters 7, 8, and 9 are designed for the application of probability in statistical decision making.

The book is also designed for a one-semester course in statistics for both undergraduate business and economics majors and MBA students. Part of Chapter 4 (the probability functions for two or more variables) and all of Chapter 5 can be omitted without loss of continuity. For MBA students the knowledge of calculus is essential for a better understanding of the probability functions being extensively used in statistical decision problems they are facing or will be facing in their practical business life.

The author is very grateful to the publishers and authors who kindly gave permission to reproduce the statistical tables presented in the Tables section in this book.

I am indebted to my family, my wife Sabah and my sons Adel, Atef, and Azmi for their patience, understanding, and encouragement during the writing of this book.

Milad A. Tawadros
South Bend, Indiana

CHAPTER

# 1

# Statistics: Its Nature

Statistics can be defined as a set of methods and theory applied to numerical or quantitative data for the purpose of making decisions under uncertainty. Statistics is applied in many areas of human activity; in business, government, industry, agriculture, the natural sciences, the social sciences, the medical research area, psychology, and in economics. These are but few of the fields where the application and development of statistical methods is widely used.

In business, especially, statistics is being commonly applied as a useful decision making tool. Since the end of World War II many quantitative methods and procedures have been developed to improve the managerial decision making capability and as the economy becomes increasingly complex it has become necessary for the information to be channeled into well designed systems. Given numerical or quantitative data, the managerial decision maker has to assemble an appropriate quantity of the highest quality of relevant information to use in decision making. Statistical analysis continue to be an integral part of the quantitative methods developed to help managers make better decisions.

Another important tool in the area of decision making is the ability to forecast the future activity of a firm. Econometric models are widely used for such forecasting. An econometric model to forecast sales is becoming a popular managerial tool to provide top management with information that can be used to make decisions concerning inventory control, production scheduling, financial, marketing, and hiring strategies for their organizations. Methods of statistical analysis play an important role in estimating the parameters of the model, testing the significance of the estimated parameters, and testing the reliability of the model as a whole.

Statistics, as defined here usually use numerical or quantitative data; however, there are situations where the available data are measured in qualitative form as in the area of marketing, psychology, and behavioral sciences. In the most common usage, statistics refer to a collection of observations that are either quantitative or qualitative. Quantitative observations are expressed in numerical data points such as closing stock prices, crime rates, sports data, and economic data concerning unemployment, income, consumption, and savings. Qualitative observations are non-numerical characteristics such as college major, sex, and occupation. A collection of observations can be referred to as a population or a sample. A population is a collection of all possible observations of the subject under study, while a sample consists only of some of the observations.

The collected data, when properly classified and summarized, can be presented in a form which is extremely valuable to researchers and decision makers. The collection, organization, graphic presentation, and calculation of measures to describe the numerical data is known as "descriptive statistics." Recently, however, the primary objective of statistics is the application of procedures and methods to the collected numerical data to help decision makers, in any field of endeavor, in making decisions under uncertainty. This branch of statistics is referred to as "statistical decision making" which includes: statistical inference, decision theory, and regression and correlation analysis. Statistical inference consists of methods applied to determine certain characteristics of a population based on information contained in a sample. Decision makers faced with alternatives may use the methods of the decision theory to select the best course of action to maximize their position. Forecasts, which are an important decision making tool, can be made through the application of regression and correlation methods.

Probability theory plays an important role in the area of statistical decision making. In statistical inference, the probability theory extends its concepts to the measurement of the chance of errors existing in the estimated characteristics of a population. Also, probability theory lends itself to both the decision theory and forecasting through the application of regression and correlation analysis.

The definition of statistics applies to both descriptive and statistical decision making. The material in this book is to cover both areas. Probability theory and probability distributions will be presented immediately after descriptive statistics in order to prepare for the discussion of statistical decision making. The remainder of this chapter and the following chapter will cover descriptive statistics. Succeeding chapters will explore probability and probability distributions. The last chapters are devoted to statistical decision making.

**Descriptive Statistics: (Frequency Distributions)**

Data collected to represent populations or samples can be organized (grouped), and presented in a frequency distribution table. This section is concerned with frequency tables and the following chapter will deal with numerical data by the calculation of different measures.

Data collected from internal or external sources could be either in the form of ungrouped data (raw data) or in the form of grouped data (presented in a frequency distribution table). Grouping raw data into a frequency distribution table will simplify the presentation of such data. When dealing with raw data, as usually is the case in business, one can simplify the presentation of such data by organizing the data into classes and frequencies and construct a frequency distribution table. To illustrate the utility of grouping raw data in a frequency distribution table, the following example is presented:

*Example 1.1:* The credit manager of a department store wants to know some information concerning the length of time of collection. Table 1-1 on page 3 shows the number of days to collect bills from 100 customers:

It is rather difficult for the credit manager to draw any meaningful information concerning the length of time to collect from these 100 customers. However, one may suggest any of the following ways to draw some conclusions:

1. arrange the data in ascending or descending order and report the lowest and highest values.
2. arrange the data in classes and frequencies.
3. calculate a measure that describes or represents the data, for instance, calculate the mean.

## TABLE 1-1

| | | | | | | | | | |
|----|----|----|----|----|----|----|----|----|----|
| 20 | 50 | 40 | 30 | 18 | 17 | 25 | 24 | 42 | 19 |
| 25 | 40 | 11 | 15 | 50 | 90 | 14 | 18 | 80 | 41 |
| 30 | 20 | 20 | 16 | 42 | 40 | 16 | 27 | 84 | 46 |
| 90 | 70 | 31 | 27 | 23 | 29 | 20 | 28 | 24 | 30 |
| 45 | 60 | 40 | 34 | 51 | 19 | 24 | 19 | 17 | 20 |
| 60 | 30 | 25 | 33 | 18 | 26 | 92 | 30 | 19 | 18 |
| 70 | 50 | 49 | 48 | 34 | 42 | 50 | 32 | 22 | 19 |
| 15 | 18 | 62 | 22 | 80 | 16 | 60 | 55 | 25 | 80 |
| 28 | 21 | 71 | 20 | 25 | 25 | 44 | 90 | 60 | 95 |
| 30 | 46 | 98 | 70 | 60 | 32 | 72 | 78 | 82 | 72 |

Reporting the lowest and highest values, which represent the shortest and longest time of collection, is not helpful because the two extreme values reflect the variation rather than describe the whole set of these data. Grouping the data into classes and frequencies is shown below. Calculation of measures to describe the data such as measures of central location, measures of dispersion, and others are presented in the following chapter.

The 100 observations in Example 1.1 can be classified into groups or classes through the application of tally sheets as follows:

## TABLE 1-2
### Tally Sheet

| Class | Tally | Frequency |
|-------|-------|-----------|
| 10–19 | ⊤⊦⊔  ⊤⊦⊔  ⊤⊦⊔  1111 | 19 |
| 20–29 | ⊤⊦⊔  ⊤⊦⊔  ⊤⊦⊔  ⊤⊦⊔  ⊤⊦⊔ | 25 |
| 30–39 | ⊤⊦⊔  ⊤⊦⊔  11 | 12 |
| 40–49 | ⊤⊦⊔  ⊤⊦⊔  1111 | 14 |
| 50–59 | ⊤⊦⊔  1 | 6 |
| 60–69 | ⊤⊦⊔  1 | 6 |
| 70–79 | ⊤⊦⊔  11 | 7 |
| 80–89 | ⊤⊦⊔ | 5 |
| 90–99 | ⊤⊦⊔  1 | 6 |
| | | 100 |

The results of grouping these data can be shown in a frequency distribution table. The frequency distribution table contains: class intervals, and frequency (f).

Assembling the data into a frequency distribution table has made it possible to simplify the presentation of the data, however, information concerning the individual observations has been lost. For example, in the class interval 50–59, we know that there are 6 observations or 6 customers who waited for 50–59 days to pay their bills, but in the absence of the raw data, we cannot determine exactly how many days elapsed for each of the six customers to pay their bills.

### Remarks About Class Intervals

1. The arrangement of groups into 10–19, 20–29, . . . , 90–99 is called class intervals. Each interval has two limits, lower limit as 10, 20, 30, . . . , 90, and upper

**TABLE 1-3**
Frequency Distribution
of collection time
(in days)

| Class | f (frequency) |
|-------|---------------|
| 10–19 | 19 |
| 20–29 | 25 |
| 30–39 | 12 |
| 40–49 | 14 |
| 50–59 | 6 |
| 60–69 | 6 |
| 70–79 | 7 |
| 80–89 | 5 |
| 90–99 | 6 |
|       | 100 |

limit as 19, 29, . . . , 99. Class intervals with unknown lower or upper limits are called open-end class intervals, e.g., less than 150 or 200 or more.

2. Class intervals may be viewed as the number of units or observations contained in the class groupings, e.g., the class interval of 40–49 is 10 units. Class intervals may or may not be equal in a frequency distribution table. While it is not a firm rule, it is advisable to use equal class intervals in the construction of frequency distribution.

3. The greater the number of class intervals in the frequency distribution table, the more detailed information about the data. There is no fixed rule to determine the number of class intervals in any frequency distribution.

4. Class intervals are either discrete or continuous depending on the type of data being grouped. If the data are discrete, as in the case of the previous example, then the class intervals contain a break:

40–49
50–59

On the other hand, if the data represents a continuous variable, then the class intervals for continuous data have no break:

40 and less than 50
50 and less than 60

5. The midpoint of a class interval is the average of the values contained in this interval. To calculate the midpoint (M) simply add the lower limit and the upper limit and divide the sum by two. For example, the midpoint for a discrete class interval: 40–49 is $(40 + 49)/2 = 44.5$, and the midpoint for a continuous class interval: 40 and less than 50 is $(40 + 50)/2 = 45$ (because less than 50 could be 49.9999 which is approximately 50).

**Graphical Representation of the Frequency Distribution**

Graphs are used to give clear visual information about the data. Many methods can be used to graphically represent the frequency distribution; for example, the bar chart, the histogram, the frequency polygon, and the smooth continuous frequency curve.

a. A bar chart: There are gaps between the class intervals.
b. A histogram: The gaps between the class intervals are closed by finding the boundaries of the class intervals, e.g., for the class interval 40–49, its boundaries are 39.5 and 49.5.
c. A frequency polygon: If we connect the midpoints of the class intervals of a histogram, this produces a frequency polygon.
d. A frequency curve: A smoothed frequency polygon is a frequency curve.

These four methods of graphical representation can be used to convert a discrete data graph to a continuous data graph. The construction of the four graphs, from a bar chart to a continuous curve of the frequency distribution of the variable, where X assumes the observations in Example 1.1, is shown below in Figures 1-1 through 1-3:

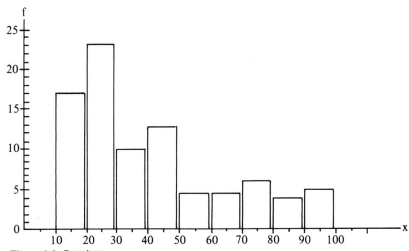

**Figure 1-1.** Bar chart.

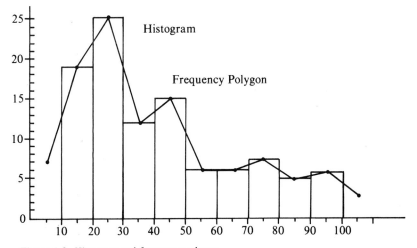

**Figure 1-2.** Histogram and frequency polygon.

**5**

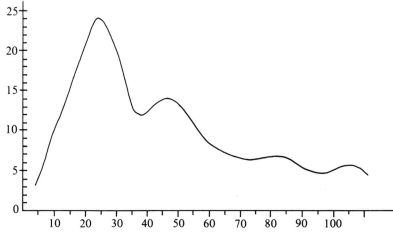

**Figure 1-3.** Frequency curve.

Cumulative and relative frequency distributions are two useful extensions of the basic frequency distribution previously presented. Cumulative and relative frequency distributions for the data of Example 1.1 are shown in the following table:

**TABLE 1-4**
Frequency Distribution
of collection time
(in days)

| Class | f (frequency) | F (cumulative frequency) | Relative frequency (R.f.) |
|---|---|---|---|
| 10–19 | 19 | 19 | 19/100 = 0.19 |
| 20–29 | 25 | 44 | 25/100 = 0.25 |
| 30–39 | 12 | 56 | 0.12 |
| 40–49 | 14 | 70 | 0.14 |
| 50–59 | 6 | 76 | 0.06 |
| 60–69 | 6 | 82 | 0.06 |
| 70–79 | 7 | 89 | 0.07 |
| 80–89 | 5 | 94 | 0.05 |
| 90–99 | 6 | 100 | 0.06 |
| | 100 | | 1.00 |

A cumulative frequency of a particular class is the sum of the frequencies of that class and those of the preceding classes. A cumulative frequency provides information concerning the total number of observations at the upper limit of each class. For example, the cumulative frequencies in the Table 1-4 of 44 and 56 shows that it took 44 customers out of 100 in this group 29 days or less to pay their bills while it took 56 customers of the same group 39 days or less to pay their bills.

A relative frequency is the ratio of the number of frequencies or observations falling in each class to the total observations. The ratios can be expressed in percentage to facilitate

**6**

comparisons of similar groups of different size. In Table 1-4, the relative frequencies of 0.19 and 0.25 indicate that 19% of the customers in this group paid their bills in a time period of 10–19 days, while 25% of the same group waited between 20–29 days to pay.

Grouping data into a frequency distribution and graphic presentation is one way to describe and simplify raw data. Another way to describe raw data is to calculate measures of different types to learn more about the characteristics of the data and the shape of the frequency distribution. Chapter 2 will continue the discussion of descriptive statistics.

EXERCISES

1.1 a. Define statistics.
   b. What is the difference between statistical inference and descriptive statistics?
   c. Differentiate between a discrete and continuous variable.

1.2 Define:
   a. Population           b. Sample           c. Parameter
   d. Statistic            e. Histogram        f. Frequency
   g. Class midpoint                              polygon

1.3 The following table shows the height of 100 football players: (in inches)

| 74 | 70 | 71 | 70 | 77 |
|----|----|----|----|----|
| 75 | 76 | 77 | 71 | 73 |
| 72 | 72 | 72 | 72 | 73 |
| 75 | 76 | 69 | 74 | 74 |
| 71 | 74 | 72 | 79 | 75 |
| 74 | 71 | 74 | 71 | 76 |
| 71 | 76 | 73 | 71 | 73 |
| 70 | 74 | 74 | 75 | 74 |
| 76 | 74 | 72 | 73 | 73 |
| 74 | 75 | 70 | 75 | 74 |
| 71 | 71 | 72 | 74 | 72 |
| 75 | 73 | 71 | 72 | 72 |
| 74 | 70 | 75 | 73 | 71 |
| 72 | 75 | 77 | 74 | 73 |
| 76 | 72 | 72 | 74 | 69 |
| 71 | 74 | 76 | 72 | 75 |
| 70 | 74 | 73 | 70 | 69 |
| 75 | 77 | 71 | 75 | 72 |
| 78 | 73 | 74 | 77 | 70 |
| 72 | 76 | 75 | 69 | 79 |

a. Construct a frequency distribution:

Class
69–70
71–72
73–74
75–76
77–78
79–80

b. Draw a bar chart

c. Draw a histogram, frequency polygon, and a frequency curve.

1.4 The following data represent High School I.Q. Scores for 110 students:

| | | | | | | | | | |
|---|---|---|---|---|---|---|---|---|---|
| 154 | 116 | 142 | 97 | 150 | 115 | 117 | 93 | 147 | 114 |
| 118 | 89 | 145 | 113 | 119 | 85 | 143 | 112 | 121 | 127 |
| 142 | 111 | 122 | 136 | 110 | 123 | 123 | 137 | 109 | 125 |
| 119 | 136 | 108 | 127 | 100 | 135 | 80 | 128 | 100 | 133 |
| 87 | 132 | 108 | 133 | 85 | 134 | 85 | 112 | 123 | 96 |
| 121 | 140 | 99 | 104 | 135 | 100 | 102 | 109 | 107 | 144 |
| 129 | 134 | 97 | 111 | 124 | 105 | 106 | 110 | 123 | 134 |
| 114 | 115 | 145 | 109 | 148 | 99 | 116 | 120 | 143 | 133 |
| 128 | 97 | 122 | 99 | 136 | 128 | 110 | 146 | 107 | 129 |
| 89 | 111 | 115 | 113 | 107 | 109 | 87 | 112 | 132 | 98 |
| 110 | 128 | 109 | 121 | 111 | 99 | 104 | 102 | 87 | 125 |

a. Group the data into the following class intervals:
   80–89, 90–99, 100–109, . . . , 150–159.

b. Draw a histogram for the grouped data.

1.5 The following table shows sales of groceries of 50 customers: (in dollars)

| | | | | |
|---|---|---|---|---|
| 25.64 | 15.10 | 30.45 | 50.34 | 17.05 |
| 34.67 | 56.06 | 20.86 | 19.05 | 8.99 |
| 56.98 | 42.11 | 11.69 | 23.56 | 31.90 |
| 12.89 | 36.56 | 44.22 | 11.75 | 9.99 |
| 33.67 | 29.06 | 15.39 | 22.91 | 22.87 |
| 35.87 | 44.98 | 50.00 | 39.11 | 34.86 |
| 18.77 | 20.90 | 38.71 | 56.70 | 34.99 |
| 52.89 | 35.87 | 25.87 | 44.00 | 29.99 |
| 21.11 | 19.03 | 19.65 | 29.06 | 28.77 |
| 34.98 | 41.09 | 27.67 | 39.06 | 51.03 |

a. Construct a frequency table for the data.

b. Draw a histogram and a frequency curve for the data.

CHAPTER

# 2

# Descriptive Statistics

As indicated in Chapter 1, statistics is the science dealing with the collection, organization, and analysis of quantitative data for decision making.

The collection, organization, and graphic presentation of numerical data help to describe and present such data into a form suitable for deriving logical conclusions. This part of descriptive statistics is presented in Chapter 1.

Analysis of data is another way to simplify quantitative data by extracting relevant information from which summarized and comprehensible numerical measures can be calculated to describe the given data. The most important measures for this purpose are: measures of location, measures of dispersion, and measures of symmetry and skewness. This is the other part of descriptive statistics which is the main topic of this chapter.

In this chapter, the three different measures of descriptive statistics are presented in the following order: Measures of location, measures of dispersion, and measures of symmetry and skewness.

### A. Measures of location

A single value can be derived for a set of data to describe the elements contained in that set or the frequency distribution representing such data. This single value is called a measure of central location or central tendency. There are three popular types, namely: averages (or means), the median, and the mode. There are four types of means: arithmetic, weighted, geometric, and harmonic. The arithmetic mean which is the most widely used measure of central location is considered in this chapter. In addition, there are other measures of location which are not central location; such as, quartiles, deciles, and percentiles. Figure 2-1 shows a summary of measures of location.

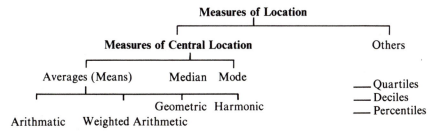

**Figure 2-1.** Summary of measures of location.

In evaluating these measures of central tendency, we shall differentiate between ungrouped data (raw data) and grouped data (frequency distribution tables). However, if we utilize computer programming to calculate the statistics of the measures of central location, it is not necessary to distinguish between ungrouped and grouped data. As a matter of fact, we do not need to develop computer programs for the grouped data at all. The computer is capable of handling an extensive number of observations without grouping the data before calculating any statistic.

A general knowledge of basic statistical methods requires, however, that the student be capable of utilizing the formulas for both ungrouped and grouped data which will be presented in this discussion.

### 1. Averages or Means

An average is a single value that describes or characterizes the elements of a set of data. As mentioned before, the discussion will be limited to the arithmetic mean because of its importance.

### Arithmetic Mean

In everyday activity you may hear or read the word "average", such as, average income, average weight, average weekly working hours, average speed, etc. The concept of "average" in ordinary usage is what statisticians call the arithmetic average or arithmetic mean. In most cases the terms "arithmetic mean, arithmetic average, mean, and average" are used interchangeably in this text.

The arithmetic mean, or simply the mean, is calculated by adding the values of all the elements of the variable (X) and dividing the total or the summation ($\Sigma X$) by the number of elements or observations (N or n) contained in that variable. Any measure being calculated using all the elements or observations contained in a population is referred to as a "parameter". Parameters are denoted by Greek letters. Any measure to be calculated for a sample is called a "statistic". A Latin letter is usually used as a symbol for a statistic. For example:

$$\mu = \frac{\Sigma X}{N}$$  The Greek letter $\mu$ (MU) is used to denote the mean of the population. $\mu$ is a parameter.

$$\overline{X} = \frac{\Sigma X}{n}$$  $\overline{X}$ ("X bar") is the symbol used to denote the mean of a sample. $\overline{X}$ is a statistic.

The arithmetic mean is affected by each observed value of the variable including the extremes. This follows from the fact that the value of each element of the distribution (or the array) of X is a part of $\Sigma X$.

To calculate the arithmetic mean, we shall differentiate between ungrouped and grouped data.

### Ungrouped Data

The formula to be used to calculate the mean for ungrouped data is as mentioned above:

$$\mu = \frac{\Sigma X}{N} \qquad \text{for a population}$$

or $\qquad \overline{X} = \frac{\Sigma X}{n} \qquad \text{for a sample}$

**10**

For example: Let $X_i$ be the weekly income per family in a city, then $X_1$, $X_2$, . . . , $X_9$, $X_{10}$ represents the weekly income for 10 families in this city. To calculate the average weekly income for these 10 families ($\overline{X}$), one must apply the previous formulas:

$$\overline{X} = \frac{\sum\limits_{i=1}^{n} X_i}{n} \qquad \text{where n} = 10$$

or $\qquad \overline{X} = \dfrac{X_1 + X_2 + \ldots + X_9 + X_{10}}{10}$

*Example 2.1:* Suppose that the weekly income of these 10 families is as shown in Table 2-1, and we want to calculate the average weekly income for these families.

### TABLE 2-1
$X_i$ (weekly income per family in dollars)

| | |
|---|---|
| $X_1$ | 140 |
| $X_2$ | 250 |
| $X_3$ | 300 |
| $X_4$ | 90 |
| $X_5$ | 180 |
| $X_6$ | 200 |
| $X_7$ | 220 |
| $X_8$ | 150 |
| $X_9$ | 180 |
| $X_{10}$ | 250 |
| | 1960 |

**Solution:**

$$\Sigma X_i = 1960$$

then $\quad \overline{X} = \dfrac{\Sigma X}{n} = \dfrac{1960}{10} = \$196.00$ (average weekly income for the ten families)

### Grouped Data

Grouped data refers to any set of observations tabulated into classes with corresponding frequencies which we call frequency distribution tables. By grouping the data, the individual value of each observation has disappeared and has been included with others falling within the same class interval.

For example: Let X be a variable representing the amount spent on gasoline by 9 drivers during a particular week. The values of X are: 10, 11, 11, 15, 16, 16, 17, 19, and 20. Grouping the values of X into three class intervals (10–14, 15–19, 20–24), a frequency distribution table is developed as shown in Table 2-2 on page 12.

In the frequency distribution (or grouped data), the first class interval (10–14) has 3 frequencies which means that there are three values of X having a magnitude between 10 and 14. In the absence of exact values of X, one can say that the 3 frequencies could be: 10, 10, 10 or 14, 14, 14 or any combination of values between 10 and 14.

## TABLE 2-2
### The Values of the Variable X, Ungrouped and Grouped

| Ungrouped | X | Grouped: Frequency distribution | |
|---|---|---|---|
| | 10 | Class | Frequency (f) |
| | 11 | | |
| | 11 | 10–14 | 3 |
| | 15 | 15–19 | 5 |
| | 16 | 20–24 | 1 |
| | 16 | | |
| | 17 | | |
| | 19 | | |
| | 20 | | |

Hence, the sum of the 3 values or frequencies contained in the first class interval might be as low as $10 + 10 + 10 = 30$ or as high as $14 + 14 + 14 = 42$ or any value between 30 and 42. Neither 30 nor 42 is an acceptable sum of the 3 frequencies; therefore, an average of both $\dfrac{30 + 42}{2} = 36$ is an appropriate value of the sum of the three values contained in the first class interval.

We arrive at the same summation by multiplying the midpoint of the interval by the frequency.

Applying this rule for the first class interval (10--14) of the frequency distribution in Table 2-2, the midpoint $(M) = \dfrac{10 + 14}{2} = 12$, the frequency $(f) = 3$, then the midpoint times the frequency or $Mf = 12 \times 3 = 36$. This rule will assist us in finding the summation of all the values of X that have been grouped where $\Sigma Mf \cong \Sigma X$. Accordingly, the general formula of the mean for grouped data becomes:

$$\overline{X} = \frac{\Sigma Mf}{\Sigma f}$$

where M = midpoint of class intervals
f = frequency of class intervals

Comparing the ungrouped formula $\overline{X} = \dfrac{\Sigma X}{n}$

with the grouped formula $\overline{X} = \dfrac{\Sigma Mf}{\Sigma f}$

we can see that: $\Sigma Mf \cong \Sigma X$

and $\Sigma f = n$

In most cases, the mean of the ungrouped values of X is not equal to the mean of the same values after being grouped because the exact values of X have been lost in the process of the grouping. To clarify this point, let us calculate the mean for the ungrouped and the grouped values of the variable X presented in Table 2-2.

### 2. Median

Although the arithmetic mean is commonly used, it is affected by extreme values in the variable. There are times when the data lends itself to open-end class intervals (100 or

*Example 2.2:*

| **Ungrouped Data** | | **Grouped Data** | | |
|---|---|---|---|---|
| X | Class | Frequency (f) | Midpoint (M) | Mf |
| 10 | 10–14 | 3 | 12 | 36 |
| 11 | 15–19 | 5 | 17 | 85 |
| 11 | 20–24 | 1 | 22 | 22 |
| 15 | | | | $\Sigma Mf = 143$ |
| 16 | | $\Sigma f = 9$ | | $\Sigma f = 9$ |
| 16 | | | | |
| 17 | | | | |
| 19 | | | | |
| 20 | | | | |

$\Sigma X = 135$
$n = 9$

$$\overline{X} = \frac{\Sigma Mf}{\Sigma f}$$

$$\overline{X} = 135/9 \qquad\qquad = 143/9$$
$$= 15 \qquad\qquad = 15.89$$

more, less than 10, etc.). In such cases it will not be possible to calculate the arithmetic mean. There is another measure of central location which may be applied—the median. The median is affected by its position in an array rather than by the value of each observation of the variable.

The median is defined as that value which divides the distribution into two equal parts: half of the values are smaller than or equal to the median, while the other half are larger than or equal to the median.

To calculate the median, we still differentiate between ungrouped and grouped data.

**Ungrouped Data**

In working with ungrouped data, the first step in locating the median is to arrange the values of X in either ascending or descending order. Determine the value which divides the array into two equal parts; this value is the median. In the following two examples the median is calculated for both an odd number of observations[1] and an even number of observations[2] of the variable X.

*Example 2.3:* (odd number of observations): Let X be the annual salaries of 7 teachers in an elementary school: $7500, $6900, $7200, $8000, $6750, $7600, $6500. Find the median salary of this group.

---

(1) A formula may be used to determine the location of the median for *odd* ungrouped numbers of observations such as:

$$\text{Med} = X_{\frac{n+1}{2}} \qquad n = \text{number of observations.}$$

(2) A formula may be used to determine the location of the median for *even* ungrouped numbers of observations:

$$\text{Med} = (X_{n/2} + X_{n/2+1})/2$$

**Solution:** To solve this problem, arrange these values either in ascending or descending order:

| X | |
|---|---|
| $6500 | 7500 |
| 6750 | 7600 |
| 6900 | 8000 |
| 7200————Median = 7200 | |

*Example 2.4* (even number of observations): Let X be the annual salaries of the first 6 teachers shown in Example 2.3: $7500, $6900, $7200, $8000, $6750, $7600. Find the median salary.

**Solution:** X is arranged in ascending order:

$6750
6900
7200
————— Median $= \dfrac{7200 + 7500}{2} = \dfrac{14700}{2} = 7350$
7500
7600
8000

## Grouped Data

Presentation of a set of data is simplified by classifying such data into a frequency distribution. To find the median for a frequency distribution (or grouped data) the following formula is applied:

$$\text{Med} = L_{med} + \frac{\frac{\Sigma f}{2} - F_{Lmed}}{f_{med}} \; i_{med}$$

where  Med = Median

$L_{med}$ = Real lower limit (or the boundary) of the class in which the median falls.

$\Sigma f$ = Sum of frequencies or the number of observations contained in the variable.

$F_{Lmed}$ = Cumulative frequencies less than the lower limit of the median class.

$f_{med}$ = Frequency of the median class.

$i_{med}$ = Class interval of the median class.

*Example 2.5:* The following is a frequency distribution of the weight of 100 females. Calculate the median weight.

| Weight (lbs.) | Frequency |
|---|---|
| 100–119 | 5 |
| 120–129 | 18 |
| 130–139 | 12 |
| 140–149 | 27 |
| 150–159 | 15 |
| 160–169 | 15 |
| 170–179 | 8 |
| $\Sigma f =$ | 100 |

**14**

**Solution:** Before applying the median formula for grouped data, we must first determine the median class, the class in which the median falls.

We know that the median is a value which divides the frequency distribution into two equal parts. Therefore, if we divide the sum of the observations by two $\frac{\Sigma f}{2}$ we know in which class interval the median will be located. In our example, $\frac{\Sigma f}{2} = 100/2 = 50$. This tells us to examine the cumulative frequencies until we find the class interval which contains the 50th observation.[3] In the example, the 50th observation falls in the class interval 140–149. This, then, is the median class.

| Weight (lbs.) | Frequency | Cumulative Frequencies (F) |
|---|---|---|
| 110–119 | 5 | 5 |
| 120–129 | 18 | 23 |
| 130–139 | 12 | 35 |
| 140–149 | 27 | 62 Median class |
| 150–159 | 15 | 77 |
| 160–169 | 15 | 92 |
| 170–179 | 8 | 100 |
| | $\Sigma f = $ 100 | |

Given this information, we may now apply the formula:

$$\text{Med} = L_{med} + \frac{\frac{\Sigma f}{2} - F_{Lmed}}{f_{med}} \, i_{med}$$

$$L_{med}^{(4)} = 139.5$$
$$\Sigma f/2 = 50$$
$$F_{Lmed} = 35 \text{ or } (5 + 18 + 12)$$
$$f_{med} = 27$$
$$i_{med} = 149.5 - 139.5 = 10$$
$$\text{Med} = 139.5 + \frac{50 - 35}{27} (10)$$
$$\text{Med} = 139.5 + \frac{15}{27} (10)$$
$$= 139.5 + 5.6$$
$$= 145.1$$

Notice that the value of the median (145.1) is contained in the median class interval between 140–149.

---

(3) It is true that we have an even number of observations (100) and hence an average of the values of the 50th and the 51st observations will be the median. However, in selecting $\Sigma f/2$ to equal $100/2 = 50$ in our example, we simply want to determine the median class. The value of the median will be determined by the formula.

(4) The lower limit of the median class is 140, the upper limit of the immediately preceding class is 139; hence, the real lower limit of the median class ($L_{med}$) is 139.5. The one unit difference between 139 and 140 is divided between the two classes to find the boundaries (real limit) or 139.5.

## 3. Mode

The mode is that value in a frequency distribution which occurs most often. In certain distributions the mode may not exist[5] or, if it exists, it may not be unique as in the case of a bimodal or trimodal distribution.[6]

**Ungrouped Data**

To find the mode for ungrouped data, it is helpful, but not essential, to arrange the values of the variable into ascending or descending order.

*Example 2.6:* The manager of a company wants to determine the mode for the number of days an employee is absent during the last year for salary decisions. The following are such data for the 18 employees in his company (in days): 8, 18, 9, 10, 18, 15, 20, 17, 20, 15, 16, 20, 15, 14, 13, 15, 20, 15. Find the mode.

**Solution:**

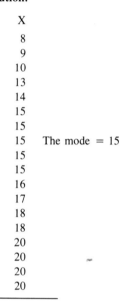

| X | |
|---|---|
| 8 | |
| 9 | |
| 10 | |
| 13 | |
| 14 | |
| 15 | |
| 15 | |
| 15 | The mode = 15 |
| 15 | |
| 15 | |
| 16 | |
| 17 | |
| 18 | |
| 18 | |
| 20 | |
| 20 | |
| 20 | |
| 20 | |

---

(5) The mode does not exist if the distribution is rectantular or if the distribution of the variable is as follows:

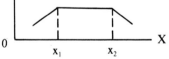

The mode is any value between $X_1$ and $X_2$.

(6) A bimodal distribution is one that has two identical peaks as shown in the following diagram:

while a trimodal distribution has three identical peaks.

### Grouped Data

The mode of the grouped data falls in the modal class. But what is the modal class? It is the class that has the highest number of frequencies.

*Example:*

| Class | Frequency | |
|-------|-----------|---|
| 10–14 | 5 | |
| 15–19 | 22 | |
| 20–24 | 18 | |
| 25–29 | 28 | This is the Modal Class. |
| 30–34 | 16 | |

The class of (25–29) is the one which has the highest frequencies; therefore, it is the modal class for this distribution.

The mode is contained in the modal class, but to determine its value for grouped data, we have to apply the following formula:

$$Mo = L_{mo} + \frac{f_{mo} - f_1}{2f_{mo} - (f_1 + f_2)} i_{mo}$$

where $Mo$ = the mode

$L_{mo}$ = real lower limit of the modal class

$f_{mo}$ = frequency of the modal class

$f_1$ = frequency of the class immediately preceding the modal class

$f_2$ = frequency of the class immediately following the modal class

$i_{mo}$ = class interval of the modal class

*Example 2.7:* The following is the age distribution of 70 students in a home economics class:

| Class | Frequency | |
|-------|-----------|---|
| 15–19 | 3 | |
| 20–24 | 10 | |
| 25–29 | 12 | |
| 30–34 | 25 | The Modal Class |
| 35–39 | 15 | |
| 40–44 | 5 | |

Find the modal age of this group.

**Solution:** Before applying the formula for grouped data to calculate the mode, first determine the modal class, which in the example is the class of (30–34).

Now apply the formula:

$$Mo = L_{mo} + \frac{f_{mo} - f_1}{2f_{mo} - (f_1 + f_2)} i_{mo}$$

$$= 29.5 + \frac{25 - 12}{2(25) - (12 + 15)} (5)$$

$$= 29.5 + \frac{13}{23} (5)$$

$$= 29.5 + 2.8$$
$$= 32.3$$

Notice that the mode[7] falls within the limits of the modal class (30–34).

### Relationship Between the Mean, the Median, and the Mode

The mean, the median, and the mode are each single values which tend to be located at the center of the distribution—hence, they name measures of central location. In a symmetrical or bell-shaped distribution, the magnitude of the mean, median, and mode will be equal and will be located at the exact center as demonstrated by the following example.

*Example 2.8:* The following shows the distribution of salaries of 50 executives in different companies. Find the relationship between the mean, the median and the mode of these salaries.

| Class | f |
|---|---|
| 30 and under 40 | 2 |
| 40 and under 50 | 6 |
| 50 and under 60 | 10 |
| 60 and under 70 | 14 |
| 70 and under 80 | 10 |
| 80 and under 90 | 6 |
| 90 and under 100 | 2 |
| | Σf = 50 |

**Solution:**

| Class | f | M | Mf | F | |
|---|---|---|---|---|---|
| 30 and under 40 | 2 | 35 | 70 | 2 | |
| 40 and under 50 | 6 | 45 | 270 | 8 | |
| 50 and under 60 | 10 | 55 | 550 | 18 | |
| 60 and under 70 | 14 | 65 | 910 | 32 | Median/Modal Class |
| 70 and under 80 | 10 | 75 | 750 | 42 | |
| 80 and under 90 | 6 | 85 | 510 | 48 | |
| 90 and under 100 | 2 | 95 | 190 | 50 | |
| | Σf = 50 | | ΣfM = 3250 | | |

(7) The value of the mode (32.3) can be interpolated graphically as shown in the diagram, representing the data of the distribution presented in Example 2.7.

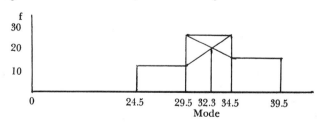

$$\overline{X} = \Sigma Mf/\Sigma f = 3250/50 = 65$$

$$Med = L_{med} + \frac{\Sigma f/2 - F_{Lmed}}{f_{med}} i_{med}$$

$$= 60 + \frac{25 - 18}{14}(10)$$

$$= 60 + 70/14 = 65$$

$$Mo = L_{mo} + \frac{f_{mo} - f_1}{2f_{mo} - (f_1 + f_2)} i_{mo}$$

$$= 60 + \frac{14 - 10}{28 - 20}(10)$$

$$= 60 + 40/8 = 65$$

From this example, we can conclude that if the distribution is symmetrical as in Figure 2-2, then the following relationship exists: $\overline{X}$ = Median = Mode. The magnitude of the three measures of central location divides the frequency distribution into two equal halves.

However, if the distribution is asymmetric, or skewed, the relationship of the three measures of location will depend upon whether the skewness is positive or negative. If the distribution is positively skewed,[8] the relationship becomes:

Mode < Median < Mean

If the distribution is negatively skewed,[9] the relationship becomes:

Mean < Median < Mode

The following two examples will serve to illustrate:

Assume that the salaries of the 50 executives are distributed as follows:

| Class | f |
|---|---|
| 30 and less than 40 | 4 |
| 40 and less than 50 | 8 |
| 50 and less than 60 | 15 |
| 60 and less than 70 | 13 |
| 70 and less than 80 | 5 |
| 80 and less than 90 | 3 |
| 90 and less than 100 | 2 |
| | $\Sigma f = 50$ |

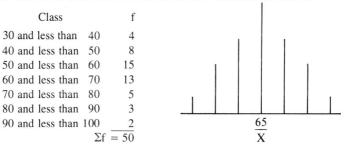

**Figure 2-2.**

Show the relationship among the mean, median, and the mode.

---

(8) A positively skewed distribution has a long tail to the right.

(9) A negatively skewed distribution has a long tail to the left.

**19**

**Solution:**

| Class | f | M | Mf | F |
|---|---|---|---|---|
| 30 and less than 40 | 4 | 35 | 140 | 4 |
| 40 and less than 50 | 8 | 45 | 360 | 12 |
| 50 and less than 60 | 15 | 55 | 825 | 27 Median/Modal Class |
| 60 and less than 70 | 13 | 65 | 845 | 40 |
| 70 and less than 80 | 5 | 75 | 375 | 45 |
| 80 and less than 90 | 3 | 85 | 255 | 48 |
| 90 and less than 100 | 2 | 95 | 190 | 50 |
| | $\Sigma f = 50$ | | $\Sigma Mf = 2990$ | |

$$\overline{X} = \frac{\Sigma Mf}{\Sigma f} = \frac{2990}{50} = 59.8$$

$$\text{Med} = L_{med} + \frac{\Sigma f/2 - F_{Lmed}}{f_{med}} i_{med}$$

$$= 50 + \frac{25 - 12}{15} (10)$$

$$= 50 + 8.7 = 58.7$$

$$\text{Mo} = L_{mo} + \frac{f_{mo} - f_1}{2f_{mo} - (f_1 + f_2)} i_{mo}$$

$$= 50 + \frac{15 - 8}{30 - 21} (10)$$

$$= 50 + 7.8 = 57.8$$

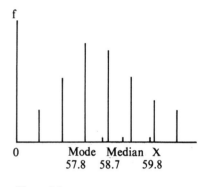

Mode   Median   X
57.8    58.7     59.8

**Figure 2-3.**

The result of computing the mean, the median, and the mode is summarized in the following double inequity:

$$57.8 < 58.7 < 59.8$$
$$\text{Mode} < \text{Median} < \overline{X}$$

A graphical presentation of the relationship between the mode, median, and mean in a positively skewed distribution is shown in Figure 2-3.

Our next example examines the same relationship in a negatively skewed distribution: Assume that the salaries of the 50 executives are distributed as follows:

| Class | f |
|---|---|
| 30 and under 40 | 2 |
| 40 and under 50 | 3 |
| 50 and under 60 | 5 |
| 60 and under 70 | 13 |
| 70 and under 80 | 15 |
| 80 and under 90 | 8 |
| 90 and under 100 | 4 |
| | $\Sigma f = 50$ |

Calculate the mean, median, and mode for this distribution and compare the results.

**20**

**Solution:**

| Class | f | M | Mf | F | |
|---|---|---|---|---|---|
| 30 and under 40 | 2 | 35 | 70 | 2 | |
| 40 and under 50 | 3 | 45 | 135 | 5 | |
| 50 and under 60 | 5 | 55 | 275 | 10 | |
| 60 and under 70 | 13 | 65 | 845 | 23 | |
| 70 and under 80 | 15 | 75 | 1125 | 38 | Median/Modal Class |
| 80 and under 90 | 8 | 85 | 680 | 46 | |
| 90 and under 100 | 4 | 95 | 380 | 50 | |
| | $\Sigma f = 50$ | | $\Sigma Mf = 3510$ | | |

$$\overline{X} = \Sigma Mf / \Sigma f = 3510/50 = 70.2$$

$$Med = L_{med} + \frac{\Sigma f/2 - F_{Lmed}}{f_{med}} \, i_{med}$$

$$= 70 + \frac{25 - 23}{15} (10)$$

$$= 70 + 1.3 = 71.3$$

$$Mo = L_{mo} + \frac{f_{mo} - f_1}{2f_{mo} - (f_1 + f_2)} \, i_{mo}$$

$$= 70 + \frac{15 - 13}{30 - 21} (10)$$

$$= 70 + 2.2 = 72.2$$

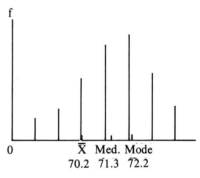

**Figure 2-4.**

Expressing the values of the mean, median and mode in a negatively skewed distribution yields the following inequality:

$$70.2 < 71.3 < 72.2$$
$$\overline{X} < \text{Median} < \text{Mode}$$

B. *Measures of Dispersion (or variation)*

Measures of central tendency alone cannot adequately represent or summarize statistical data because they do not provide any information concerning the spread of the actual observations. Two sets of data having the same value of central tendency measure (e.g., same average) do not necessarily exhibit the same dispersion of the actual or observed data about that central tendency measure. For example, the income of two groups having the same average cannot indicate that the two groups are of comparable status unless the income of each group has the same spread or dispersion about the average. The difference in the spread of the actual observations about the mean for the two groups may indicate otherwise as shown in the following example:

|  | Group A | Group B |
|---|---|---|
|  | $ | $ |
|  | 10,000 | 8,000 |
|  | 12,000 | 5,000 |
|  | 18,000 | 6,000 |
|  | 15,000 | 20,000 |
|  | 10,000 | 30,000 |
|  | 13,000 | 9,000 |
|  | 78,000 | 78,000 |
| Average | 13,000 | 13,000 |

Graphically the dispersion of the observed values about the mean for each group is as follows:

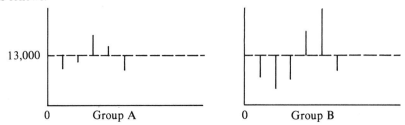

Measures of dispersion provide information concerning the variability of the data about a central tendency measure. These measures are useful in evaluating the significance of the central tendency measure as a representative value of the statistical data. The degree of dispersion determines the effectiveness of a central tendency measure to represent the data. This fact can be proved by inspecting the graphs of group A and group B. The first graph shows that the data are clustered about the mean which indicates that the average of $13,000 is a good representative of the income of group A, while in the case of group B the situation is completely different where the mean cannot be viewed as a good representative measure for the income of this group.

Measures of dispersion may be measured in two basic ways: in terms of distance between two particular observations, e.g., range and interquartile range, or in terms of deviations about the mean, e.g., mean deviation and standard deviation.

The interquartile range which is the difference between the first and third quartile, or ($Q_3 - Q_5$) has limited application in business, therefore only the other three types, namely, the range, the mean deviation, and the standard deviation, are presented in this section. Each measure is defined and calculated for ungrouped as well as grouped data.

1. *Range*

The range is very simple to calculate, but it is less reliable than other measures of variation because it depends on two extreme values: the highest and the lowest. As a matter of fact, the range is the difference between the highest and the lowest values.

For ungrouped data, the range is defined as:

Range = Highest value − Lowest value

*Example 2.9:* Let X be the annual income of 10 families: 17,000; 10,000; 20,000; 5,000; 6,000; 9,500; 15,000; 4,000; 16,000; 25,000.

**Solution:**

Range = 25,000 − 4,000 = $21,000

or    The range is $21,000.

## Grouped Data

For grouped data, the range is the difference between the lower limit of the first class and the upper limit of the last class.

*Example 2.10:* Assume that the income of 50 families are distributed as follows:

| Class | Frequency |
|-------|-----------|
| 3,000– 7,999 | 10 |
| 8,000– 12,999 | 4 |
| 13,000– 17,999 | 18 |
| 18,000– 22,999 | 11 |
| 23,000– 27,999 | 7 |

Range = 27,999 − 3,000 = $24,999

or    The range is $24,999.

### 2. *Mean Deviation*

Deviations have been defined before as the differences between observed data and its mean. For example:[10]

$$X - \overline{X} = \text{deviation of X from its mean}$$

For every observation there exists a deviation that can be positive, negative, or zero. A measure of dispersion is a single value that represents these deviations. In other words, a measure of dispersion is an average of these deviations and can be calculated by the following formula:

$$\frac{\text{Total deviations}}{\text{Number of observations (or deviations)}}$$

This formula cannot be used to calculate the mean deviation because the total deviations divided by the number of observations or: $\frac{\Sigma(X-\overline{X})}{n}$ equals zero.[11] Therefore, summation of absolute values of the deviations, $\Sigma |X - \overline{X}|$, is used to calculate the mean deviation as follows:

## Ungrouped Data

$$\text{Mean Deviation} = \frac{\Sigma |X - \overline{X}|}{n}$$

---

(10) Deviation or variation of the variable $\overline{X}$ from its mean (X) is defined as: $X - \overline{X}$ or $x$.

(11) $\frac{\Sigma(X - \overline{X})}{n}$ is called the first moment measured about the mean and its value is always zero.

*Example 2.11:* Let X be the time in minutes it takes 15 secretaries to type a particular letter:

$$X: 13, 12, 5, 10, 6, 8, 14, 6, 12, 10, 11, 11, 12, 9, 11.$$

Calculate the mean deviation.

**Solution:**

1. Find the mean: $\overline{X} = \dfrac{\Sigma X}{n}$

$$= \frac{150}{15} = 10$$

2. Find $|X - \overline{X}|$ as shown below:

| X | $|X - \overline{X}|$ |
|---|---|
| 13 | 3 |
| 12 | 2 |
| 5 | 5 |
| 10 | 0 |
| 6 | 4 |
| 8 | 2 |
| 14 | 4 |
| 6 | 4 |
| 12 | 2 |
| 10 | 0 |
| 11 | 1 |
| 11 | 1 |
| 12 | 2 |
| 9 | 1 |
| 11 | 1 |
| 150 | 32 |

Mean Deviation $= \dfrac{32}{15} = 2.133$

**Grouped Data**

Mean deviation can be calculated for grouped data (frequency distributions) by applying the following formula:

$$\text{Mean Deviation} = \frac{\Sigma |M - \overline{X}| f}{\Sigma f} \qquad M = \text{class midpoint}$$

*Example 2.12:* The following is a distribution of weights for 100 females. Calculate mean deviation.

**24**

| Weight (lbs) | Frequency |
|---|---|
| 110–119 | 5 |
| 120–129 | 18 |
| 130–139 | 12 |
| 140–149 | 27 |
| 150–159 | 15 |
| 160–169 | 16 |
| 170–179 | 7 |
| $\Sigma f =$ | 100 |

**Solution:**

1. Calculate the mean of X: $\overline{X} = \dfrac{\Sigma Mf}{\Sigma f} = \dfrac{14500}{100} = 145.0$

| Weight (lbs) | Frequency | M | Mf |
|---|---|---|---|
| 110–119 | 5 | 114.5 | 572.5 |
| 120–129 | 18 | 124.5 | 2241.0 |
| 130–139 | 12 | 134.5 | 1614.0 |
| 140–149 | 27 | 144.5 | 3901.5 |
| 150–159 | 15 | 154.5 | 2317.5 |
| 160–169 | 16 | 164.5 | 2632.0 |
| 170–179 | 7 | 174.5 | 1221.5 |
| $\Sigma f =$ 100 | | $\Sigma Mf =$ | 14500.0 |

2. Find $\Sigma |M - \overline{X}| f$ as follows:

| Class | f | M | $|M - \overline{X}|$ | $|M - \overline{X}| f$ |
|---|---|---|---|---|
| 110–119 | 5 | 114.5 | 30.5 | 152.5 |
| 120–129 | 18 | 124.5 | 20.5 | 369.0 |
| 130–139 | 12 | 134.5 | 10.5 | 126.0 |
| 140–149 | 27 | 144.5 | .5 | 13.5 |
| 150–159 | 15 | 154.5 | 9.5 | 142.5 |
| 160–169 | 16 | 164.5 | 19.5 | 312.0 |
| 170–179 | 7 | 174.5 | 29.5 | 206.5 |
| | 100 | | | 1322.0 |

Mean Deviation $= \dfrac{1322.0}{100} = 13.220$

The magnitude of the mean and the mean deviation together can give a better description of the data because the mean deviation shows the variability of the data around the mean.

3. *Standard Deviation*

The standard deviation is the most important measure of dispersion. It is similar to the mean deviation where deviations are measured from the mean. As mentioned before, the summation of these deviations, $\Sigma(X - \overline{X})$, is equal to zero. That is why, in the calculation of the mean deviation, the absolute values of the deviations are used and in case of standard deviation, deviations of the variable X from its mean $(X - \overline{X})$ are squared $(X - \overline{X})^2$, and the average squared deviations[12], $\dfrac{\Sigma(X - \overline{X})^2}{n}$, is called the vari-

ance. The standard deviation is the non-negative root of the variance: $\sqrt{\dfrac{\Sigma(X - \overline{X})^2}{n}}$

The standard deviation for a population is denoted by the lower case of the Greek letter sigma $(\sigma)$, while the standard deviation of a sample is denoted by small s. Accordingly, the standard deviation formulas are as follows:

$$\text{For the Population:} \quad \sigma = \sqrt{\frac{\Sigma(X - \mu)^2}{N}}$$

$$\text{And for the Sample:} \quad s = \sqrt{\frac{\Sigma(X - \overline{X})^2}{n}}$$

The sample standard deviation s being used as a measure of dispersion present in a sample may be calculated using the above formula. However, the sample standard deviation, as presented in later chapter, is also used to estimate the standard deviation of the population from the sample was drawn. In this case, in order to derive an unbiased estimator for the standard deviation of the population, $n - 1$ is being used instead of n in the above formula to become:

$$s = \sqrt{\frac{\Sigma(X - \overline{X})^2}{n - 1}}$$

This formula can serve to calculate the sample standard deviation. However, it is easier to use the following one which is derived from the above formula:

$$s = \sqrt{\frac{\Sigma X^2}{n - 1} - \frac{(\Sigma X)^2}{n(n - 1)}}$$

Examples to illustrate the calculation of sample standard deviation for ungrouped as well as grouped data are presented below.

**Ungrouped Data**

To calculate the standard deviation for ungrouped data, the following formula can be used:

$$s = \sqrt{\frac{\Sigma X^2}{n - 1} - \frac{(\Sigma X)^2}{n(n - 1)}}$$

*Example 2.13:* Let X be the bi-weekly salaries (in dollars) of ten teachers in high school.

X: 250, 230, 300, 410, 260, 250, 300, 300, 410, 300.

Calculate the mean and standard deviation for this group.

---

(12) $\dfrac{\Sigma(X - \overline{X})^2}{n}$ is called the second moment about the mean which is the variance.

**26**

**Solution:**

1. Calculate the mean: $\overline{X} = \dfrac{\Sigma X}{n} = \dfrac{3010}{10} = \$301.00$

2. Calculate the standard deviation:

This is a sample, therefore, the formula to be used is:

$$s = \sqrt{(\Sigma X^2)/(n-1) - (\Sigma X)^2/(n)(n-1)}$$

| X | $X^2$ |
|------|--------|
| 250 | 62500 |
| 230 | 52900 |
| 300 | 90000 |
| 410 | 168100 |
| 260 | 67600 |
| 250 | 62500 |
| 300 | 90000 |
| 300 | 90000 |
| 410 | 168100 |
| 300 | 90000 |
| 3010 | 941700 |

$$s = \sqrt{\dfrac{941700}{10-1} - \dfrac{(3010)^2}{(10)(9)}}$$

$$= 62.97$$

**Grouped Data**

For grouped data, the standard deviation can be calculated by using the following formula:

$$s = \sqrt{\dfrac{\Sigma M^2 f}{n-1} - \dfrac{(\Sigma Mf)^2}{n(n-1)}}$$

where $n = \Sigma f$   and   M = class midpoint

*Example 2.14:* The following is a frequency distribution of the monthly salaries in dollars of 125 instructors at a state college:

| Salaries ($) | No. of Instructors |
|---------------------|--------------------|
| 800 and under 1000 | 10 |
| 1000 and under 1200 | 20 |
| 1200 and under 1400 | 40 |
| 1400 and under 1600 | 22 |
| 1600 and under 1800 | 18 |
| 1800 and under 2000 | 5 |
| 2000 and under 2200 | 5 |
| 2200 and under 2400 | 3 |
| 2400 and under 2600 | 2 |
| | 125 |

Calculate the mean and the standard deviation.

**Solution:**

1. The mean is a part of the general formula: $\overline{X} = \dfrac{\Sigma Mf}{\Sigma f}$

2. Calculate the standard deviation by applying the general formula:

$$s = \sqrt{(\Sigma M^2 f)/(n - 1) - (\Sigma Mf)^2/(n)(n - 1)}$$

| Class | f | M | Mf | $M^2f = M \times Mf$ |
|---|---|---|---|---|
| 800– 1000 | 10 | 900 | 9000 | 8100000 |
| 1000– 1200 | 20 | 1100 | 22000 | 24200000 |
| 1200– 1400 | 40 | 1300 | 52000 | 67600000 |
| 1400– 1600 | 22 | 1500 | 33000 | 49500000 |
| 1600– 1800 | 18 | 1700 | 30600 | 52020000 |
| 1800– 2000 | 5 | 1900 | 9500 | 18050000 |
| 2000– 2200 | 5 | 2100 | 10500 | 22050000 |
| 2200– 2400 | 3 | 2300 | 6900 | 15870000 |
| 2400– 2600 | 2 | 2500 | 5000 | 12500000 |
| | 125 | | 178500 | 269890000 |

$$\overline{X} = \frac{\Sigma Mf}{\Sigma f} = \frac{178500}{125} = 1428.00$$

$$s = \sqrt{\frac{269890000}{125 - 1} - \frac{(178500)^2}{125(124)}}$$

$$= \$347.71$$

The use of $\mu$ and $\sigma$:

The mean of the population $\mu$ is a constant that tends to fall in the center of the data. It is being used as a representative value for all observations contained in the population. The standard deviation $\sigma$ is a measure of variability that measures the dispersion of the observed values about the mean.

Most of the populations are normally distributed. The mean $\mu$ divides the normal distribution curve into two symmetrical halves. Knowing the magnitude of the standard deviation $\sigma$ one will be able to determine percentages of observations that fall in any interval constructed by deviating of $k\sigma$ from the mean ($k = 1, 2, 3, \ldots$). For example, a deviation of one sigma (or $k = 1$) from the mean in both directions will create an interval $\mu \pm 1\sigma$ that contains 68.26% of the observations in the population. A deviation of $2\sigma$ from $\mu$ will create an interval of $\mu \pm 2\sigma$ that contains 95.44% of the observed values in the population. This topic will be discussed in more detail in a later chapter.

C. *Measures of Skewness and Kurtosis:*

Measures of central location and of dispersion provide information concerning the average value of a distribution and the variability of the data about the mean, but nothing about the skewness or the peakedness of that frequency distribution.

a. Measures of Skewness:

It is often important to know whether a frequency distribution is symmetrical (normal) or non-symmetrical (skewed). This can be determined by calculating the coefficient of skewness. The popular formula of skewness[13] developed by the famous statistician, Karl Pearson is:

$$Sk_p = \frac{\text{Mean} - \text{Mode}}{\text{Standard Deviation}} = \frac{\overline{X} - \text{Mode}}{s}$$

However, if the mode does not exist or is not unique the following formula may be used:

$$Sk_p = \frac{3(\text{Mean} - \text{Median})}{\text{Standard Deviation}} = \frac{3(X - \text{Med})}{s}$$

The magnitude of the skewness coefficient indicates the degree of skewness. If the coefficient is zero, then the frequency distribution is symmetrical; if it is different than zero, then it is skewed. The sign associated with the magnitude of the skewness coefficient refers to the type of skewness. The positive sign indicates that the distribution is positively skewed, or it has a tail to the right; the negative sign indicates a negatively skewed distribution with a tail to the left.

b. Measures of Kurtosis:

Measures of kurtosis provide information concerning the peakedness of a frequency distribution. The most important coefficient of kurtosis is Beta two ($\beta_2$) or sometimes called Alpha four ($\alpha_4$). The magnitude of $\beta_2$ determines one of three types of distribution with different degrees of peakedness. If $\beta_2 = 3$, then we have a Mesokurtic distribution; if $\beta_2 > 3$, the distribution is Leptokurtic; and if $\beta_2 < 3$, the distribution becomes Platykurtic as shown in Figure 2-5.

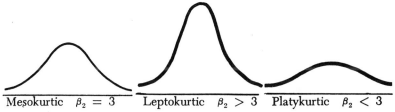

Mesokurtic $\beta_2 = 3$    Leptokurtic $\beta_2 > 3$    Platykurtic $\beta_2 < 3$

**Figure 2-5.** Three distributions with different degrees of peakedness.

---

(13) The third moment, $\dfrac{\Sigma(X - \overline{X})^3}{n}$, may be used as an absolute skewness measure. A more valuable skewness measure is the relative skewness measure, which is the ratio of the 3rd moment squared to the 2nd moment cubed; it is called Beta one.

$$\beta_1 = \frac{[\Sigma(X - \overline{X})^3/n]_2}{[\Sigma(X - \overline{X})^2/n]^3} \quad \text{or} \quad \sqrt{\beta_1}, \text{ which is called } \alpha_3.$$

Nonetheless, the sign of the third moment indicates the type of skewness.

The formula of $\beta_2$ or $\alpha_4$ is as follows:[14]

| Ungrouped Data | Grouped Data |
|---|---|
| $\beta_2 = \dfrac{\Sigma(X - \bar{X})^4/n}{[\Sigma(X - \bar{X})^2/n]^2}$ | $\beta_2 = \dfrac{\Sigma f(M - \bar{X})^4/\Sigma f}{[\Sigma f(M - \bar{X})^2/\Sigma f]^2}$ |

## EXERCISES

2.1 Differentiate among the measures of descriptive statistics: measures of location, measures of dispersion, and measures of skewness.

2.2 a. Define:
   The arithmetic mean
   The median
   The mode
   b. Describe the relationship existing among the mean, the median, and the mode.

2.3 a. Define:
   The range
   The variance
   The standard deviation
   b. What is the interpretation of the standard deviation?

2.4 The following table shows the acreage of 26 private camp grounds in Indiana:

| 48 | 80 | 136 | 65 | 100 |
|---|---|---|---|---|
| 50 | 80 | 160 | 258 | 70 |
| 54 | 80 | 160 | 114 | 71 |
| 55 | 83 | 161 | 120 | 74 |
| 60 | 96 | 180 | 75 | 121 |
| 125 | | | | |

Calculate:
1. The arithmetic mean
2. The median
3. The mode
4. The standard deviation

2.5 a. Construct a frequency distribution for the data in Exercise 2.4. Compute the mean, the median, the mode, and the standard deviation.
   b. Compare the results of Exercise 2.4 and 2.5 and state the reason for discrepancies if they exist.

2.6 Calculate the mean, the median, the mean deviation, and the standard deviation for the following distribution:

---

(14) The coefficient of peakedness, $\beta_2$ or $\alpha_4$, is based on the Fourth Moment: $\dfrac{\Sigma(X - \bar{X})^4}{n}$. As a matter of fact, $\beta_2$ is the ratio of the fourth moment to the second moment squared.

Average Hourly Wage of Production Workers
in 100 Metropolitan Areas
(in Dollars)

| Class | Frequency (f) |
|---|---|
| 2.00 and less than 2.25 | 1 |
| 2.25 and less than 2.50 | 6 |
| 2.50 and less than 2.75 | 5 |
| 2.75 and less than 3.00 | 11 |
| 3.00 and less than 3.25 | 21 |
| 3.25 and less than 3.50 | 22 |
| 3.50 and less than 3.75 | 15 |
| 3.75 and less than 4.00 | 11 |
| 4.00 and less than 4.25 | 6 |
| 4.25 and less than 4.50 | 2 |
| | 100 |

2.7 The following are two frequency distributions of the weight (in pounds) and the height (in inches) of 110 football players:

| Weight (in pounds) | | Height (in inches) | |
|---|---|---|---|
| Class | f | Class | f |
| 166–175 | 9 | 68–69 | 5 |
| 176–185 | 19 | 70–71 | 17 |
| 186–195 | 14 | 72–73 | 30 |
| 196–205 | 9 | 74–75 | 37 |
| 206–215 | 13 | 76–77 | 17 |
| 216–225 | 12 | 78–79 | 4 |
| 226–235 | 11 | | 110 |
| 236–245 | 8 | | |
| 246–255 | 7 | | |
| 256–265 | 8 | | |
| | 110 | | |

Compare the two frequency distributions.

2.8 The following frequency distribution represents the number of home runs hit in both National and American Leagues that occurred during a fifty-year period:

| Class | Frequency |
|---|---|
| 10–19 | 1 |
| 20–29 | 10 |
| 30–39 | 31 |
| 40–49 | 45 |
| 50–59 | 11 |
| 60–69 | 2 |
| | 100 |

As a consultant, what do you recommend as bases to rate a new player?

# 3

# Probability

Traditionally, statistics has been looked upon as the science dealing with the collection, organization, analysis, and interpretation of quantitative data. This phase of statistics has been covered in the previous chapter.

Today, statistics is concerned with decision making. The role of statistics in decision making is to help the businessman to draw conclusions about the unknown future. Uncertainty is a fact of life that businessmen face in making their decisions. Decisions concerning inventory, production, and sales are just few of many decisions to be made for the uncertain future. Probability theory lends itself to the statistical methods used to analyze the uncertain phenomena. Such analysis will guide the decision maker to implement the right course of action. On one hand, a businessman can use his past experience to express his judgment by assigning probabilities to each possible event that might affect the outcome of his decision. Moreover, the decision maker can use these probabilities together with other economic information to improve his decision-making process.

On the other hand, many of the business phenomena are of repetitive nature. Probability theory can provide an appropriate mathematical model, or probability distribution, to describe and interpret such observed phenomena. These mathematical models are constructed to simulate actual situations. Conclusions drawn can be reliable to the extent that the model is a good approximation to the real phenomenon.

Even though probability theory is an important branch of pure mathematics, it has its vital role in statistical methods used for decision making. The roots of probability lie in a simple mathematical theory of games of chance. In such games, the outcome of any trial cannot be accurately predicted. For example, if a coin is tossed many times under the same conditions, one cannot predict with accuracy the outcome of any toss. It is true that the possible outcomes of tossing a coin is known in advance, it is either Head or Tail. Nonetheless, this information will not change the fact that if the experiment is repeated under the same conditions, it is not possible to tell whether the outcome of the next trial is a Head or a Tail. These experiments are called random phenomena. The outcome of a random phenomenon is called the random event.

Probability theory has many notations from the set theory. Set theory is the main subject of any finite mathematical course, but a review of some of the set theory concepts will pave the road to understand the theory of probability.

### Review of Set Theory

A Set: is a collection of objects.

For example: Let A be a set of the outcomes of tossing a coin, then the elements contained in the set A are: Head, and Tail, and the set A is listed as follows:

$$A = \{H, T\} \quad \text{where}^{(1)} \quad H \epsilon A, \quad T \epsilon A$$

Another example: if B is set, elements of this set are the outcome of rolling a die, then the set B is:

$$B = \{1, 2, 3, 4, 5, 6\}$$

**A Subset:** (or event) is a part of a set.

For example: if B and $B_1$ are two sets:

$$B = \{1, 2, 3, 4, 5, 6\}$$
$$\text{and} \quad B_1 = \{2, 5\}$$

$B_1$ is called a subset of B and can be written in the following form:

$$B_1 \subset B \quad \text{or} \quad B \supset B_1$$

$B_1$ is a subset of B if every element in $B_1$ is an element in B but not vice versa. (Figure 3-1)

**Figure 3-1.** $B_1$ is
a subset of B.

Two sets, A and C, are said to be equal if and only if every element in A is also an element in C and vice versa, or if A is a subset of C, and C is a subset of A:

$$A = C \quad \text{if} \quad A \subset C \quad \text{and} \quad C \subset A.$$

For example: if $A = \{2, 4, 6, 8\}$ and $C = \{8, 6, 2, 4\}$ then $A = C$

**Union and intersection:**

The union and intersection are two notations associated with sets. $\cup$ is the symbol to denote the union of two or more sets, e.g., $A \cup B$ (A union B), $A \cup B \cup C$ (A union B union C) and so on. $\cap$ is used to denote the intersection of two or more sets.

*Example 3.1:* Let A and B be two sets:

$$A = \{a, b, c, d, f\}$$
$$\text{and} \quad B = \{1, 2, a, b, c\}$$

then $A \cup B$ is a new set, say C, that contains all the elements in A OR B:

$$C = \{1, 2, a, b, c, d, f\}; \quad C = A \cup B$$

The $A \cup B$ is shown graphically in Figure 3-2a.

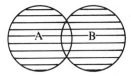

**Figure 3-2a.** The shaded
area is $A \cup B$.

---

(1) $\epsilon$ is a symbol used to denote "the element of" or the member contained in the set.

On the other hand, the A ∩ B, is a set, say D, that contains the elements that are in A AND B:

$$D = \{a, b, c\}; \quad D = A \cap B$$

Figure 3-2b. Shows the intersection of the two sets.

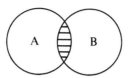

**Figure 3-2b.** The shaded area is A ∩ B.

*Example 3.2:* Let E and F to be two sets:

$$E = \{1, 2, 3\}$$
and $F = \{a, b, c, d\}$
then $E \cup F = \{1, 2, 3, a, b, c, d\}$
and $E \cap F = \phi$, $\phi$ is the null set[2]

### Joint and disjoint sets:

In the first example, A and B are said to be joint or partially overlapping sets, while in the second example, E and F are disjoint or mutually exclusive sets. As a general rule, if the intersection of two or more sets is the empty set, then these sets are mutually exclusive, mathematically:

if $A_i \cap A_j = \phi$ for $i \neq j$
then A's are mutually exclusive sets.

### Complement set:

For every set there is a complement, e.g., if A is a set then $A^c$ (or $A'$) is the complement of A that contains all the elements in the space or universe which are **not** in A.

*Example 3.3:* If S, a sample space, has these elements:

$$S = \{a, b, c, d, e, f, g\}$$
and let $A = \{b, e, d\}$
then $A^c = \{a, c, f, g\}$

This is shown graphically in Figure 3-3.

**Figure 3-3.** The unshaded area of S is $A^c$.

---

(2) $\phi$, 0, or ⌣ are symbols used to denote the null, void, or empty set. A null set is one that has no elements.

In this example:

$$A \cup A^c = S$$
$$\text{and } A \cap A^c = \phi$$

Also, de Morgan's laws show the connection between complement sets and the two notations of $\cup$ and $\cap$ as follows:

$$(A \cup B)^c = A^c \cap B^c$$
$$\text{and } (A \cap B)^c = A^c \cup B^c$$

The de Morgan's laws can be expanded to n sets or events.

The three important algebraic laws: commutative, associative, and distributive, are applicable to the union and the intersection of the events A, B, and C as follows:

Commutative law:

$$A \cup B = B \cup A \qquad \text{and} \qquad AB = BA$$

Associative law:

$$A \cup (B \cup C) = (A \cup B) \cup C \qquad A(BC) = (AB)C$$

Distributive law:

$$A(B \cup C) = AB \cup AC \qquad A \cup (BC) = (A \cup B)(A \cup C)$$

This brief review of the set theory and its concepts will help to better understand the probability theory.


## Probability

Probability of an event is a value between zero and one assigned to measure the degree of uncertainty of the occurrence of the event. If it is certain that the event will occur, then the probability of this event equals one. This means that we are sure 100% that this event will occur. On the contrary, if the event never occurs, then its probability equals zero. For example, the probability that it snows in July in the Midwest is zero, while the probability it snows in January is that region is one. On the other hand, the event that it snows on a particular day in January in the Midwest cannot be predicted with certainty; however, the event is more likely to occur, therefore its probability falls between one and zero. Also, the event that it rains in January in the Midwest is unlikely to occur, but it can occur; therefore the probability of this event falls between zero and one. To assign one or zero as the probability of an event is easier than to assign values in between.

There are three approaches that help to assess the probability of any event between the two extreme values: one and zero. The three approaches or concepts of probability assessments are:

1. Theoretical  2. Experimental  3. Subjective

### 1. Theoretical assessments of probability:

This approach is also called the classical, equiprobable, or the equally likely outcome approach. This approach is useful in assigning probabilities of events involving games of chance. Probabilities determined by this method are based on a prior concept. For example, if a fair coin is tossed under the same conditions, the possible outcomes are either

**36**

Head or Tail; the chances of a Head occurring are equally likely to those of a Tail. Therefore, one may assign 1/2 as a probability value for a Head or a Tail, or: P(H) = P(T) = 1/2. Another example, the possible outcomes of rolling a die are: 1, 2, 3, 4, 5, 6. If the die is fair and the process of rolling this die is uniform, then there is no reason to favor one outcome over the other. Therefore, the probability of each event occurring is equal to 1/6.

If the condition of equally likely outcomes does not apply for any reason, then the experimental approach to assess probabilities will be more appropriate to use.

### 2. **Experimental assessment of probabilities:**

The theoretical approach to assess probabilities is limited to events that can assume symmetrical probabilities, such as tossing a fair coin or rolling a fair die under the same conditions. Neither of these two criteria can exist, e.g., one cannot be sure that the die or the coin is well balanced or fair or that uniform or consistent tossing will be performed. Difficulties encountered in the selection of a criterion for the symmetry or equal possibility of outcomes led to the foundation of the experimental method.

The experimental approach is very useful to assess probabilities for the outcome of business and economic phenomena. Business and economic phenomena are not symmetrical and in order to analyze any business event we need to gather data from an experiment that deals with this event. For example, to determine the percentage of defectives in the production of a particular machine, an experiment or series of experiments should take place before arriving at any decision about this problem. The more observations (trials) to be included in the experiment the closer the decision to reality. In this example, the ratio of the number of defectives to the number of items produced included in the experiment, indicates the relative frequency of the occurrence of the event.

In general, if a random experiment is repeated n times, then the experimental assessment of the probability of an event A (denoted by P(A)) is the limiting value of relative frequency of A as to the number of trials, such as:

$$P(A) = \lim_{n \to \infty} \frac{f}{n}; \quad f = \text{frequency of the event A}$$

and n = number of observations or trials.

As the number of trials n gets larger and larger, the relative frequency $\frac{f}{n}$ reaches a stable limit which is used to assess the probability of A.

### 3. **Subjective assessment of probabilities:**

Subjective or personalistic probability is an approach used frequently in assigning probabilities for business events where experiments are either very costly or cannot be conducted. For example, the probabilities assigned to selling a new product can be determined by the decision maker (say the head of Marketing Research Division) based on his own judgment. Such assessment of probabilities reflect the experience, the attitude, and the values of the decision maker and the information available to him. These factors can differ from one decision maker to another, and even for the same decision maker probabilities assigned can differ from one time to another time depending on the availability of more information.

Subjective approach of assigning probabilities is very useful and it is frequently applied in business and economic decision making.

## Axioms of Probability

Given a random phenomenon whose possible outcomes are the events or the points on a sampling description space, S, a non-negative value between zero and one inclusive is the probability of any of these events. The property of the probabilities can be described by the following axioms:

Axiom 1.     $0 \leqslant P(A_i) \leqslant 1$

This axiom is stating that the probability assigned to any event of $A_i$ is a non-negative value that falls between zero and one inclusive.

Axiom 2.     $P(S) = 1$,   S = sample space

The second axiom of probability is concerned with the summation of probabilities assigned to all the elements or events contained in the sample space (or the set). If $A_i \in$ S, then $\Sigma P(A_i) = P(S) = 1$.

Axiom 3.     $P(A_1 \cup A_2 \cup A_3 \ldots \cup A_n) = P(A_1) + P(A_2) + P(A_3) + \ldots + P(A_n)$ provided that the subsets $A_i$ are mutually exclusive events or that $A_i \cap A_j = \phi$; for $i \neq j$

This axiom is considered the basis for the addition theorem of probability which will be presented later in this chapter.

The three axioms of probability are illustrated in the following example:

*Example 3.4:*   Let S be the sample space that contains all the possible outcomes of rolling two fair dice for 36 times under the same conditions. The 36 possible outcomes are:

| 1,1 | 1,2 | 1,3 | 1,4 | 1,5 | 1,6 |
| 2,1 | 2,2 | 2,3 | 2,4 | 2,5 | 2,6 |
| 3,1 | 3,2 | 3,3 | 3,4 | 3,5 | 3,6 |
| 4,1 | 4,2 | 4,3 | 4,4 | 4,5 | 4,6 |
| 5,1 | 5,2 | 5,3 | 5,4 | 5,5 | 5,6 |
| 6,1 | 6,2 | 6,3 | 6,4 | 6,5 | 6,6 |

The sample space, S, contains 36 events:
$$S = \{(1,1),(2,1),(3,1), \ldots ,(5,6),(6,6)\}$$
or    $$S = \{A_1, A_2, A_3, \ldots , A_{35}, A_{36}\}$$

The probability of each event of $A_i$ equals 1/36, a non-negative value between 0 and 1 inclusive, which satisfies Axiom 1.

The probability of all the 36 elements of S, is one, or $\Sigma P(A_i) = P(S) = 1$ which is consistent with Axiom 2.

On the other hand, events $A_1, A_2, \ldots , A_{36}$ are mutually exclusive and:

$$P(A_1 \cup A_2 \cup A_3 \cup \ldots \cup A_{36}) = 1/36 + 1/36 + 1/36 + \ldots + 1/36$$
$$= P(A_1) + P(A_2) + P(A_3) + \ldots + P(A_{36})$$

The probability of the union of $A_1$ events is an obvious illustration of Axiom 3.

## Theorems of Probability

There are many probability theorems based on the three axioms presented above. Some of these theorems will be stated without proof to be a good exercise for the reader to apply the 3 axioms in proving them.

Theorem 1.   $P(A) + P(A^c) = 1$
Theorem 2.   $P(\phi) = 0$
Theorem 3.   $P(A_1) \le P(A_2)$   , $A_1 \subset A_2$
Theorem 4.   $P(A_1) \le 1$   , $A \subset S$
Theorem 5.   The addition theorem
Theorem 6.   The multiplication theorem
Theorem 7.   The Bayesian theorem

Theorems 5, 6, and 7 will be presented in detail.

## The Addition Theorem

To apply this theorem, one has to differentiate between mutually exclusive and non-mutually exclusive events. Event: $A_1$ and $A_2$ are said to be mutually exclusive if the two events cannot occur together, otherwise they are non-mutually exclusive. For example: In tossing a coin, either a Head or a Tail occurs but one cannot expect both of them to occur at the same time in one toss. Therefore, the events of the occurrence of a Head or that of a Tail are considered to be mutually exclusive events.

On the other hand, a card being drawn from a deck of 52 cards can be a face card, a spade card, or a face card and a spade card at the same time. Therefore, the events of drawing a face card or a spade card are considered to be non-mutually exclusive events.

### Addition of mutually exclusive events:

If $A_1$ and $A_2$ are mutually exclusive events, then the probability of $A_1$ or $A_2$ is:

$$P(A_1 \cup A_2) = P(A_1) + P(A_2)$$

*Example 3.5:*   What is the probability of one or five occurring when rolling a fair die?

**Solution:**   let $A_1$ be the event of one occurring
                   and $A_2$ be the event of five occurring

$A_1$ and $A_2$ are mutually exclusive events because in one throw of a die either one or five can occur. Therefore,

$$P(A_5 \text{ OR } A_2) = P(A_5) + P(A_2)$$
$$= 1/6 + 1/6 = 1/3$$

Notice that the addition theorem for mutually exclusive events is an application of Axiom 3.

### Addition of non-mutually exclusive events:

If $B_1$ and $B_6$ are not-mutually exclusive events, then the probability of their union, or $P(B_1 \cup B_2)$ is:

$$P(B_1 \cup B_2) = P(B_1) + P(B_2) - P(B_1 \cap B_2)$$

This is illustrated graphically in Figure 3-4:

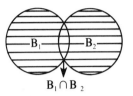

$$B_1 \cap B_2$$

**Figure 3-4.** The shaded area
is $P(B_1 \cup B_2)$.

*Example 3.6:* What is the probability of drawing one card from a deck of 52 cards to be either a face card or a spade card?

    **Solution:**   Let $B_1$ to be the event of drawing a face card
               and $B_2$ to be the event of drawing a spade card
         then   $B_1$, $B_2$ elements and probabilities are:
               $B_1 = \{J_S, Q_S, K_S, J_H, Q_H, K_H, J_D, Q_D, K_D, J_C, Q_C, K_C\}$
               $P(B_1) = 12/52$
               $B_2 = \{J_S, Q_S, K_S, 1_S, 2_S, 3_S, 4_S, 5_S, 6_S, 7_S, 8_S, 9_S, 10_S\}$
               $P(B_2) = 13/52$

$B_1$ and $B_2$ are not mutually exclusive events then $P(B_1 \cup B_2) = P(B_1) + P(B_2) - P(B_1 B_2)$. Values of $P(B_1)$ and $P(B_2)$ have been calculated and we need to find $P(B_1 B_2)$. $B_1 B_2$ consists of the elements that are in $B_1$ and $B_2$; $J_S$, $Q_S$, $K_S$. Hence the $P(B_1 B_2) = 3/52$.

    Accordingly, $P(B_1 \cup B_2) = P(B_1) + P(B_2) - P(B_1 B_2)$
                                 $= 12/52 + 13/52 - 3/52$
                                 $= 22/52$

## The Multiplication Theorem

This theorem differentiates between independent and dependent events. Two or more events are said to be independent if neither affects the occurrence of the other. In other words, events $A_1$ and $A_2$ are considered to be independent events if and only if the occurrence of $A_1$ does not depend on the occurrence of $A_2$ or vice versa, the occurrence of event $A_2$ does not depend on the occurrence of $A_1$.

### Multiplication of independent events:

If $A_1$ and $A_2$ are independent events, then the joint probability of $A_1$ AND $A_2$, or the $P(A_1 \cap A_2)$ is:

$$P(A_1 \cap A_2) = P(A_1) \cdot P(A_2)$$
or     $$P(A_2 \cap A_1) = P(A_2) \cdot P(A_1)$$

*Example 3.7:* An urn contains 6 white balls, and two black balls; two balls are drawn simultaneously from this urn with replacement. Find the probability that the two balls will be white.

**Solution:**

Drawing the balls with replacement means that the event of drawing the first ball in no way affects the event of drawing the second ball, or that the two events are independent.

Let $W_1$ be the event of drawing the first ball to be white

and $W_2$ be the event of drawing the second ball to be white

$P(W_1) = 6/8$, the probability of drawing a white ball

and $\quad P(W_2) = 6/8$, is the same as $P(A_1)$ because the first ball has been replaced in the urn before drawing the second ball

then $\quad P(W_1 W_2) = P(W_1) \cdot P(W_2)$

$$= 6/8 \cdot 6/8$$
$$= 36/64 = 9/16$$

A tree diagram showing the different values of joint probabilities for the independent events in the previous example is shown in Figure 3-5.

### Multiplication of dependent events:

Event $A_2$ is said to depend on event $A_1$ if the occurrence of $A_1$ affects the occurrence of $A_2$. The joint probability of the two dependent events $A_1$ AND $A_2$ is:

$$P(A_1 \cap A_2) = P(A_1) \cdot P(A_2/A_1)$$

the probability of the second event given (denoted by / ) or knowing the outcome of event $A_1$; $P(A_2/A_1)$ is known as the conditional probability.

*Example 3.8:* An urn contains 8 balls, 6 white and 2 black. Two balls are drawn simultaneously **without** replacement from the urn. Find the probability of drawing the first ball to be white and the second ball to be black.

**Solution:**

Drawing two balls without replacement indicates that the probability of drawing the second ball depends on the outcome of the first draw. If the first ball drawn is white, then 7 balls are left in the urn: 5 white and 2 black, while drawing a black ball in the first draw will leave 7 balls: 6 white and one black.

In the example: the assumption is that a white ball has been drawn in the first draw, event $W_1$, then the $P(W_1) = 6/8$. The second ball drawn is assumed to be black, event $B_2$. The probability of drawing the second ball, given that the first ball is white, $P(B_2/W_1)$, equals 2/7. The joint probability, $P(W_1 \cap B_2)$, is:

$$P(W_1 B_2) = P(W_1) \cdot P(B_2/W_1)$$
$$= 6/8 \cdot 2/7$$
$$= 3/14$$

The $P(B_2/W_1)$ is called the conditional probability which can be derived from this formula as:

$$P(B_2/W_1) = \frac{P(W_1 B_2)}{P(W_1)}$$

A tree diagram to show the different joint probabilities for different assumptions of drawing the balls is shown in Figure 3-6.

**41**

Joint probabilities

$6/8 \cdot 6/8 = 9/16$

$6/8 \cdot 2/8 = 3/16$
$6/8 \cdot 2/8 = 3/16$

$2/8 \cdot 2/8 = 1/16$

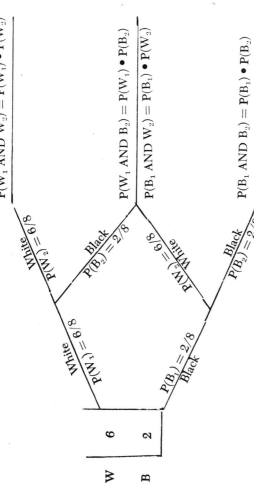

$P(W_1 \text{ AND } W_2) = P(W_1) \cdot P(W_2)$

$P(W_1 \text{ AND } B_2) = P(W_1) \cdot P(B_2)$
$P(B_1 \text{ AND } W_2) = P(B_1) \cdot P(W_2)$

$P(B_1 \text{ AND } B_2) = P(B_1) \cdot P(B_2)$

White
$P(W_2) = 6/8$

Black
$P(B_2) = 2/8$

White
$P(W_2) = 6/8$

White
$P(W_1) = 6/8$

Black
$P(B_2) = 2/8$

Black
$P(B_1) = 2/8$

W 6

B 2

**Figure 3-5.** A tree diagram for joint probabilities of independent events.

42

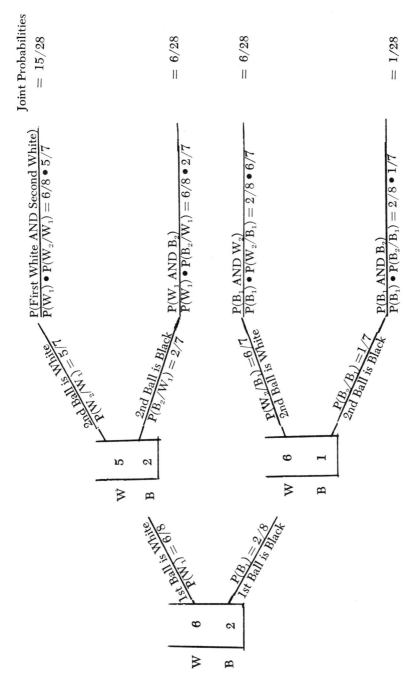

**Figure 3-6.** A tree diagram to calculate the joint probabilities for dependent events.

43

The addition theorem looks upon the events as mutually exclusive and non-mutually exclusive, while the multiplication theorem is concerned with independent and dependent events. It is not necessarily true that mutually exclusive events are independent or that non-mutually exclusive events are dependent. The following illustration is presented to show the relationship that exists between mutually exclusive and independent events:

Let $A_1$ be the event that Mr. X, a chess player, wins.

and $A_2$ be the event that Mr. Y, another chess player, wins.

a. If Mr. X is playing Mr. Y

then $A_1$ and $A_2$ are mutually exclusive but **not** independent events

b. If Mr. X is playing Mrs. Y

and Mr. Y is playing Mrs. X

then $A_1$ and $A_2$ may be considered as independent events but **not** mutually exclusive events.

### The Bayesian Theorem

The theorem has been developed by the Reverend Thomas Bayes, an English minister who was interested in making inductive inferences about the hypothesis given the occurrence of an event based on this hypothesis. The theorem is a systematic method to revise subjective probabilities assigned to the outcome of the hypothesis by the decision maker. Such revision is based on the availability of more relevant empirical data about an event which is part of the hypothesis.

Bayesian theorem is a direct application of the conditional probability and can be derived as follows:

Let $H_1$, $H_2$, $H_3$, . . . , $H_n$ be n mutually exclusive and exhaustive events in the sample space, S, and let E be an event; that is E $\subset$ S.

Probabilities of $H_1$, $H_2$, . . . , $H_n$ are assigned by the subjective approach and called the prior probabilities.

An experiment is conducted to gather data about the event E, and the conditional probabilities of E given $H_1$, or $P(E/H_1)$, are referred to as the likelihood probabilities.

The main goal of the Bayesian theorem is to revise the subjective (or prior probabilities) by providing the posterior probabilities, $P(H_1/E)$ which in turn can be revised as more relevant empirical data becomes available.

A graphical presentation of the Bayesian theorem in Figure 3-7, where S, the sample space is partitioned into $H_1$, $H_2$, . . . , $H_n$. The event E is shown as a subset of S.

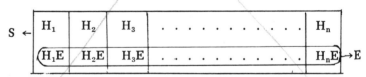

**Figure 3-7.** Event E is a subset of the partitioned sample space S.

The assumptions stated before are: $P(H_i)$, and $P(E/H_i)$ are known. Also, the joint events of $H_i$ and E, ($H_i \cap E$), are mutually exclusive. Having these assumptions in mind, the posterior probabilities, $P(H_i/E)$, can be derived as follows:

$$E \qquad = (H_1 \cap E) \cup (H_2 \cap E) \cup (H_3 \cap E) \cup \ldots \cup (H_n \cap E)$$

$$P(E) \qquad = P(H_1 \cap E) + P(H_2 \cap E) + P(H_3 \cap E) + \ldots + P(H_n \cap E);$$
$$H_1 \cap E \text{ are mutually exclusive}$$

$$P(E/H_1) \quad = \frac{P(E \cap H_1)}{P(H_1)} = \frac{P(H_1 \cap E)}{P(H_1)}; \ H_1 \cap E = E \cap H_1$$

$$P(H_1 \cap E) = P(H_1) \cdot P(E/H_1)$$

$$P(E) \qquad = P(H_1) \cdot P(E/H_1) + P(H_2) \cdot P(E/H_2)$$
$$+ \ldots + P(H_n) \cdot P(E/H_n)$$
$$= \Sigma P(H_i) \cdot P(E/H_i)$$

The posterior probability to be figured out is:

$$P(H_1/E) = \frac{P(H_1 \cap E)}{P(E)}$$

$$= \frac{P(H_1) \cdot P(E/H_1)}{\Sigma P(H_i) \cdot P(E/H_i)}$$

In general, $P(H_i/E) = \dfrac{P(H_i) \cdot P(E/H_i)}{\Sigma P(H_i) \cdot P(E/H_i)}$ $\qquad i = 1, 2, \ldots, n$

*Example 3.9:* Let $H_1$, $H_2$, $H_3$ be the output produced by Machine I, II, and III. Machine I produces 35%, Machine II produces 20%, and Machine III produces 45% of the output. Percentage of defectives produced by Machine I, II, III is 2%, 1.5%, 1% respectively.

An item has been selected randomly for inspection and found to be defective. What is the probability that this item has been produced by Machines I, II, or III?

**Solution:**

1. Prior Probabilities: $P(H_1) = .35$, $P(H_2) = .20$, $P(H_3) = .45$, indicate the probabilities of producing defective as well as non-defective items by Machine I, II, and III.

2. The event of selecting an item for the purpose of inspection, is the event E. The item was found to be defective.

3. Conditional probabilities (or likelihood) $P(E/H_i)$ refer to the probability of a defective item being produced by Machine I, II, or III:

    $$P(E/H_1) = 2\%, \quad P(E/H_2) = 1.5\%, \quad P(E/H_3) = 1\%$$

4. From the information we have the posterior probability for each machine, $P(H_i/E)$, indicates the probability that the defective item has been produced by Machine I, II, or III.

The following table shows the calculations of the posterior probabilities:

| Event | Prior Prob. $P(H_i)$ | Likelihood × Prob. $P(E/H_i)$ | Joint = Prob. $P(H_i \cap E)$ | Posterior Prob. $P(H_i/E)$ |
|---|---|---|---|---|
| $H_1$: Output of Machine I | .35 | .020 | .0070 | $\dfrac{.0070}{.0145} = .48$ |
| $H_2$: Output of Machine II | .20 | .015 | .0030 | $\dfrac{.0030}{.0145} = .21$ |
| $H_3$: Output of Machine III | .45 | .010 | .0045 | $\dfrac{.0045}{.0145} = .31$ |
| | 1.00 | | $P(E) = .0145$ | 1.00 |

The probability of an item (defective or nondefective) to be produced by Machine I, II, III is shown by the prior probabilities: .35, .20, .45 respectively. If an item is selected randomly and found to be defective, then the probability that this defective item belongs to Machine I, II, or III is shown by the posterior probabilities: .48, .21, .31 respectively.

Machine III, producing 45% of the total output, has the lowest percentage of defective output (1%), which explains why the probability that an item selected randomly and found to be defective belongs to Machine III is lower (.31) than that of Machine I (.48).

Bayesian Theorem is very helpful as a decision making tool especially when dealing with small sample size. More of the Bayesian theorem application to decision theory is introduced later in this book.

## Counting Methods

To assign probabilities, one needs to know the possible outcomes of the event. In some cases, it is very easy to enumerate the possible outcomes of the event such as tossing a coin, or rolling a die. If the event has few stage experiments then the tree diagram method can be used to determine the probability of the possible outcomes as shown in Figure 3-5, and Figure 3-6. In these two Figures, the tree diagram method has been used to calculate the joint probabilities for independent and dependent events.

The more stages of an experiment and the more possibilities for each stage, the more difficult to use the tree diagram method to enumerate the possible outcomes of the complex event. In these cases other counting methods based on algebraic formulas are applied. The most popular algebraic methods used for counting the possible outcomes of an experiment or event are as follows:

1. If an event $E_1$ can result in $n_1$ outcome, and event $E_2$ can result in $n_2$ outcomes, then the joint events ($E_1$ and $E_2$) can result in $n_1 n_2$ outcomes.

This can be extended to more than two events. For example, a woman has 15 dresses, 10 shoes, 6 hats: How many combinations of attire can she wear?

$E_1$ = event of having a dress, its outcome $n_1$ = 15
$E_2$ = event of having a shoe, its outcome $n_2$ = 10
$E_3$ = event of having a hat, its outcome $n_3$ = 6

The number of combinations of attire she can wear or the outcome of the joint events

$(E_1, E_2,$ and $E_3) = n_1 \cdot n_6 \cdot n_3$
$$= 15 \cdot 10 \cdot 6 = 900.$$

2. *Permutations:*

Permutation is arrangement of n different elements of a set. Let S be a finite set with distinct elements:

$$S = \{e_1, e_2, e_3, \ldots, e_n\}$$

The number of distinct **arrangements** that can be formed from the n elements contained in S, using X of them at a time, $X \leq n$, is called "the number of permutations of n objects taken X at a time." Using the notation of permutation, the number of arrangements can be calculated by:

$$_nP_x \text{ or } (n)x$$

where [3] $\quad _nP_x = \dfrac{n!}{(n-x)!}$

*Example 3.10:* if $S = \{a,b,c,d,e,f\}$, then the number of permutations of the 6 letters of S taken 2 at a time is:

$$_6P_2 = \dfrac{6!}{(6-2)!} = \dfrac{6!}{4!} = 6 \cdot 5 = 30$$

The 30 permutations are:

| ab | ba | bc | cb | cd | dc | de | ed | ef | fe |
|----|----|----|----|----|----|----|----|----|----|
| ac | ca | bd | db | ce | ec | df | fd |  |  |
| ad | da | be | eb | cf | fc |  |  |  |  |
| ae | ea | bf | fb |  |  |  |  |  |  |
| af | fa |  |  |  |  |  |  |  |  |

In the previous example if we are interested to find the number of permutations of the 6 letters taken 6 at a time, then:

$$_6P_6 = \dfrac{6!}{(6-6)!} = \dfrac{6!}{0!} = 6!$$

In general, the number of permutations of the n elements of set S, taken all of them at a time, $_nP_n$, equals n!

If the set S contains of $n_1$ elements that are alike, $n_2$ elements that are alike but different kind, and so on, or simply the set S is partitioned into r subgroups that are alike, then the number of permutations of the n elements of this partitioned set is:

$$\dfrac{n!}{n_1! n_2! \ldots n_r!} \qquad , n = n_1 + n_2 + \ldots + n_r$$

---

(3) ! is the symbol used for factorial.

This permutation is called the multinomial coefficient. For example, the number of permutations of a bridge deck be partitioned in four hands each of size 13 is:

$$n = n_1 + n_2 + n_3 + n_4$$
$$52 = 13 + 13 + 13 + 13$$

Then the number of permutations $= \dfrac{52!}{13!13!13!13!}$

### 3. Combinations:

The calculation of permutations or arrangements is concerned with the order of the different elements or objects, the arrangement ab is different than the arrangement ba. If the order does not matter, then ab is not different than ba, and that ab and ba are considered to be a subset that contained the two elements of a and b. The number of subsets of size x to be formed from the n elements of S is called the number of combinations and is denoted by:

$$\binom{n}{x} \qquad \text{or } _nC_x \qquad \text{or } C_x^n$$

where $_nC_x$ is the number of combinations of n different elements of S, taken X at a time, is

$$\binom{n}{x} = \frac{n!}{x!(n - x)!}$$

The number of combinations $_nC_x$ is known as the binomial coefficient. In example 3.10 where S is a set of 6 elements:

$$S = \{a,b,c,d,e,f\}$$

the number of combinations of the 6 letters contained in S taken 2 letters at a time is:

$$\binom{6}{2} = \frac{6!}{2!(6 - 2)!} = \frac{6!}{2!4!} = 15$$

The 15 subsets of combinations are:

| | | | | |
|----|----|----|----|----|
| ab | bc | cd | de | ef |
| ac | bd | ce | df | |
| ad | be | cf | | |
| ae | bf | | | |
| af | | | | |

Comparing the formulas used to find the number of permutations and the number of combinations, one can express one formula in terms of the other:

$$_nP_x = x!\,_nC_x$$

and $\quad _nC_x = \dfrac{1}{x!}\,_nP_x$

In example 3.10 of the 6 letters taken 2 at a time, the number of permutations can be figured out by using the number of combinations time x! ($= 2!$):

**48**

$_6C_2 = 15$, then $_6P_2 = 2!(15) = 30$. Also the number of combinations can be calculated from the number of permutations:

$$_6P_2 = 30 \text{ and } x! = 2!, \text{ then } _6C_2 = \frac{1}{2!}\ 30 = 15.$$

EXERCISES

3.1   a.  What is the role of the probability theory in statistics?
       b.  Differentiate among the following probability assessments:
            Theoretical
            Experimental
            Subjective

3.2   Define:
       A set
       A subset
       Equal sets
       Union of two or more sets
       Intersection of two or more sets
       Empty set
       Mutually exclusive sets
       Independent sets

3.3   Prove the following probability theorems:
       a.  $P(A) + P(A^c) = 1$
       b.  $P(\phi) = 0$
       c.  $P(A_1) \leq P(A_6)$   , $A_1 \subset A_2$
       d.  $P(A_1) \leq 1$   , $A_1 \subset S$

3.4   a.  Three balls are drawn with replacement from a box containing 10 balls, of which 6 are white and 4 are red.
            Find the probability:
            1.  P (all the three balls will be white)
            2.  P (all the three balls will be the same color)
            3.  P (at least one is red)
       b.  Solve part a assuming that the balls are drawn without replacement.
            1.  P (all the three balls will be white)
            2.  P (all the three balls will be the same color)
            3.  P (at least one is red)

3.5   a.  If one card is drawn from a regular deck of 52 cards, what is the probability it will be a heart or a face card?
       b.  What is the probability of an odd number appearing in a single toss of a die?
       c.  Toss two fair coins. What is the probability of:
            1.  P (2 heads)
            2.  P (at least one head)
       d.  Two fair dice are tossed. What is the probability that the sum of the dice will be either 6 or 12?

3.6 X and Y are members of a football team of 25 players. After a winning game, all of the 25 players throw their helmets in the air. Each player starts to pick up randomly one of the helmets. What is the probability that:
. a. X will get his own helmet.
b. X and Y will get their own helmets.
c. At least one, either X or Y, will get his own helmet.

3.7 a. An instructor grades each student as follows: A die is cast; if one occurs, the student receives A grade. If two or three occur, the student receives B grade; otherwise the student receives C grade.
If 15 students are graded according to the above scheme, what is the probability of 4 A's, 5 B's, and 6 C's?
b. The student body of a certain college is composed of 55% men and 45% women. 80% of the men and 58% of the women are democrats.
1. What is the probability that a student who is a democrat is a man?
2. What is the probability that a student who is a democrat is a woman?

3.8 a. A man tosses 2 fair coins; what is the probability that he has tossed 2 tails, given that he has tossed at least one tail?
b. A man tosses 2 fair dice. What is the probability that the sum of the two dice will be 5 given that the sum is odd?

3.9 a. Find
$(5)_2$
$(5)^2$
$5!$
$$\binom{5}{2}$$
b. 1. Four politicians met at a party. How many hand shakes are exchanged if each politician shakes hands with every other politician once?
2. A restaurant menu lists 5 soups, 9 meat dishes, 6 desserts and 4 beverages. An ordered meal consists of all these components. How many customers can order different meals?
3. Five of 12 girls are to be selected to form a discussion team. In how many ways can the team be made?

3.10 What is the practical significance of the Bayesian theorem?

3.11 Urn I contains 7 white and 4 red balls, while urn II contains 5 white and 3 red balls. An urn is selected at random, and a ball is drawn from it. Given that the ball drawn is white, what is the probability that urn II was selected?

3.12 Mr. X, Mr. Y, and Mr. Z are the only production workers in a small workshop. Mr. X produces 30% and Mr. Y produces 25% of the total production. The previous records showed that the percentage of defective units produced by X, Y, and Z is 2%, 4%, 1.5% respectively. A customer received a defective unit. What is the probability that it has been produced by Z?

# 4

# Probability Functions of Random Variables

The preceding chapter discussed the fundamental concepts of probability. In this chapter, those probability concepts will be used to construct mathematical models known as probability functions. The mathematical models may be viewed as a good approximation of empirical frequency distributions presented in chapters one and two. As we shall see in later chapters, a test ($\chi^2$ − test) can be conducted to prove the validity of using a probability function as a reasonable approximation of a frequency distribution. One can describe the frequency distribution of any variable by a mathematical model called a probability function and then use this probability function to make decisions concerning the variable. For example, the frequency distribution of a defective product from a particular machine can be approximated by a probability function (in this case a binomial probability function). This probability function can be used to make probability statements concerning the occurrence of the number of defects produced by this machine. Also, decisions concerning the performance of this machine relative to a similar machine of a different brand may be made through the use of the probability functions.

A probability function is a special mathematical function that depicts the relationship between observed numerical values of a random variable and their probabilities. The probability function can be used to generate probability distributions, however, most statisticians use probability function, probability distribution, or probability models to denote the same concept.

Probability functions are of two types—discrete and continuous—and can be constructed for one random variable or more than one. This chapter will present probability functions of a discrete and continuous type for one and two random variables. It seems appropriate at this point to review functions and random variables in order to pave the way for a clear understanding of probability functions.

## Functions

Given two sets: $X = \{D\}$ and $Y = \{R\}$ where the elements of X are called the *domain* and the elements of Y are called the *range,* a function from X into Y is defined as a set of ordered pairs (D,R) in which each *element or elements* in X is paired with one and only one element in Y.

*Example:* Let X and Y be two sets having the following elements:

$X = \{2,4,6,8\}$
$Y = \{1,2,3,4\}$

Each element in X is paired with one and only one element in Y, therefore, the set containing those ordered pairs is called a function, as shown below:

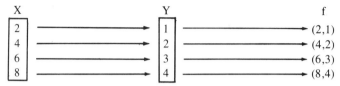

where f is a function whose elements are (2,1), (4,2), (6,3), and (8,4) where 2,4,6, and 8, the values of X, are the elements of the domain of the function f, while 1,2,3, and 4, the values of Y, are the elements of the range of the function f. Any function can be expressed by an algebraic equation. For example, the f function can be written in equation form as follows: $Y = f(X)$

or $\quad Y = 1/2X \qquad X = 2,4,6,8$

or $\quad f(X) = 1/2X \qquad X = 2,4,6,8$

A graphical representation of the f function is shown in Figure 4-1:

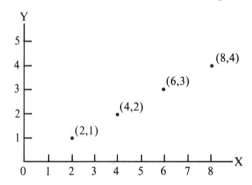

**Figure 4-1.** Function of one variable.

There are single variable functions and multiple variable functions. We have briefly reviewed the definition of functions of one variable. However, there are situations, as in bivariate probability distributions, where functions of two variables have been applied. Therefore, the functions of two variables is defined below:

A function of two variables is made up of three sets: X, Y, and Z. The domain of this function is the product of the elements of the two sets (X,Y) and its range is the elements of the third set (Z). The function becomes:

$Z = f(X,Y)$

A graphical representation of this function is shown in Figure 4-2:

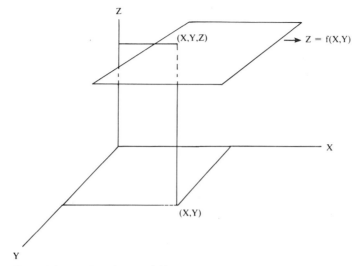

**Figure 4-2.** Function of two variables.

### Discrete and Continuous Functions

Functions can be classified as discrete or continuous depending on the domain of the function. If the domain consists of values of finite or countable numbers, then the function is described as discrete. If the domain of the function assumes numerical values on a continuous scale or infinite values contained in an interval, then this function is classified as continuous. For example, if

$$f(X) = 2X \qquad X = 0,1,2,3$$

the domain of the function $f(X)$ consists of finite integers, therefore, $f(X)$ is a discrete function. A graphical representation of this discrete function is shown in Figure 4-3:

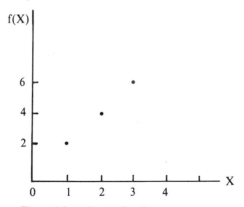

**Figure 4-3.** A discrete function.

**53**

On the other hand, if

$$f(X) = 2X \qquad\qquad 0 < X < 3$$

then the domain of X contains infinite values that fall between zero and three and the function f(X) is a continuous function. One can view this situation as a spinning wheel that starts at zero and stops at three and includes all values between 0 and 3 on a continuous basis. This continuous function is graphically represented in Figure 4-4:

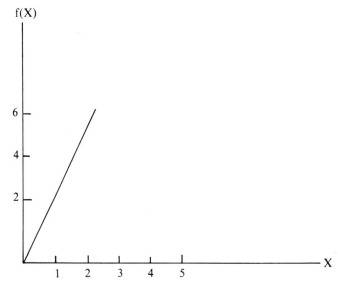

**Figure 4-4.** A continuous function.

This brief review of functions will enhance the understanding of probability functions because a probability function is a special case of mathematical functions. Probability functions show the relationship between the observed numerical values of random variables and their probabilities.

**Random Variables**

As mentioned in the preceding chapter, the set of points of possible outcome of an experiment is called the sample space of the experiment. If the outcome of an experiment is determined by chance, then it is called a random experiment. The variable of a random experiment is a random variable. Therefore, a random variable can be defined as a valued variable determined by the outcome of a random experiment and defined on a sample space.

To illustrate the relationship between the sample space and the random variable, consider the experiment of randomly selecting two units produced by Machine A. The selection is done one unit at a time and with replacement. The defective unit is referred to by D while N is used to denote a nondefective unit. The possible outcomes of this random experiment constitute the sample space, S, while the random variable X denotes the number of nondefective (or defective) units in the selection process. The sample space that shows the possible outcomes and the random variable X that assumes the outcome of this experiment can be expressed as follows:

| | | | | |
|---|---|---|---|---|
| Sample space | S = | { (NN), | (ND), (DN), | (DD) } |
| Random Variable | X = | 2N | 1N | 0N |
| | = | 2, | 1, | 0 |

This relationship can be represented in a diagram as shown in Figure 4-5:

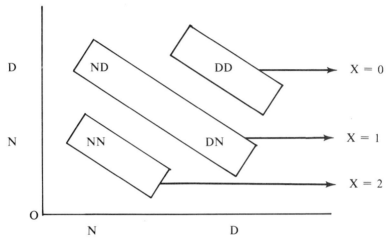

**Figure 4-5.** Sample space and random variable X for selecting two units from the production of Machine A.

Another example follows: Let X be a random variable that denotes the sum of points obtained when rolling two dice. The sample space for rolling two dice consists of the following 36 sample points:

$$S = \{(1,1), (1,2), (1,3), (1,4), (1,5), (1,6),$$
$$(2,1), (2,2), (2,3), (2,4), (2,5), (2,6),$$
$$(3,1), (3,2), (3,3), (3,4), (3,5), (3,6),$$
$$(4,1), (4,2), (4,3), (4,4), (4,5), (4,6),$$
$$(5,1), (5,2), (5,3), (5,4), (5,5), (5,6),$$
$$(6,1), (6,2), (6,3), (6,4), (6,5), (6,6) \}$$

The random variable X assumes integral values from 2 to 12 because the smallest sum when rolling two dice is $1 + 1$ or two and the largest sum is $6 + 6$ or twelve. The values that X may assume are 2, 3, 4, 5, 6, 7, 8, 9, 10, 11, and 12.

A graphical representation of the sample space and the random variable for rolling two dice is shown in Figure 4-6.

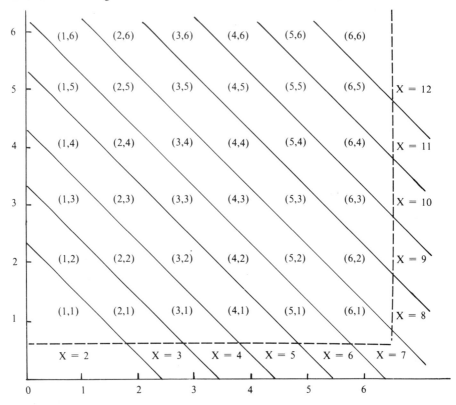

**Figure 4-6.** Sample space and the random variable X for rolling two dice.

Many events that occur in the business world can be viewed as random because the outcome depends on chance. For example, sales for a period of time cannot be predicted with complete accuracy because sales depend on some uncontrollable factors.

Random variables can be classified as discrete or continuous depending on the phenomena they represent. Random variables described above are of a discrete nature because they assumed a finite number of distinct values. On the other hand, a random variable that assumes values on a continuous scale in some interval is called a continuous random variable. Variables involving time are of continuous nature.

This introduction of functions and random variables aids in understanding the concept of probability functions and probability distribution, because a probability function is a mathematical model of the observed numerical values of random variables and their probabilities.

## Probability Functions

As mentioned before, a probability function f(X) is a special function. Its domain is the mathematical values assumed by the random variable X, and the elements of its range are the probabilities associated with those numerical values of the random variable.

A probability function is usually presented in tabular form and this table contains the value of the random variable X and the probabilities of the values that X may assume denoted by P(X). The following is an illustration of the probability function of selecting two units from Machine A (assume that the machine produces 5% defective units).

The sample space: $S = $ (NN), (ND), (DN), (DD)  
The random variable: $X = $ 2, 1, 0 (number of nondefectives)  
The Probabilities: $P(X) = $ 0.9025, 0.0950, 0.0025

The probability function of selecting two units produced by Machine A can be arranged in a table form as a distribution of the random variable X that assumes the number of nondefective units obtained in the experiment. The distribution is shown in Table 4-1:

**TABLE 4-1**
Probability Distribution of Selecting Two Units
of the Production of Machine A

| X (nondefective units) | P(X) |
|---|---|
| 0 | 0.0025 |
| 1 | 0.0950 |
| 2 | 0.9025 |
| | 1.0000 |

A graphical representation of the probability function of selecting two units produced by Machine A is shown in Figure 4-7:

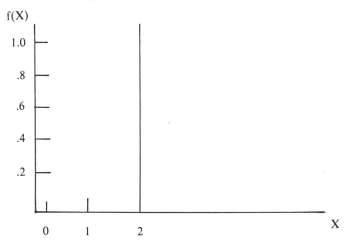

**Figure 4-7.** Probability function of selecting two units from the production of Machine A.

Probability functions are mathematical models formulated to describe the relationship between the observed numerical values of a random variable and their corresponding probabilities. Since random variables are either discrete or continuous, the probability function is either discrete or continuous. Discrete and continuous probability functions for one variable will be presented in the next section followed by discrete and continuous joint probability functions.

1. *Probability Functions for One Random Variable:*

   A. Discrete Variables:

   Most variables in business and economics are discrete such as sales, production, rate of investment, orders, accounts receivable, etc. Some of these discrete random variables describe the outcome of one variable, while others describe the outcome of two or more variables. In this section a discrete probability function of one random variable, f(X), is discussed. A discrete probability function must satisfy two properties:

$$1. \ 0 \leqslant f(X) \leqslant 1 \qquad \text{for all X}$$

This property implies that the probability of the different values of X are non-negative and fall between 0 and 1 inclusive. This is consistent with axiom I of the probability theory.

$$2. \quad \sum_{\text{all X}} f(X) = 1$$

This implies that the summation of all probabilities of X, equals one. This property agrees with the second axiom of probability:

$$P(S) = 1, \text{ since } P(S) = \sum P(X_1) = \sum_{\text{all X}} f(X)$$

The previous example, selecting two units produced by Machine A, is a discrete random experiment where the random variable X assumed a finite number of distinct values. The probability function for that random variable is discrete and satisfies the two properties stated above. Another example of a probability function for a discrete random variable is presented below.

*Example 4.1:* A car dealer was able to derive the following probability distribution of random variable X, where X denotes the number of cars sold per week.

**TABLE 4-2**
Probability Distribution of X

| $X_i$ (number of cars sold per week) | $P(X_i)$ |
|---|---|
| 1 | 0.04 |
| 2 | 0.08 |
| 3 | 0.12 |
| 4 | 0.16 |
| 5 | 0.18 |
| 6 | 0.16 |
| 7 | 0.14 |
| 8 | 0.12 |
| | 1.00 |

**58**

From the probability distribution table, a discrete probability function of car sales per week can be derived and stated as follows:

$$f(X) = X/25 \qquad \text{for } X = 1,2,3,4$$
$$= (14 - X)/50 \qquad X = 5,6,7,8$$
$$= 0 \qquad \text{otherwise}$$

From the probability function f(X), and its two properties, one can find the probability of any value that X assumes.

For example:

$$P(X = 4) = f(X = 4) = 4/25 = 0.16,$$

which indicates that the probability of selling four cars during any week is 0.16.

$$P(X = 7) = f(X = 7) = (14 - 7)/50 = 0.14,$$

this means that the probability of selling seven cars per week is 0.14.

$$P(4 \leqslant X \leqslant 6) = \sum_{X=4}^{6} f(X) = f(X = 4) + f(X = 5) + f(X = 6)$$
$$= 4/25 + (14 - 5)/50 + (14 - 6)/50$$
$$= 0.16 + 0.18 + 0.16$$
$$= 0.50$$

This result indicates that the chances are 50% to sell either four, five or six cars during any given week.

In general, if the probability function is discrete, then the $P(a \leqslant X \leqslant a) = P(X = a) = f(X = a)$, while the $P(a \leqslant X \leqslant b) = \sum_{X=a}^{b} P(X) = f(X = a) + f(X = a + 1) + \ldots + f(X = b - 1) + f(X = b)$.

## Cumulative Probability Function: F(X)

A function closely related to the probability function f(X) is the cumulative probability function F(X) that provides the probability of those values of the random variable less than or equal to the specific value of X.

The discrete cumulative probability function can be defined as:

$$F(X) = \sum_{t \leqslant x} f(t)$$

where the summation occurs over all the values of the random variable X that are less than or equal to a particular value of X, referred to by t. For example, the cumulative probability function F(X) for the random variable X, where X refers to the number of cars sold per week is presented in the following table:

**TABLE 4-3**
Cumulative Probability Function
for the Random Variable X

| $X_i$ | $f(X_i)$ | $F(X_i)$ |
|-------|----------|----------|
| 1 | 0.04 | 0.04 |
| 2 | 0.08 | 0.12 |
| 3 | 0.12 | 0.24 |
| 4 | 0.16 | 0.40 |
| 5 | 0.18 | 0.58 |
| 6 | 0.16 | 0.74 |
| 7 | 0.14 | 0.88 |
| 8 | 0.12 | 1.00 |

From the cumulative probability function constructed for the car sales, cumulative probabilies can be calculated. For example, the cumulative probability of $X \leqslant 5$ is:

$$P(X \leqslant 5) = F(X = 5) = 0.58,$$

or the probability of selling five cars or less is 0.58.

The cumulative probability of $X \leqslant 3$ is 0.24 (from the table) means that the chances of the selling three cars or less is 24%.

Graphical representations for the probability function f(X) and the cumulative probability function F(X) for car sales are shown in Figure 4-8.

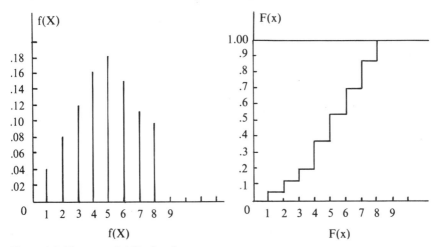

**Figure 4-8.** Discrete probability function.

B. Continuous Variables:

As mentioned before, a random variable is a function of a random phenomenon. If the random phenomenon is of a continuous nature, then the random variable will be continuous. For example, waiting for a bus or waiting in line at the supermarket can be considered to be continuous and the random variable, X, representing the continuous phenomenon is a continuous random variable.

**60**

## Probability Density Function (p.d.f.)

The probability function formulated for the continuous random variable is a continuous function and is called the probability density function. The probability density function must satisfy two properties:

1. $0 \leq \int_a^b f(X) \, dx \leq 1$

2. $\int_{\text{all } X} f(X) \, dx = 1$

These two properties are similar to those of the discrete probability function except that in the continuous case integral calculus is used to find the probability of X between any two values in the domain of X. The first property of the continuous probability function means that the probability of any interval in the domain of X must be positive and have a magnitude of between zero and one, inclusive. The second property implies that the sum of all the probabilities of the interval in the domain of X is equal to one. In other words, the area under the probability curve and covering the whole domain of the function must be equal to unity.

An example of a probability function of the continuous type is shown in the following example.

*Example 4.2:* The following p.d.f. shows the time elapsed between arrivals at a check out counter of a supermarket during weekends (in minutes):

$$
\begin{aligned}
f(X) &= 0 & &\text{for } X < 0 \\
&= 0.1(X + 1) & &0 \leq X \leq 1 \\
&= 0.4(X - 1/2) & &1 \leq X \leq 2 \\
&= 0.3(3 - X) & &2 \leq X \leq 3 \\
&= 0.2(4 - X) & &3 \leq X \leq 4 \\
&= 0.1 & &4 \leq X \leq 6 \\
&= 0 & &\text{elsewhere}
\end{aligned}
$$

Prove that this function satisfies the two properties of a probability density function.

**Solution:** Let us examine $f(X)$ and prove that it is a p.d.f. This means that the function should satisfy the two properties;

1. $0 \leq \int_a^b f(X) \, dx \leq 1$   or   $0 \leq P(a \leq X \leq b) \leq 1$     $f(X)$

$f(X) = 0$ $\qquad\qquad$ $X < 0$ $\qquad$ $P(X < 0)$ $\qquad\qquad\qquad\qquad$ $= 0.0$

$\quad = 0.1(X + 1)$ $\quad$ $0 \leq X \leq 1$ $\quad$ $P(0 \leq X \leq 1) = \int_0^1 0.1(X + 1)dx$ $\;= 0.15$

$\quad = 0.4(X - 1/2)$ $\;$ $1 \leq X \leq 2$ $\quad$ $P(1 \leq X \leq 2) = \int_1^2 0.4(X - 1/2)dx = 0.40$

$\quad = 0.3(3 - X)$ $\quad$ $2 \leq X \leq 3$ $\quad$ $P(2 \leq X \leq 3) = \int_2^3 0.3(3 - X)dx$ $\;= 0.15$

$\quad = 0.2(4 - X)$ $\quad$ $3 \leq X \leq 4$ $\quad$ $P(3 \leq X \leq 4) = \int_3^4 0.2(4 - X)dx$ $\;= 0.10$

$\quad = 0.1$ $\qquad\qquad$ $4 \leq X \leq 6$ $\quad$ $P(4 \leq X \leq 6) = \int_4^6 0.1dx$ $\qquad\quad$ $= 0.20$

$\quad = 0$ $\qquad\qquad\quad$ elsewhere $\quad$ $P(X > 6)$ $\qquad\qquad\qquad\qquad$ $= 0$

**61**

The first property of a probability density function has been satisfied because the $f(X)$ for any interval within its domain is non-negative.

2. $\int\limits_{\text{all } X} f(X)dx = 1$

For this $f(X)$, the second property can be proved by:

$$\int\limits_0^6 f(X)dx = \int\limits_0^1 0.1(X + 1)dx + \int\limits_1^2 0.4(X - 1/2)dx + \int\limits_2^3 0.3(3 - X)dx$$

$$+ \int\limits_3^4 0.2(4 - X)dx + \int\limits_4^6 0.1dx$$

$$= 0.15 + 0.04 + 0.15 + 0.10 + 0.20 = 1.00$$

In the case of the continuous functions, the value of the probability function for any interval may be obtained by evaluating the integral of the p.d.f. over the desired interval, for example:

$$P(0 \leq X \leq 1) = \int\limits_0^1 f(X)dx = \int\limits_0^1 0.1(X + 1)dx = 0.1\left(\frac{X^2}{2} + X + C\right)\Big|_0^1$$

$$= 0.1(1/2 + 1 + C) - 0.1(0/2 + 0 + C) = 0.15$$

$$P(0 \leq X \leq 1/2) = \int\limits_0^{1/2} f(X)dx = \int\limits_0^{1/2} 0.1(X + 1)dx = 0.1\left(\frac{X^2}{2} + X + C\right)\Big|_0^{12}$$

$$= 0.1(1/8 + 1/2 + C) - (0 + C) = 0.0625$$

$$P(1.5 \leq X \leq 1.5) = \int\limits_{1.5}^{1.5} f(X)dx = \int\limits_{1.5}^{1.5} 0.4(X - 1/2)dx$$

$$= 0.4\left(\frac{X_2}{2} - \frac{X}{2} + C\right)\Big|_{1.5}^{1.5} = 0$$

The last example, $P(1.5 \leq X \leq 1.5)$, shows that the probability of any point on a continuous probability function is equal to zero. Mathematically, $P(a \leq X \leq a) = \int_a^a f(X)\, dx = 0$, if the probability function is continuous.

### Cumulative Probability Function: F(X)

The cumulative probability function is related to the p.d.f. and is defined as:

$$F(X) = \int\limits_{t \leq x} f(t)\, dt$$

For a given t value, the cumulative probability is the total area under the probability density function for the interval to the left of X. The values derived from the cumulative probability function have the same meaning for discrete and continuous random variables.

The cumulative function F(X) for *Example 4.2* is:

$$f(X) = 0 \qquad\qquad X < 0 \qquad 0.0 \qquad\qquad P(X < 0) = 0.0$$

$$= 0.1(X + 1) \quad 0 \leqslant X \leqslant 1 \quad 0.15 \qquad \int_0^1 f(X)dx = P(X \leqslant 1) = 0.15$$

$$= 0.4(X - 1/2) \; 1 \leqslant X \leqslant 2 \quad 0.40 \qquad \int_0^2 f(X)dx = P(X \leqslant 2) = 0.55$$

$$= 0.3(3 - X) \quad 2 \leqslant X \leqslant 3 \quad 0.15 \qquad \int_0^3 f(X)dx = P(X \leqslant 3) = 0.70$$

$$= 0.2(4 - X) \quad 3 \leqslant X \leqslant 4 \quad 0.10 \qquad \int_0^4 f(X)dx = P(X \leqslant 4) = 0.80$$

$$= 0.1 \qquad\qquad 4 \leqslant X \leqslant 6 \quad 0.20 \qquad \int_0^6 f(X)dx = P(X \leqslant 6) = 1.00$$

$$= 0 \qquad\qquad\quad \text{elsewhere}$$

A graphical presentation of f(X) and F(X) is shown in Figure 4-9:

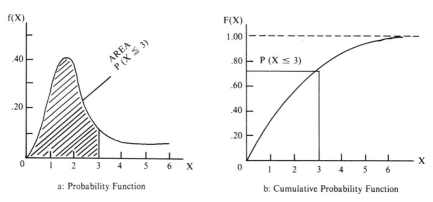

a: Probability Function          b: Cumulative Probability Function

**Figure 4-9.** Continuous probability functions.

## 2. *Probability Functions For More Than One Random Variable:*

In the preceding section, probability functions of discrete and continuous random variables have been considered. However, in business many problems require the use of several random variables rather than one random variable. For example, the probability of a certain car with a certain type of brake stopping within a specified distance on various road surfaces (blacktop, concrete, gravel, etc.) and different surface condition (ice, water, dry, snow), involves several random variables. The probability function of those random variables is a joint function known as a multivariate probability function. Another example of two random variables is the number of cars to be sold during a certain period of time (day, week, or month), and the number of customers visiting the show room. In this example let X be a random variable that shows the number of cars sold, and Y, another random variable, the number of customers visiting the show room. f(X,Y) is the joint probability of the two random variables X and Y. This bivariate probability function can be represented graphically on a three-dimension diagram. The joint probability function, bivariate or multivariate, must satisfy the same two properties of any probability function.

Probability functions formulated to describe more than one random variable are called Joint Probability functions. If X and Y are two random variables defined as functions on the same sample space, then their joint probability function f(X,Y) describes the simultaneous probability of the two variables. Joint probability functions are either discrete or continuous.

A. Discrete Joint Probability Functions:

The joint probability function that describes two or more discrete random variables is a discrete function. Let X and Y be two random variables where X refers to the number of cars per family and Y denotes the number of working members in a family. The joint probability distribution of the two variables is presented in Table 4-4:

### TABLE 4-4
Joint Probability Distribution

| X / Y | 1 | 2 | Total |
|-------|------|------|-------|
| 1 | 0.15 | 0.30 | 0.45 |
| 2 | 0.20 | 0.35 | 0.55 |
|   | 0.35 | 0.65 | 1.00 |

The values of the joint probability function f(X,Y) are shown in the body of Table 4-4. For example, the joint probability of having one car and one working member in the family is $P(X = 1) \cap P(Y = 1)$ which can be derived from the table as the value of $f(X = 1, Y = 1) = 0.15$. Also, the probability of a family owning two cars and one working is the joint probability $f(X = 2, Y = 1)$ equals to 0.30.

The joint probability function, similar to other probability functions, must satisfy the following two criteria:

1. $0 \leqslant f(X,Y) \leqslant 1$

2. $\sum_x \sum_y f(X,Y) = 1$

Inspect the values of the joint probability function listed in Table 4-4 to see that they are nonnegative and that they are between one and zero, inclusive. This will satisfy the requirement of the first criterion. For the second criterion, note that the total is equal to one.

Another example of a joint probability function of two variables, expressed in equation form, follows:

*Example 4.3:* Let X be a random variable representing the number of children in a family, and Y, another random variable, representing the number of TV sets owned by this family. The bivariate probability function for these two variables f(X,Y) is expressed in the equation:

$$f(X,Y) = \frac{X + Y + 1}{42} \qquad \begin{array}{l} \text{for } X = 0,1,2,3 \\ \text{and } Y = 0,1,2 \end{array}$$

**64**

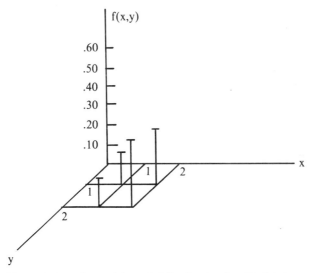

**Figure 4-10.** A discrete joint probability function for table 4-4.

Then the joint probabilities, $P(X \cap Y)$ are calculated in Table 4-5:

**TABLE 4-5**
Joint Probabilities of $f(X,Y)$

| X <br> Y | 0 | 1 | 2 | 3 | f(Y) |
|---|---|---|---|---|---|
| 0 | 1/42 | 2/42 | 3/42 | 4/42 | 10/42 |
| 1 | 2/42 | 3/42 | 4/42 | 5/42 | 14/42 |
| 2 | 3/42 | 4/42 | 5/42 | 6/42 | 18/42 |
| f(X) | 6/42 | 9/42 | 12/42 | 15/42 | 1 |

The $f(X,Y)$ must satisfy the following two criteria:

1. $0 \le f(X,Y) \le 1$         for all ordered pairs of X and Y
2. $\sum_x \sum_y f(X,Y) = 1$

From the joint probability function a probability function for each random variable is derived. This probability function is called a marginal probability function.

**Marginal Probability Function: (Discrete)**

Marginal probability function is simply the function that describes the probabilities of a single, random variable. Accordingly, $f(X)$, and $f(Y)$ are the two marginal probability functions involved in the joint probability function $f(X,Y)$.

**65**

The marginal probability function, f(X) or f(Y), can be derived from the joint probability function f(X,Y) as follows:

$$f(X) = \sum_{\text{all } Y} f(X,Y)$$

and

$$f(Y) = \sum_{\text{all } X} f(X,Y)$$

*Example 4.4:* From Example 4.3, calculate the marginal probability functions f(X) and f(Y) from the joint probability function f(X,Y):

$$f(X,Y) = \frac{X + Y + 1}{42} \qquad\qquad X = 0,1,2,3; \quad Y = 0,1,2$$

**Solution:**

(A) $f(X) = \sum_{Y=0}^{2} \dfrac{X + Y + 1}{42}$

$$= \frac{X + 0 + 1}{42} + \frac{X + 1 + 1}{42} + \frac{X + 2 + 1}{42}$$

$$= \frac{X + 1 + X + 2 + X + 3}{42}$$

Hence,

$$f(X) = \frac{3X + 6}{42} \qquad\qquad X = 0,1,2,3$$

and the probabilities of X are calculated from f(X) as follows:

$$f(X=0) = \frac{0 + 6}{42} = 6/42$$

$$f(X=1) = \frac{3 + 6}{42} = 9/42$$

$$f(X=2) = \frac{6 + 6}{42} = 12/42$$

$$f(X=3) = \frac{9 + 6}{42} = 15/42$$

These values are the sum of the columns of the joint probability function in Table 4-5.

(B) $f(Y) = \sum_{X=0}^{3} \dfrac{X + Y + 1}{42}$

$$= \frac{0 + Y + 1}{42} + \frac{1 + Y + 1}{42} + \frac{2 + Y + 1}{42} + \frac{3 + Y + 1}{42}$$

$$= \frac{Y + 1 + Y + 2 + Y + 3 + Y + 4}{42}$$

Hence,

$$f(Y) = \frac{4Y + 10}{42} \qquad\qquad Y = 0,1,2$$

66

and the probabilities of Y can be derived from f(Y) as follows:

$$f(Y=0) = \frac{0 + 10}{42} = \frac{10}{42}$$

$$f(Y=1) = \frac{4 + 10}{42} = \frac{14}{42}$$

$$f(Y=2) = \frac{8 + 10}{42} = \frac{18}{42}$$

These values are the sum of the rows of the joint probabilities in Table 4-5.

## Conditional Probability Functions: (Discrete)

Marginal probability functions determine the probability of observed numerical values of a single random variable without regard to any other variable. However, in some cases it is important to know the probability of a single variable given information concerning values of another variable (or several variables). For example, the probability of the variable Y given information about the variables $X_1$, $X_2$, . . . , $X_n$. The probability of Y given X is called the conditional probability. The preceding chapter introduced the conditional probability concept of events. The conditional probability function is an extension of the conditional probability concept of the two events A and B, where A and B are two dependent events such that the outcome of one depends on the outcome of the other. In other words, if the outcome of the event A depends on the outcome of the event B, then the conditional probability of event A, given event B, is defined as follows:

$$P(A/B) = \frac{P(A \cap B)}{P(B)} \quad ; P(B) > 0$$

Likewise, if event B depends on event A, then the conditional probability of event B, given event A, becomes:

$$P(B/A) = \frac{P(A \cap B)}{P(A)} \quad ; P(A) > 0$$

In general, for two dependent events, the conditional probability is equal to a ratio of the joint probability of the two events to the marginal probability or:

$$\text{Conditional probability} = \frac{\text{Joint Probability}}{\text{Marginal Probability}}$$

Applying the same concept of conditional probability of two events to two random variables, X and Y, where one depends on the other, then the conditional probability function of the random variable X, given the random variable Y, becomes:

$$f(X/Y) = \frac{f(X,Y)}{f(Y)}$$

and the conditional probability of Y, given X, is:

$$f(Y/X) = \frac{f(X,Y)}{f(X)}$$

where f(X,Y) is the joint probability function of the two random variables X and Y, while f(X) and f(Y) are the marginal probability functions for X and Y, respectively. Both the

**67**

joint probability function and the marginal probability function have been considered in this chapter.

The conditional probability function must satisfy the same two properties of any probability function:

1. $0 \leq f(X/Y) \leq 1$     or     $0 \leq f(Y/X) \leq 1$

2. $\sum_x f(X/Y = y_i) = 1$     or     $\sum_y f(Y/X = x_i) = 1$    ;    $i = 1, 2, \ldots, n$

The following is an example of the conditional probability function of two random variables:

*Example 4.5:*    For the function in Example 4.4, calculate the conditional probability function for X, given Y, and for Y, given X.

**Solution:**    From Example 4.4, the joint and marginal probability functions for the two random variables X and Y are as follows:

a. The joint probability function of X and Y:

$$f(X,Y) = \frac{X + Y + 1}{42} \qquad X = 0,1,2,3 \quad ; \quad Y = 0,1,2$$

b. The marginal probability of X:

$$f(X) = \frac{3X + 6}{42} \qquad X = 0,1,2,3$$

c. The marginal probability of Y:

$$f(Y) = \frac{4Y + 10}{42} \qquad Y = 0,1,2$$

A. The conditional probability function of X given Y is:

$$f(X/Y) = \frac{f(X,Y)}{f(Y)}$$

$$= \frac{(X + Y + 1)/42}{(4Y + 10)/42}$$

$$= \frac{X + Y + 1}{4Y + 10} \qquad X = 0,1,2,3 \quad ; \quad Y = 0,1,2$$

B. The conditional probability function of Y given X is:

$$f(Y/X) = \frac{f(X,Y)}{f(X)}$$

$$= \frac{(X + Y + 1)/42}{(3X + 6)/42}$$

$$= \frac{X + Y + 1}{3X + 6} \qquad X = 0,1,2,3 \quad ; \quad Y = 0,1,2$$

**Statistical Independence of Random Variables**

Conditional probability functions assume that one random variable depends on another random variable (or several variables). However, not all random variables are interdependent; many random variables are independent. Statistical independence, rather than interdependence, of random variables is vital in statistics. For example, the theory of sampling, which assumes that successive observations are statistically independent. This eliminates the chance of transmitting errors in measurement from one observation to the next.

The concepts of independence and interdependence of events have been considered in the preceding chapter. Nonetheless, a review of those concepts with two events will be very useful for the study of two random variables. A and B are two independent events such that the outcome of the event A does not affect the outcome of the event B, and vice versa. In this case:

$$P(A/B) = P(A)$$

or $\quad P(B/A) = P(B)$

which means that, in general, two events are said to be statistically independent if their conditional probability equals their marginal probability. On the other hand, if the conditional probability of the two events differs from their marginal probability as:

$$P(A/B) \neq P(A)$$

or $\quad P(B/A) \neq P(B)$

then the two events are said to be interdependent.

We can extend the concepts of independence and interdependence of two events to random variables. To determine whether or not two random variables are statistically independent, one may compare the value of the marginal and conditional probabilities of the random variable and if the two values are not different, then the two random variables are statistically independent. If the two values of the conditional and marginal probabilities are different, then the two random variables are interdependent.

With two random variables there are two cases: $f(X/Y)$ and $f(Y/X)$:

1. X and Y are two random variables, their conditional probability function of X, given Y, follows:

$$f(X/Y) = \frac{f(X,Y)}{f(Y)}$$

if X and Y are statistically independent, then

$$f(X,Y) = f(X)f(Y)$$

and hence, $f(X/Y) = \dfrac{f(X)\ f(Y)}{f(Y)} = f(X)$

This shows that the two random variables are statistically independent if the conditional probability equals the marginal probability. On the other hand, if the conditional probability does not equal the marginal probability then the two random variables are said to be statistically interdependent.

2. The same analysis is true with the conditional probability of Y, given X.

*Example 4.6:* Determine whether the two random variables in Example 4.5 are statistically independent or interdependent.

**Solution:**

$$P(X=0/Y=0) \neq P(X=0) \qquad \text{and} \qquad P(Y=1/X=0) \neq P(Y=1)$$
$$\text{(because)}$$
$$P(X=0/Y=0) = 1/10 \qquad\qquad P(Y=1/X=0) = 2/6$$
$$\text{(and)}$$
$$P(X=0) \qquad = 6/42 \qquad\qquad P(Y=1) \qquad = 14/42$$

then, X and Y are statistically interdependent.

B. Continuous Joint Probability Functions:

If the random variables are continuous, then the joint probability function for two or more variables is continuous.

Let X and Y be two continuous random variables; their joint probability function is:

$$\int_x \int_y f(X,Y) dy \; dx$$

This function can be represented graphically by a surface in three dimensions as shown in Figure 4-11:

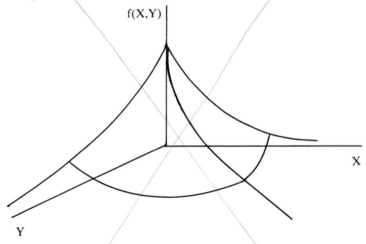

**Figure 4-11.** A graphical representation of a joint probability function of two continuous random variables.

A continuous probability function is a p.d.f. (probability density function) where the value of a probability corresponding to any set of its domain is calculated by integrating $f(X,Y)$ over that subset of the domain.

The probability function for two continuous random variables is the same as the joint probability function for two discrete variables except that integrals are used in the continuous case to derive both conditional and marginal probability density functions.

The continuous joint probability function must satisfy the two properties of any continuous probability function:

1. $f(X,Y) \geqslant 0$

2. $\int_x \int_y f(X,Y) \; dy \; dx = 1$

**70**

The following example illustrates a joint probability function for two continuous random variables:

*Example 4.7:* Let X and Y be two continuous random variables, where X denotes the number of days needed for a patient to recover from a surgical operation, while Y refers to the number of hours spent in the surgical room. The joint probability function of X and Y is:

$$f(X,Y) = (1/18) (X + Y - 2) \quad \text{for } 1 < X < 4;$$
$$\text{and } 0 < Y < 3$$

Two examples to evaluate the joint probability for the two intervals:

a. $1 < X < 2$ and $0 < Y < 1$

b. $2 < X < 3$ and $1 < Y < 2$

are presented below:

a. For $1 < X < 2$ and $0 < Y < 1$, the joint probability is:

$$\int_1^2 \int_0^1 1/18(X + Y - 2)dy\,dx = \int_1^2 [\int_0^1 1/18(X + Y - 2)dy]dx$$

$$= \int_1^2 \left[ 1/18\left( XY + \frac{Y^2}{2} - 2Y \right) \Big|_0^1 \right] dx$$

$$= \int_1^2 [1/18\{X(1) + 1/2 - 2(1)\} - 0]dx$$

$$= \int_1^2 [1/18(X - 3/2)]dx$$

$$= 1/18(1/2X^2 - 3/2X) \Big|_1^2$$

$$= 1/18[(4/2 - 3/2(2)) - (1/2 - 3/2)]$$

$$= 1/18(-1 + 1)$$

$$= 0$$

b. For $2 < X < 3$ and $1 < Y < 2$, the joint probability is:

$$\int_2^3 \int_1^2 1/18(X + Y - 2)dy\,dx = \int_2^3 [\int_1^2 1/18(X + Y - 2)dy]dx$$

$$= \int_2^3 \left[ 1/18\left( XY + \frac{Y^2}{2} - 2Y \right) \Big|_1^2 \right] dx$$

$$= \int_2^3 1/18\left\{ \left( 2X + \frac{4}{2} - 4 \right) - (X + 1/2 - 2) \right\}dx$$

$$= \int_2^3 1/18(X - 1/2)dx$$

$$= 1/18(1/2X^2 - 1/2X) \Big|_2^3$$

$$= 1/18\{(9/2 - 3/2) - (4/2 - 1)\}$$
$$= 1/18(3 - 1)$$
$$= 2/18$$

## Marginal Probability Function: (continuous)

The marginal probability function f(X) or f(Y) can be derived from the joint probability function in the same way as in the discrete case.

where $f(X) = \int_y f(X,Y)\, dy$

and $f(Y) = \int_x f(X,Y)\, dx$

The marginal probability function for the random variable X and for the random variable Y has been derived from the joint probability function of Example 4.6 as shown in Example 4.8:

*Example 4.8:* For the joint probability function f(X,Y) in Example 4.6 derive the marginal probability of X and Y.

**Solution:** The joint probability function of X and Y is:

$$f(X,Y) = (1/18)(X + Y - 2) \qquad 1 < X < 4 \;\; ; \;\; 0 < Y < 3$$

1. The marginal probability function of X is derived as follows:

$$f(X) = \int_0^3 (1/18)(X + Y - 2)dy$$
$$= (1/18)[(XY + 1/2Y^2 - 2Y)]\Big|_0^3$$
$$= 1/18[(3X - 4.5 + 6) - (0)]$$
$$= 1/18(3X - 3/2) \qquad\qquad 1 < X < 4$$

2. The marginal probability function of Y is derived as follows:

$$f(Y) = \int_1^4 1/18(X + Y - 2)dx$$
$$= 1/18[(1/2X^2 + XY - 2X)]\Big|_1^4$$
$$= 1/18[(8 + 4Y - 8) - (1/2 + Y - 2)]$$
$$= 1/18(3Y + 1.5) \qquad\qquad 0 < Y < 3$$

## Conditional Probability

In the case of continuous dependent random variables, the conditional probability can be calculated by dividing the continuous joint probability by the continuous marginal probability.

$$f(X/Y) = \frac{f(X,Y)}{f(Y)}$$

and $f(Y/X) = \dfrac{f(X,Y)}{f(X)}$

**72**

The two conditional probability functions for X, given Y, and for Y, given X, have been calculated from the joint probability function and the marginal probability functions of Examples 4.7 and 4.8.

This is an example of calculations for the conditional probability for X, given Y, where $1 < X < 2$ and $0 < Y < 1$:

$$f(1 < X < 2 \mid 0 < Y < 1) = \frac{\int_1^2 \int_0^1 1/18(X + Y - 2)dy\, dx}{\int_0^1 f(Y)dy} = 0$$

For the conditional probability of Y, given X, the following example is presented:

$$f(1 < Y < 2 \mid 1 < X < 2) = \frac{\int_1^2 \int_1^2 1/18(X + Y - 2)dy\, dx}{\int_1^2 f(X)dx} = \frac{1/18}{3/18} = 1/3$$

The two continuous random variables X and Y are statistically interdependent because their conditional probability differ from their marginal probability as shown in the following examples:

a.  $f(1 < X < 2 \mid 1 < Y < 2)$ $= 0$
    while $f(1 < X < 2)$ $= 3/18$

b.  $f(1 < Y < 2 \mid 3 < X < 4)$ $= 9/18$
    while $f(1 < Y < 2)$ $= 6/18$

EXERCISES

4.1  Define:
    A random variable
    Probability density function
    Cumulative probability function
    Joint probability function
    Conditional probability function

4.2  a. What are the two properties that a probability function must possess?
    b. Differentiate between discrete and continuous random variables. Sustain your answer graphically.

4.3  The following is a probability function:

$$f(X) = \frac{3 - X}{6} \qquad \text{for } X = 0,1,2$$
$$= 0 \qquad \text{elsewhere}$$

    a. Find $P(X = 2)$, $P(X = 1)$.
    b. Construct the cumulative probability function $F(X)$.
    c. Graph $f(X)$ and $F(X)$.

4.4 Graph the following probability function:
$$f(X) = 1/40 \, (X^2 + X) \qquad \text{for } X = 1,2,3,4$$
$$= 0 \qquad\qquad\qquad \text{elsewhere}$$
a. Find $P(X = 3)$, $P(X = 4)$
b. Find $P(X \geqslant 2)$, $P(X > 4)$, $P(X \leqslant 3)$

4.5 Given the following probability function:

$$f(X) = \binom{5}{X} (1/4)^X (3/4)^{5-X} \qquad X = 0,1,2,3,4,5$$

$$= 0 \qquad\qquad\qquad \text{elsewhere}$$

a. Find $P(X = 3)$, $P(X \leqslant 2)$, $P(X \geqslant 4)$
b. Graph the function.
c. Calculate $F(X)$ and graph the function.

4.6 The following is a probability function of X:

$$f(X) = \frac{\binom{6}{X}\binom{4}{4-X}}{\binom{10}{4}} \qquad X = 0,1,2,3,4$$

$$= 0 \qquad\qquad \text{elsewhere}$$

Calculate:
a. $P(X = 2)$, $P(X = 3)$
b. $P(X \leqslant 1)$, $P(X > 3)$
c. $P(X > 4)$

4.7 The following is a probability density function:
$$f(X) = 3/2X^2 \qquad\qquad \text{for } 0 \leqslant X \leqslant 1$$
$$= 3/2(6X - 2X^2 - 3) \qquad 1 \leqslant X \leqslant 2$$
$$= 3/2(X^2 - 6X + 9) \qquad 2 \leqslant X \leqslant 3$$
$$= 0 \qquad\qquad\qquad \text{elsewhere}$$
Calculate: $P(1.5 \leqslant X \leqslant 2.5)$, $P(2.5 \leqslant X)$, $P(X = 2)$
Graph the function.

4.8 Prove that the following function is a p.d.f.:
$$f(X) = 1/11(X^2 - X + 2) \qquad 0 \leqslant X \leqslant 1$$
$$= 2 - X \qquad\qquad\qquad 1 \leqslant X \leqslant 2$$
$$= 1/50(X^2 - 1) \qquad\qquad 2 \leqslant X \leqslant 4$$
$$= 0 \qquad\qquad\qquad\qquad \text{elsewhere}$$

4.9 Prove that the following is a p.d.f.:
$$f(X) = 3/2(X^2 - 3X + 2) \qquad 0 \leqslant X \leqslant 2$$
$$= 0 \qquad\qquad\qquad\qquad \text{elsewhere}$$
and if it is a p.d.f., calculate:
$$P(X \leqslant 1), \, P(1.5 \leqslant X \leqslant 2.5)$$

4.10 The following is a joint probability function for X and Y:

$$f(X,Y) = \frac{X - Y + 3}{30} \qquad \begin{array}{l} \text{for } X = 0,1,2 \\ \text{and } Y = 0,1,2,3 \end{array}$$

a. Construct a table to show the different probabilities of each pair of values of X and Y.
b. Derive f(X) and f(Y)
c. Calculate: $P(Y = 1 \,/\, X = 2)$, $P(X = 1 \,/\, Y = 0)$

4.11 Prove that the following is a joint probability function:

$$f(X,Y) = \frac{4Y - X^2}{57} \qquad \begin{array}{l} \text{for } X = 0,1,2 \\ \text{and } Y = 0,1,2,3 \end{array}$$

If this function proved to be a joint probability function, calculate:
a. $P(X = 2)$, $P(X = 1)$
b. $P(Y = 1)$, $P(Y \geqslant 2)$
c. $P(X = 0 \,/\, Y = 3)$, $P(Y = 2 \,/\, X = 2)$

4.12 Prove that the following function is a joint probability distribution function for X and Y:

$$f(X,Y) = 1/36(X + Y) \qquad \begin{array}{l} 0 \leqslant X \leqslant 3 \\ \text{and } 1 \leqslant Y \leqslant 4 \end{array}$$

If it is proved that it is a joint probability distribution function, find:
a. f(X)
b. f(Y)
c. $P(1 \leqslant X \leqslant 2)$ and $(1 \leqslant Y \leqslant 3)$
d. $P(0 \leqslant X \leqslant 1)/(2 \leqslant Y \leqslant 3)$
e. $P(3 \leqslant Y \leqslant 4)/(0 \leqslant X \leqslant 1)$

# CHAPTER
# 5
# Expected Value and Variance of Random Variables

In Chapter 2, it is found that the mean, as a measure of central location, and the variance as a measure of dispersion, together can provide enough information to describe the data of a deterministic frequency distribution.

In a probabilistic situation, where the outcome of the event is uncertain, the expected value (or the mathematical expectation) and the variance of a probability model are corresponding to the mean and the variance in a deterministic situation. The expected value, or the mean, of a random variable indicates the value of the variable that occurs "on the average," while the variance of a random variable shows the spread of the values of the variable about its mean.

The expected value concept is used extensively in making decisions. The expected pay off, and the expected opportunity loss are a few to mention the application of the expected value concept in making decisions.

Properties of the expected value and the variance of random variables are presented in this chapter and followed by the different methods to calculate these two concepts for discrete as well as continuous random variables.

**Properties of the expected value of a random variable:**

a. $E(C) = C$

b. $E(X+C) = E(X) + C$   $\left.\right\}$ $C$ = constant

c. $E(CX) = CE(X)$

d. $E(X+Y+Z) = E(X) + E(Y) + E(Z)$

e. $E(X \cdot Y) = \sum_X \sum_Y (X \cdot Y) f(X,Y)$

$\qquad = E(X)E(Y) + E[(X - E(X))(Y - E(Y))]$

**Properties of the variance of a random variable:**

a. $Var(C) = 0$

b. $Var(X+C) = Var(X)$   $\left.\right\}$ $C$ = constant

c. $Var(CX) = C^2 Var(X)$

d. $Var(X+Y) = Var(X) + 2 Cov(X,Y) + Var(Y)$

$\qquad = Var(X) + 2E[(X-E(X))(Y-E(Y))] + Var(Y)$

e. If $X_1, X_2, \ldots X_n$ are independent random variables, then

$\qquad Var(X_1 + X_2 + \ldots + X_n) = Var(X_1) + Var(X_2) + \ldots + Var(X_n)$

**Expected value and variance of discrete random variables:**

Let X be a discrete random variable, the expected value of X is:

$$E(X) = \sum_{i=1}^{n} X_i f(X_i)$$

and the variance of X is:

$$Var(X) = E[(X - E(X)]^2$$
$$= \sum_{x}[X - E(X)]^2 f(X)$$
$$= E(X^2) - [E(X)]^2$$

where $E(X^2) = \sum X^2 f(X)$

The expected value of X is the sum of the variable X weighted by its probabilities, and the variance of X is equal to the mean square of X, $E(X^2)$, minus its square mean, $[E(X)]^2$. The standard deviation of X is the square root of the variance.

*Example 5.1:* Find the E(X), Var(X), and the standard deviation of X, where X is a discrete random variable of the possible outcomes of rolling two dice 36 times; X and f(X) are as follows:

**Solution:**

| X | f(X) | Xf(X) | $X^2$ | $X^2f(X)$ | $[X - E(X)]^2f(X)$ | |
|---|------|-------|-------|-----------|------------------|---|
| 2 | 1/36 | 2/36 | 4 | 4/36 | $(2 - 7)^2(1/36) =$ | 25/36 |
| 3 | 2/36 | 6/36 | 9 | 18/36 | $(3 - 7)^2(2/36) =$ | 32/36 |
| 4 | 3/36 | 12/36 | 16 | 48/36 | $(4 - 7)^2(3/36) =$ | 27/36 |
| 5 | 4/36 | 20/36 | 25 | 100/36 | = | 16/36 |
| 6 | 5/36 | 30/26 | 36 | 180/36 | = | 5/36 |
| 7 | 6/36 | 42/36 | 49 | 294/36 | = | 0 |
| 8 | 5/36 | 40/36 | 64 | 320/36 | = | 5/36 |
| 9 | 4/36 | 36/36 | 81 | 324/36 | = | 16/36 |
| 10 | 3/36 | 30/36 | 100 | 300/36 | = | 27/36 |
| 11 | 2/36 | 22/36 | 121 | 242/36 | = | 32/36 |
| 12 | 1/36 | 12/36 | 144 | 144/36 | = | 25/36 |
| | | 252/36 | | 1974/36 | | 210/36 |
| | | $\sum Xf(X)$ | | $\sum X^2f(X)$ | | Var(X) |
| | | E(X) = 7 | | $E(X^2) = 54.83$ | | |

$$Var(X) = E(X)^2 - [E(X)]^2$$
$$= 54.83 - 7^2$$
$$= 54.83 - 49 = 5.83$$

standard deviation $(X) = \sqrt{5.83} = 2.41$

In the case of having two probability distributions, X and Y, the expected value of the joint probability distribution, XY, is:

$$E(XY) = E(X)E(Y) + E[(X - E(X))(Y - E(Y))]$$

*Example 5.2:* This example is based on the joint probability function defined in Example 4.3, and its two marginal probability functions derived in Example 4.4. 4.3(B):

$$f(X,Y) = \frac{X + Y + 1}{42} \qquad ; \quad \begin{array}{l} X = 0,1,2,3 \\ Y = 0,1,2 \end{array}$$

$$f(X) = \frac{3X + 6}{42} \qquad ; \quad X = 0,1,2,3$$

$$f(Y) = \frac{4Y + 10}{42} \qquad ; \quad Y = 0,1,2$$

Find the E(XY).

**Solution:** To solve for E(XY), one needs to calculate E(X), E(Y), and E[(X − E(X))(Y − E(Y))].

$$E(X) = \Sigma Xf(X) = \sum_{x=0}^{3} X\left(\frac{3X + 6}{42}\right)$$

$$= 0 + 1\left(\frac{3 + 6}{42}\right) + 2\left(\frac{6 + 6}{42}\right) + 3\left(\frac{9 + 6}{42}\right)$$

$$= 0 + 9/42 + 24/42 + 45/42$$

$$= 78/42 = 1.8571$$

$$E(Y) = \Sigma Yf(Y) = \sum_{Y=0}^{2} Y\left(\frac{4Y + 10}{42}\right)$$

$$= 0 + 1\left(\frac{4 + 10}{42}\right) + 2\left(\frac{8 + 10}{42}\right)$$

$$= 0 + 14/42 + 36/42$$

$$= 50/42 = 1.1905$$

$E[(X − E(X))(Y − E(Y))] = -0.0681$ as calculated in Table 5-1.

The E(XY) can be calculated from f(X,Y) shown in Table 4-5. The formula to be used is:

$$E(XY) = \underset{x\ y}{\Sigma\Sigma} \ (XY)f(X,Y)$$

Table 5-2 shows the steps to find E(XY).

$$E(XY) = 90/42 = 2.1429$$

The variance of the sum of the two random variables X and Y can be calculated by the following formula:

$$Var(X + Y) = Var(X) + Var(Y) + 2Cov(X,Y)$$

**TABLE 5-1**

| | (1)<br>X − E(X) | | (2)<br>Y − E(Y) | | (3)<br>f(X,Y) | (4) = (1) × (2) × (3)<br>(X − E(X))(Y − E(Y))·f(X,Y)<br>(−1.8571)(−1.1905)(.0238) |
|---|---|---|---|---|---|---|
| 0 − 1.8571 = −1.8571 | | 0 − 1.1905 = − 1.1905 | | 1/42 = .0238 | = .0526 |
| 1 − 1.8571 = − .8571 | | 0 − 1.1905 = − 1.1905 | | 2/42 = .0476 | = .0486 |
| 2 = .1429 | | = − 1.1905 | | 3/42 = .0714 | = −.0122 |
| 3 = 1.1429 | | = − 1.1905 | | 4/42 = .0952 | = −.1296 |
| 0 = −1.8571 | | 1 = − .1905 | | 2/42 = .0476 | = .0168 |
| 1 = − .8571 | | 1 = − .1905 | | 3/42 = .0714 | = .0117 |
| 2 = .1429 | | 1 = − .1905 | | 4/42 = .0952 | = −.0026 |
| 3 = 1.1429 | | 1 = − .1905 | | 5/42 = .1190 | = −.0259 |
| 0 = −1.8571 | | 2 = .8095 | | 3/42 = .0714 | = −.1074 |
| 1 = − .8571 | | 2 = .8095 | | 4/42 = .0952 | = −.0661 |
| 2 = .1429 | | 2 = .8095 | | 5/42 = .1190 | = .0138 |
| 3 = 1.1429 | | 2 = .8095 | | 6/42 = .1429 | = .1322 |
| | | | | | −0.0681 |

$$E(XY) = (1.8571)(1.1905) + (-0.0681)$$
$$= 2.2109 - 0.0681$$
$$= 2.1428$$

TABLE 5-2

| $\diagdown$ X<br>Y $\diagdown$ | 0 | 1 | 2 | 3 | $\diagdown$ X<br>Y $\diagdown$ | 0 | 1 | 2 | 3 |
|---|---|---|---|---|---|---|---|---|---|
| 0 | 1/42 | 2/42 | 3/42 | 4/42 | 0 | 0 | 0 | 0 | 0 |
| 1 | 2/42 | 3/42 | 4/42 | 5/42 | 1 | 0 | 3/42 | 8/42 | 15/42 | 26/42 |
| 2 | 3/42 | 4/42 | 5/42 | 6/42 | 2 | 0 | 8/42 | 20/42 | 36/42 | 64/42 |
| | | | | | | 0 | 11/42 | 28/42 | 51/42 | 90/42 |

$$\underset{X\,Y}{\sum\sum}\,(XY)f(X,Y)$$

The variance of $(X + Y)$ of Example 5.2 is calculated from the functions defined the two random variables:

$$f(X) = \frac{3X + 6}{42} \qquad\qquad X = 0,1,2,3$$

and $\quad f(Y) = \dfrac{4Y + 10}{42} \qquad\qquad Y = 0,1,2$

$Var(X) = E(X^2) - [E(X)]^2$, and the
$Var(Y) = E(Y^2) - [E(Y)]^2$.

The values of $E(X)$, and $E(Y)$ are already available in the solution of Example 5.2.

$$E(X^2) = \Sigma X^2 f(X) \qquad\qquad\qquad E(Y^2) = \Sigma Y^2 f(X)$$

$$= \sum_{x=0}^{3} X^2\left(\frac{3X + 6}{42}\right) \qquad\qquad = \sum_{y=0}^{2} Y^2\left(\frac{4Y + 10}{42}\right)$$

$$= 0 + 1(9/42) + 4(12/42) + 9(15/42) \qquad = 0 + 1(14/42) + 4(18/42)$$

$$= 4.5714 \qquad\qquad\qquad\qquad\qquad = 2.0476$$

$$Var(X) = E(X^2) - [E(X)]^2 \qquad Var(Y) = E(Y^2) - [E(Y)]^2$$

$$= 4.5714 - (1.8571)^2 \qquad\qquad = 2.0476 - (1.1905)^2$$

$$= 4.5714 - 3.4488 \qquad\qquad\quad = 2.0476 - 1.4173$$

$$= 1.1226 \qquad\qquad\qquad\qquad = 0.6303$$

$$Cov(X,Y) \quad = E[(X - E(X))\,(Y - E(Y))]$$

$$= -0.0681$$

$$Var(X + Y) = Var(X) + Var(Y) + 2Cov(X,Y)$$

$$= 1.1226 + 0.6303 + 2(-0.0681)$$

$$= 1.6167$$

**Expected value and variance of continuous random variables**

Let X be a continuous random variable; the expected value of X is:

$$E(X) = \int\limits_{\text{all } x} Xf(X)dx$$

and the variance of X is:

$$
\begin{aligned}
Var(X) &= \int [X - E(X)]^2 f(X)dx \\
&= \int X^2 f(X)dx - [\int Xf(X)dx]^2 \\
&= E(X^2) - [E(X)]^2
\end{aligned}
$$

*Example 5.3:*  Find the E(X), and the Var(X) of the following probability function:

$$
\begin{aligned}
f(X) &= 0.1\,(X + 1) & 0 \leqslant X \leqslant 1 \\
&= 0.4\,(X - 1/2) & 1 \leqslant X \leqslant 2 \\
&= 0.3\,(3 - X) & 2 \leqslant X \leqslant 3 \\
&= 0.2\,(4 - X) & 3 \leqslant X \leqslant 4 \\
&= 0.1 & 4 \leqslant X \leqslant 6
\end{aligned}
$$

**Solution:**

$$E(X) = \int Xf(X)dx$$

$$= \int\limits_{0}^{1} X[0.1(X + 1)]dx + \int\limits_{1}^{2} X[0.4(X - 1/2)]dx + \int\limits_{2}^{3} X[0.3(3 - X)]dx +$$

$$\int\limits_{3}^{4} X[0.2(4 - X)]dx + \int\limits_{4}^{6} X(0.1)dx$$

$$= \int\limits_{0}^{1} 0.1(X^2 + X)dx + \int\limits_{1}^{2} 0.4(X^2 - 1/2X)dx + \int\limits_{2}^{3} 0.3(3X - X^2)dx +$$

$$\int\limits_{3}^{4} 0.2(4X - X^2)dx + \int\limits_{4}^{6} (0.1X)dx$$

$$= 0.083 + 0.633 + 0.351 + 0.333 + 1.0$$

$$= 2.40$$

$$\int\limits_{0}^{1} 0.1(X^2 + X)dx \quad = 0.1\left(\frac{X^3}{3} + \frac{X^2}{2}\right)\Big|_{0}^{1} = 0.1(1/3 + 1/2) = 0.083$$

$$\int\limits_{1}^{2} 0.4(X^2 - 1/2X)dx = 0.4\left(\frac{X^3}{3} - \frac{X^2}{4}\right)\Big|_{1}^{2}$$

$$= 0.4[(8/3 - 4/4) - (1/3 - 1/4)] = 0.633$$

$$\int\limits_{2}^{3} 0.3(3X - X^2)dx \quad = 0.3\left(\frac{3X^2}{2} - \frac{X^3}{3}\right)\Big|_{2}^{3}$$

$$= 0.3[(27/2 - 27/3) - (12/2 - 8/3)] = 0.351$$

$$\int_3^4 0.2(4X - X^2)dx = 0.2 \left( \frac{4X^2}{2} - \frac{X^3}{3} \right) \Big|_3^4$$

$$= 0.2[(64/2 - 64/3) - (36/2 - 27/3)] = 0.333$$

$$\int_4^6 (0.1X)dx = \frac{0.1X^2}{2} \Big|_4^6 = (1.8 - 0.8) = 1.0$$

$$Var(X) = E(X^2) - [E(X)]^2$$

To use this formula to calculate the variance of X, one needs to find the value of $E(X^2)$ because the value of $E(X)$ is already calculated and equals 2.40.

$$E(X^2) = \int X^2 f(X)dx$$

$$= \int_0^1 X^2[0.1(X + 1)]dx + \int_1^2 X^2[0.4(X - 1/2)]dx + \int_2^3 X^2[0.3(3 - X)]dx +$$

$$\int_3^4 X^2[0.2(4 - X)dx] + \int_4^6 X^2(0.1)dx$$

$$= \int_0^1 0.1(X^3 + X^2)dx + \int_1^2 0.4(X^3 - 1/2X^2)dx + \int_2^3 0.3(3X^2 - X^3)dx +$$

$$\int_3^4 0.2(4X^2 - X^3)dx + \int_4^6 0.1(X^2)dx$$

$$= 0.058 + 1.034 + 0.825 + 1.117 + 5.066$$

$$= 8.10$$

$$Var(X) = 8.10 - (2.4)^2$$

$$= 8.10 - 5.76 = 2.34$$

$$\int_0^1 0.1(X^3 + X^2)dx = 0.1 \left( \frac{X^4}{4} + \frac{X^3}{3} \right) \Big|_0^1 = 0.1(1/4 + 1/3) = .058$$

$$\int_1^2 0.4(X^3 - 1/2X^2)dx = 0.4 \left( \frac{X^4}{4} - \frac{X^3}{6} \right) \Big|_1^2$$

$$= 0.4[(4 - 4/3) - (1/4 - 1/6)] = 1.034$$

$$\int_2^3 0.3(3X^2 - X^3)dx = 0.3 \left( X^3 - \frac{X^4}{4} \right) \Big|_2^3$$

$$= 0.3[(27 - 81/4) - (8 - 4)] = 0.825$$

$$\int_3^4 0.2(4X^2 - X^3)dx = 0.2 \left( \frac{4X^3}{3} - \frac{X^4}{4} \right) \Big|_3^4$$

$$= 0.2[(256/3 - 64) - (36 - 81/4)] = 1.117$$

$$\int_4^6 0.1(X^2)dx = 0.1 \left( \frac{X^3}{3} \right) \Big|_4^6 = 0.1 \left( \frac{216 - 64}{3} \right) = 5.066$$

The expected value of a continuous joint probability of X and Y is:

$$E(XY) = \iint_{XY} XY \, f(X,Y)dy \, dx$$

*Example 5.4:* From Example 4.7, X and Y are continuous random variables; their joint probability function is:

$$\int_{1}^{4}\int_{0}^{3} 1/18(X + Y - 2)dy \, dx$$

Calculate the E(XY).
**Solution:**

$$E(XY) = \int_{1}^{4}\int_{0}^{3} 1/18[XY(X + Y - 2)]dy \, dx$$

$$= \int_{1}^{4}\int_{0}^{3} 1/18(X^2Y + XY^2 - 2XY)dy \, dx$$

$$= \int_{1}^{4} [1/18(1/2X^2Y^2 + 1/3XY^3 - XY^2)\Big|_{0}^{3}]dx$$

$$= \int_{1}^{4} [1/18(9/2X^2 + 9X - 9X) - 0]dx$$

$$= \int_{1}^{4} (1/18(9/2X^2)dx$$

$$= \int_{1}^{4} 1/4X^2dx$$

$$= 1/12X^3 \Big|_{1}^{4}$$

$$= 1/12(64 - 1) = 63/12 = 5.25$$

The expected value concept is vital to decision theory. The course of action that yields the highest expected pay off or the lowest expected opportunity loss will be followed. The expected pay off values as well as the opportunity loss values are different terms for the expected value of random variables. Decision theory will be examined in later chapters in this book.


EXERCISES

5.1 a. For the probability function of Exercise 4.3, calculate: E(X), Var(X), and $\sigma(X)$.
   b. Find out the expected value, the variance, and the standard deviation for the probability function in Exercise 4.4.

5.2 For the p.d.f. in Exercise 4.7 calculate: E(X), Var(X), and $\sigma(X)$.

5.3 For the joint probability in Exercise 4.10 calculate: E(XY), Var(X + Y).

5.4 For the joint probability in Exercise 4.12 calculate: E(XY).

# CHAPTER

# 6

# Probability Distributions

A probability function that depicts the relationship between observed numerical values of a random variable and their probabilities is defined as a special mathematical function satisfying specific properties. Probability functions of both discrete and continuous random variables, their mathematical expectations, and their variances presented in the previous two chapters can be applied to any probability function. However, there are specific probability functions that play a vital role in applied statistics. These functions, known as probability distributions, are the topic of this chapter. A probability distribution is a probability law which the outcome of a random phenomenon may obey. It is used to calculate the probability of any event that is a part of the random phenomenon. For example, tossing a coin n times is a random phenomenon obeying a probability law that allows a probability distribution (binomial distribution) to be devised in order to calculate the probabilities of the outcome of the n trials, e.g., probability of 0, 1, 2 . . . , n heads (or tails) to occur.

Probability distributions are either discrete or continuous depending on the type of random variables they describe. Discrete probability distributions and continuous probability distributions are presented in this chapter.

A. Discrete Probability Distributions:

Three probability distributions representing discrete random variables are considered. These distributions are:

1. Bionomial
2. Hypergeometric
3. Poisson

Probability function, expected value, and variance of these probability distributions are presented below.

## 1. *Binomial Probability Distribution:*

A random phenomenon that has only two outcomes such as true or false, success or failure, heads or tails, . . . , etc., can be described by a binomial probability distribution. The binomial distribution is widely used in business, the social sciences and other areas where statistical applications involve phenomenon that have only two complementary non-numerical outcomes such as: defective or nondefective, yes or no, male or female, like or do not like, . . . etc. The binomial distribution is based on the Bernoulli[1] probability law:

---

(1) Named after the Swedish mathematician, James Bernoulli (1645–1705).

$$P(X) = p \qquad \text{for } X = 1 \quad \text{(refers to the success of the outcome)}$$
$$\phantom{P(X)} = 1 - p = q \qquad X = 0 \quad \text{(refers to the failure of the outcome)}$$
$$\phantom{P(X)} = 0 \qquad\qquad \text{elsewhere}$$

where p is the probability of success and $1 - p$ or q stands for the probability of failure. The Bernoulli probability law is devised from a sequence of Bernoulli trials where each trial is independent and results in one of two possible mutually exclusive outcomes: success or failure.

If the independent trials are repeated n times, the number of successes X are described to be independent; and therefore, the probability of success p remains constant from trial to trial. Such random phenomenon obeys a binomial probability function:

$$f(X) = {}_nC_x p^x (1 - p)^{n-x}$$
$$\text{or} \qquad f(X) = {}_nC_x p^x q^{n-x} \qquad\qquad x = 0,1,2, \ldots , n$$
$$\phantom{f(X)} = 0 \qquad\qquad\qquad \text{elsewhere}$$

where  n = number of trials

x = number of successes

p = probability of success

q = probability of failure

*Example 6.1:*   Toss a fair coin twice. Find the probability of: 0,1, and 2 heads to occur.

**Solution:**

number of trials $= n = 2$

number of successes for a head to occur $= x = 0,1,2$

probability of success for a head to occur $= p = 1/2$

The possible outcomes of tossing a fair coin twice are: TT, TH, HT, HH. The probabilities of 0,1, and 2 heads to occur can be calculated as follows:

| Possible outcome | x | | | P(x) | |
|---|---|---|---|---|---|
| TT | 0 | $q \cdot q$ | $= q^2$ | $(1/2 \cdot 1/2)$ | $= 1/4$   $P(x = 0)$ |
| TH⎫<br>HT⎭ | 1 | $\begin{cases} q \cdot p \\ p \cdot q \end{cases} = 2qp$ | | $2(1/2 \cdot 1/2)$ | $= 1/2$   $P(x = 1)$ |
| HH | 2 | $p \cdot p$ | $= p^2$ | $(1/2 \cdot 1/2)$ | $= 1/4$   $P(x = 2)$ |
| | | $q^2 + 2qp + p^2$ | $= (q + p)^2$ | | $= (1/2 + 1/2)^2 = 1$ |

The same results can be found by applying the Binomial distribution formula:

$$f(x) \qquad = {}_nC_x p^x q^{n-x} \qquad\qquad x = 0,1,2$$
$$P(x = 0) = {}_2C_0 \, (1/2)^0 (1/2)^{2-0} = 1/4$$
$$P(x = 1) = {}_2C_1 \, (1/2)^1 (1/2)^{2-1} = 1/2$$
$$P(x = 2) = {}_2C_2 \, (1/2)^2 (1/2)^{2-2} = 1/4$$
$$\Sigma_n C_x p^x q^{n-x} \qquad\qquad = 1 = (p + q)^n$$

One can utilize the Binomial distribution table in the Appendix section of this book to arrive at the probabilities calculated above. In using the tables one needs to know the value of n and p to locate the probabilities of x = 0,1,2, . . . , n. The maximum value of p used in the table is 0.5. This may cause a problem if the value of p is greater than 0.5. In this situation use the magnitude of $(1 - p)$ which should be less than 0.5. The only change is to switch the values of x such that the value of x = 0 for p (p is greater than 0.5) will equal the value of x = n for $(1 - p)$. The following is an example for this situation:

*Example 6.2:* A couple is planning to have 4 children. It has been determined that the probability of having a boy for this couple is 0.52. Assuming no multiple births occur, what is the probability of having—0,1,2,3, or 4 boys in this family?

**Solution:** Birth is a random phenomenon that has only two outcomes—a boy or a girl. The binomial distribution can be used to calculate the desired probabilities.

n = 4

p = Probability of success to have a boy = 0.52

x = number of successes to have a boy = 0,1,2,3,4

We cannot use the binomial distribution table because the probability of success, p, is greater than 0.50. One has to switch the problem by setting p to denote the probability of success to have a girl. The new p will assume a value of $(1 - 0.52)$ or 0.48. To use the binomial table with the new magnitude of p (p = 0.48), one should switch the values assigned to x from x = 0,1,2,3,4 to x = 4,3,2,1,0. For example, if the family is planning to have four children, then the probability of success to have no boys, P (x = 0) where p = 0.52, is the same as the probability of success to have four girls, P (x = 4) where p = 0.48. Also, the probability of success to have one boy and three girls or P (x = 1) where p = 0.52 is the same as the probability of success to have three girls and one boy or P (x = 3) where p = 0.48, and so on. Accordingly, the probabilities derived from the binomial distribution table for this problem are as follows: (use n = 4 and p = 0.48)

P (success to have no boys and four girls) = P (x = 4) = 0.0531

P (success to have one boy and three girls) = P (x = 3) = 0.2300

P (success to have two boys and two girls) = P (x = 2) = 0.3738

P (success to have three boys and one girl) = P (x = 1) = 0.2700

P (success to have four boys and no girls) = P (x = 0) = 0.0731

The probabilities calculated for this example can be depicted in a diagram, (Figure 6-1) where the horizontal axis represents the number of successes of having a girl and the vertical axis shows the probability values assigned to each value of x.

A cumulative binomial distribution F(x) provides the probability of those values of the random variable less than or equal to a specific value of x. For example, in Table 6-1 which presents the cumulative binomial distribution for example 6.2, F(x = 2) refers to the probability of having two or less girls in the family, and F(x = 3) indicates the probability of having no more than 3 girls in the family and so on.

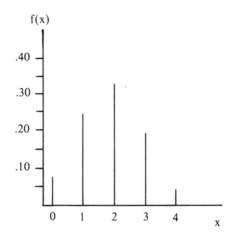

**Figure 6-1.** Binomial distribution.
(n = 4 and p = 0.48)

**TABLE 6-1**

| x | P(x) or f(x) | F(x) |
|---|---|---|
| 0 | 0.0731 | 0.0731 |
| 1 | 0.2700 | 0.3431 |
| 2 | 0.3738 | 0.7169 |
| 3 | 0.2300 | 0.9469 |
| 4 | 0.0531 | 1.0000 |

A graphic presentation of this cumulative binomial distribution is shown in Figure 6-2.

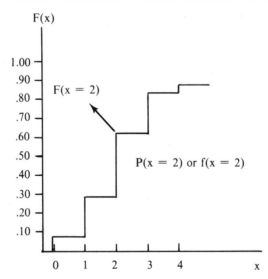

**Figure 6-2.** Cumulative binomial distribution.
(n = 4 and p = 0.48)

**Expected value of the binomial distribution:**

$$E(X) = \sum_{x=0}^{n} Xf(X)$$

$$= \sum_{x=0}^{n} X \binom{n}{x} p^x q^{n-x}$$

$$= \sum_{x=0}^{n} X \frac{n!}{x!(n-x)!} p^x q^{n-x}$$

$$= \sum_{x=1}^{n} \frac{n(n-1)!}{(x-1)!(n-x)!} p(p^{x-1} q^{n-x})$$

$$= \sum_{x=1}^{n} np \frac{(n-1)!}{(x-1)!(n-x)!} p^{x-1} q^{n-x}$$

$$= np \sum_{x=1}^{n} \binom{n-1}{x-1} p^{x-1} q^{n-x}$$

$$= np(p+q)^{n-1}$$

$$= np(1)^{n-1}$$

$$= np$$

**Variance of the binomial distribution:**

$$\sigma^2(X) = E(X^2) - [E(X)]^2$$

$$E(X^2) = \sum_{x=0}^{n} X^2 f(X)$$

$$= \sum_{x=0}^{n} X^2 \frac{n!}{x!(n-x)!} p^x q^{n-x}$$

Let $\quad X^2 = x(x-1) + x$

Then $\quad E(X^2) = \sum_{x=2}^{n} X(x-1)\frac{n!}{x!(n-x)!} p^x q^{n-x} + \sum_{x=0}^{n} X \,_nC_x p^x q^{n-x}$

$$= n(n-1)p^2 \Sigma \binom{n-2}{x-2} p^{x-2} q^{n-x} + E(X)$$

$$= n(n-1)p^2(p+q)^{n-2} + np$$

$$= n(n-1)p^2 + np$$

$$= n^2 p^2 - np^2 + np$$

$$= n^2 p^2 + np(1-p)$$

$$= n^2 p^2 + npq$$

Then $\quad \sigma^2(X) = E(X^2) - [E(X)]^2$

$$= n^2 p^2 + npq - (np)^2$$

$$= n^2 p^2 + npq - n^2 p^2$$

$$= npq$$

*Example 6.3:* From past experience it is known that machine A produces units that are consistently 5% defective. A sample of size 20 is inspected. What is the probability that:

    a. none to be defective
    b. at least one to be defective
    c. not more than two to be defective.

**Solution:** $n = 20$   $p =$ the probability of success to find defectives $= .05$

    a. $P(\text{none defectives}) = P(X = 0) = {}_{20}C_0(.05)^0(.95)^{20}$

    b. $P(\text{at least one defective}) = P(X \geqslant 1)$

$$= P(X = 1) + P(X = 2) + \ldots + P(X = 20)$$

$$\text{or} \quad = 1 - P(X = 0)$$

$$= 1 - {}_{20}C_0(.05)^0(.95)^{20}$$

    c. $P(\text{not more than two defectives}) = P(X \leqslant 2)$

$$= P(X = 0) + P(X = 1) + P(X = 2)$$

$$= {}_{20}C_0(.05)^0(.95)^{20} + {}_{20}C_1(.05)^1(.95)^{19} + {}_{20}C_2(.05)^2(.95)^{18}$$

*Example 6.4:* Find the expected value, the variance, and the standard deviation for the data presented in Example 6.3.

    **Solution:**

        Expected Value of a binomial variable $= np$

        Variance of a binomial variable $= npq$

        Standard deviation of a binomial variable $= \sqrt{npq}$

        $E(X) \quad = np \quad = (20)(.05) = 1$

        $\text{Var}(X) = npq \quad = (20)(.05)(.95) = .95$

        $\sigma(X) \quad = \sqrt{npq} = \sqrt{.95} = .975$

The expected value $(E(X) = 1)$ indicates that samples of size 20 selected from the production of machine A yields on the average one defective item.

### 2. Hypergeometric Probability Distribution:

In the binomial distribution, the drawings are with replacement. This is consistent with the assumption of independent trials and having infinite population or sample space.

If the sample space is finite, and drawings are without replacement, then the trials are not independent. Accordingly, the probability of success (p) will change from one trial to another.

*Example 6.5:* A box contains 10 balls, 6 white and 4 blue. Two balls are selected from the box simultaneously. What is the probability that:

    a. the two balls are white.
    b. the two balls are different colors.

**Solution:**

A. With Replacement:

   a. P(two white balls) = (0.6)(0.6)
$$= {}_2C_2(0.6)^2(0.4)^0$$

   b. P(one white and one blue) = [(0.6)(0.4)] $\times$ 2
$$= {}_2C_1(0.6)^1(0.4)^1$$

   Notice that the probability of success to draw a white ball (p) did not change in the 2 trials.

B. Without Replacement:

   a. P(two white balls) = (6/10)(5/9) = 30/90
   P(the 1st ball to be white) = 6/10
   And it is white.

   $\left|\begin{matrix}6\\4\end{matrix}\right|\left|\begin{matrix}5\\4\end{matrix}\right|$

   P(the 2nd ball to be white) = 5/9      10    9

   Notice that one probability of success to obtain a white ball has changed from (6/10) in the first trial to (5/9) in the second trial.

   b. P(1 white and 1 blue) = (6/10)(4/9) + (4/10)(6/9) $=\dfrac{48}{90}$

   These results can be calculated by using the following hypergeometric probability distribution:

$$f(x) = \frac{\dbinom{Np}{x}\dbinom{Nq}{n-x}}{\dbinom{N}{n}} \quad \text{for } x = 0,1,2,\ldots,n$$

$$= 0 \qquad \text{elsewhere}$$

where $Np + Nq = N$

*Example 6.6:* Solve Example 6.5 by using the hypergeometric distribution.

**Solution:** (without Replacement)

   N = 10       n = 2       Np = number of white balls = 6

                                 Nq = number of non-white balls = 4

   a. P(two balls to be white) $= \dfrac{\dbinom{6}{2}\dbinom{4}{0}}{\dbinom{10}{2}} = \dfrac{15}{45}$

   b. P(one white and one blue) $= \dfrac{\dbinom{6}{1}\dbinom{4}{1}}{\dbinom{10}{2}} = \dfrac{24}{45}$

**Expected value of the hypergeometric distribution:**

$$E(X) = \sum_{x=0}^{n} Xf(X)$$

$$= \sum_{x=0}^{n} X \frac{\dbinom{Np}{x} \dbinom{Nq}{n-x}}{\dbinom{N}{n}}$$

$$= \frac{1}{\dbinom{N}{n}} \sum_{x=1}^{n} X \dbinom{Np}{x} \dbinom{Nq}{n-x}$$

$$= np$$

**Variance of the hypergeometric distribution:**

$$\sigma^2 = E(X^2) - [E(X)]^2$$
$$= E[X(X-1)] + E(X) - [E(X)]^2$$
$$= npq\,\frac{N-n}{N-1}$$

The expected value of the hypergeometric distribution is the same as the expected value of the binomial distribution ($\mu = np$), while the variance differs:

$$\sigma^2 \text{ (binomial)} = npq$$

$$\sigma^2 \text{ (hypergeometric)} = npq\,\frac{N-n}{N-1}$$

$\dfrac{N-n}{N-1}$ is called the correction factor for finite population, if N, the number of observations in the population, is very large, or as $N \to \infty$, the limit $\dfrac{N-n}{N-1} \to 1$. This means that the variance of the hypergeometric distribution of a very large population sampled without replacement is very close to the variance of the same population sampled with replacement. In general, a binomial distribution can be used as a good approximation to a hypergeometric distribution of a very large population.

### 3. *Poisson Distribution*

Many statistical applications involve events occurring over an interval of time such as numbers of telephone calls that come through a switchboard, number of accidents arriving at an emergency room, numbers of cars arriving at a gas station, numbers of customers going through a supermarket check line, numbers of malfunctions or defects, and demand for a product or service. The occurrences of these events can be described in terms of a discrete random variable that assumes values 0,1,2, and so forth. The Poisson[2] function has been used to provide probabilities for the number of events occurring in a time inter-

---

(2) Named after a French mathematician and physicist, Simeon Poisson (1781–1840).

val. Such probabilities can help the decision maker to design a plan or facility to minimize idle resources and provide high quality service to the customers. For example, a decision maker may need to determine the appropriate number of attendants in a gas station to serve any customer within a reasonable time. Also, the Poisson distribution is a suitable approximation of the binomial distribution when n is large and p is small.[3] There are similarities between the Poisson and the binomial distributions. The binomial distribution provides probabilities for the number of events of a particular success occurring in n independent trials of a random experiment, while the Poisson distribution provides these probabilities for undefined ($\infty$) independent trials. In Poisson distribution the number of events occurring in an interval of time is described as independent of what occurred or will occur in another interval of time.

As stated before, the binomial distribution has its limitation. If $n \to \infty$, and p is small, then the calculation of the P(X) becomes more complicated. For example, if n = 100, and p = 0.05, then:

$$P(X) = {}_{100}C_x(.05)^x(.95)^{100-x} \qquad x = 0,1,2, \ldots ,100$$

Poisson probability distribution is used to approximate a binomial distribution of a large number of trials (large n). The Poisson probability distribution is derived from the binomial probability distribution as follows:

$$f(X) = \lim_{n \to \infty} {}_nC_x p^x (1 - p)^{n-x}$$

if $\qquad np = \lambda \quad$ or $\quad p = \dfrac{\lambda}{n}$

then $\qquad f(X) = \lim_{n \to \infty} {}_nC_x \left(\dfrac{\lambda}{n}\right)^x \left(1 - \dfrac{\lambda}{n}\right)^{n-x}$

$$= e^{-\lambda}\dfrac{\lambda^x}{x!}$$

**Poisson probability function:**

$$f(X) = e^{-\lambda}\dfrac{\lambda^x}{x!} \qquad x = 0,1,2, \ldots$$

$$= 0 \qquad \text{elsewhere}$$

This function is a probability function and satisfies its conditions:

1.  $f(X) \geq 0$ \qquad (probabilities of X are non-negative)
2.  $\sum\limits_{\text{all x}} f(X) = 1$

$$\sum\limits_{\text{all x}} f(X) = \sum_{x=0}^{\infty} e^{-\lambda}\dfrac{\lambda^x}{x!} = e^{-\lambda} \sum_{x=0}^{\infty} \dfrac{\lambda^x}{x!}$$

$\sum\limits_{x=0}^{\infty} \dfrac{\lambda^x}{x!}$ a Maclaurin series which is a special case of Taylor series.

---

(3) As a rule of thumb when $p \leq 0.05$ and $n \geq 20$.

$$= \left[ 1 + \frac{\lambda}{1!} + \frac{\lambda^2}{2!} + \frac{\lambda^3}{3!} + \ldots + \frac{\lambda^n}{n!} \right]$$

$$= e^\lambda$$

then, $\quad \sum_{\text{all x}} f(X) = e^{-\lambda} e^\lambda$

$$= e^{-\lambda + \lambda}$$

$$= e^0$$

$$= 1$$

### Expected value of the Poisson distribution:

$$E(X) = X \Sigma f(X) = \sum_{x=0}^{\infty} X e^{-\lambda} \frac{\lambda^x}{x!}$$

$$= \sum_{x=1}^{\infty} X e^{-\lambda} \frac{\lambda \, \lambda^{x-1}}{x(x-1)!}$$

$$= \lambda e^{-\lambda} \sum_{x=1}^{\infty} \frac{\lambda^{x-1}}{(x-1)!}$$

let $\quad n = x - 1 \qquad = \lambda e^{-\lambda} \sum_{n=0}^{\infty} \frac{\lambda^n}{n!}$

$$= \lambda e^{-\lambda} e^\lambda$$

$$= \lambda$$

One of the features of the expected value is to remain constant for a particular probability function. However, in many statistical applications the expected value of the Poisson distribution $\lambda$ does not stay constant over an extended duration. For example, in many guessing situations rates of random arrivals may change with the time of day, season, or other factors.

### Variance of the Poisson distribution:

$$\sigma^2(X) = E(X^2) - [E(X)]^2$$

$$= \lambda^2 + \lambda - \lambda^2$$

$$= \lambda$$

Expected value of the Poisson distribution = variance = $\lambda$ = np. The standard deviation of the Poisson distribution = $\sqrt{\lambda}$.

*Example 6.7:* Past experience showed the number of defectives produced by Machine A to follow a Poisson Probability Distribution. The percentage of defectives is 2.5%. In a particular shift, the machine produced 200 parts.

a. What is the probability of these conditions?
    1. all produced parts are non-defective.
    2. at least one defective.
    3. not more than two defectives.
b. Find the E(X), Var(X), and $\sigma(X)$.

**94**

**Solution:**

$n = 200$    $p = 0.025$    $\lambda = (200)(0.025) = 5$    $x$ = number of defectives

a.  1.  $P(X = 0) = e^{-5}\dfrac{5^0}{0!} = e^{-5}$

    2.  $P(X \geq 1) = 1 - P(X = 0) = 1 - e^{-5}$

    3.  $P(X \leq 2) = P(X = 0) + P(X = 1) + P(X = 2)$

$$= e^{-5} + e^{-5}\frac{5}{1!} + e^{-5}\frac{5^2}{2!}$$

One may use the Poisson distribution table to find the desired probabilities. The value of $\lambda$ is the only information needed to locate such probabilities. In the previous example where $\lambda = 5$ the probabilities derived from the table are presented in Table 6-1:

**TABLE 6-2**
Poisson Probabilities for $\lambda = 5$

| $X$ | $P(X)$ | |
|-----|--------|---|
| 0 | 0.0067 | $P(X = 0)$ |
| 1 | 0.0337 | |
| 2 | 0.0842 | $P(X \leq 2)$ |
| 3 | 0.1404 | |
| 4 | 0.1755 | |
| 5 | 0.1755 | |
| 6 | 0.1462 | |
| 7 | 0.1044 | |
| 8 | 0.0653 | |
| 9 | 0.0363 | $P(X \geq 1)$ |
| 10 | 0.0181 | |
| 11 | 0.0082 | |
| 12 | 0.0034 | |
| 13 | 0.0013 | |
| 14 | 0.0005 | |
| 15 | 0.0002 | |

a.  1.  P(all produced parts to be non-defective) $= P(X = 0) = 0.0067$

    2.  P(at least one defective) $= P(X = 1) + P(X = 2) + \ldots + P(X = 15)$

$$= .0337 + \ldots + 0.0002$$

or    $= 1 - P(X = 0)$

$$= 1 - 0.0067 = 0.9933$$

    3.  P(not more than two defectives) $= P(X = 0) + P(X = 1) + P(X = 2)$

$$= 0.0067 + 0.0337 + 0.0842 = 0.1246$$

b.  $E(X) = Var(X) = \lambda = 5$

    $\sigma(X) = \sqrt{5}$

## B. Continuous Probability Distributions:

There are many continuous probability distributions. However, only the following are considered:

1. Exponential probability distribution.
2. Normal probability distribution.
3. Sampling probability distribution.

### 1. *Exponential Probability Distribution:*

The exponential probability distribution has an important role in business. It can be used to describe random phenomena that are spread over intervals of time; e.g., the time intervals between accidents. Also, it describes the lengths of waiting time. These components of the waiting theory are used by management and operation researchers to make decisions about the maximum number of beds in a hospital, number of barbers in a barbershop, number of telephone operators or lines in a particular area, and so on.

The exponential probability distribution is a special case of the gamma probability distribution with $\alpha = 1$.[4] The exponential distribution, sometimes called the negative exponential distribution, can be derived from the Poisson probability distribution.

**Probability Density Function:**

$$f(x) = \frac{1}{\beta} e^{-x/\beta} \qquad 0 < x < \infty$$

$$= 0 \qquad \text{elsewhere}$$

where $1/\beta$ is the average number of successes per interval equals to $\lambda$, the mean of the Poisson distribution.

A graphic presentation of the exponential probability distribution is shown in Figure 6.3.

Expected value and variance for the exponential probability distribution are:

$$E(X) = \beta$$
$$Var(X) = \beta^2$$
$$\sigma(X) = \sqrt{\beta^2}$$
$$= \beta$$

---

(4) Gamma probability distribution is:

$$f(x) = c\, x^{\alpha-1}\, c^{-x/\beta} \qquad \text{for } x > 0$$
$$= 0 \qquad \text{for } x \leq 0$$

where $\alpha > 0$, $\beta > 0$

and C is a constant equal to the value that makes the $\int_{-\infty}^{\infty} f(x)dx = 1$

C is defined as $\dfrac{1}{\Gamma(\alpha)\beta^{\alpha}}$ for $0 < x < \infty$

**96**

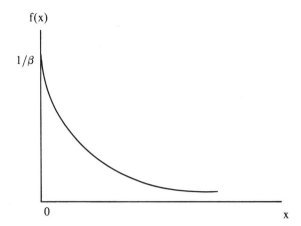

**Figure 6-3.** Exponential probability distribution.

The following is an example to show the calculation of probabilities for a phenomenon that follows the exponential probability function:

*Example 6.8:* In an emergency room of a large hospital the operators receive an average of two phone calls per minute. From past experience it has been established that the number of phone calls per minute follows a Poisson distribution.

Find the probability of:

a. No calls within the next two minutes following the previous call.

b. No calls within the next one minute.

c. The length of time until the first call received is one-half minute.

**Solution:** We are dealing with a continuous variable X, the length of time. So the Poisson distribution should be approximated by a continuous function, the Exponential distribution:

$$f(x) = \frac{1}{\beta}e^{-x/\beta} \qquad 0 < x < \infty$$

$$= 0 \qquad \text{elsewhere}$$

$$1/\beta = \lambda = 2 \quad \text{(average of 2 calls/minute)}$$

a. P(no calls within the next 2 minutes) = $P(2 < X < \infty)$

$P(2 < X < \infty) = 1 - P(0 < X < 2)$
which represents the shaded area in
the diagram

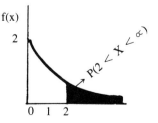

$$P(0 < X < 2) = \int_0^2 2e^{-2x}\,dx$$

$$= -e^{-2x}\Big|_0^2$$

$$= 1 - e^{-4} = .9817$$

$$P(2 < X < \infty) = 1 - .9817 = .0183$$

**97**

b. P(no calls within one minute) = P(1 < X < ∝)

$$P(1 < X < \propto) = 1 - P(0 < X < 1)$$

$$P(0 < X < 1) = \int_0^1 2e^{-2x}\, dx$$

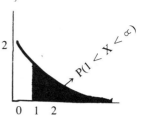

$$= -e^{-2x}\Big|_0^1$$

$$= 1 - e^{-2} = .8647$$

$$P(1 < X < \propto) = 1 - .8647 = .1353$$

c. P(the length of time until the first call received is one-half minute)
   = P(0 < X < 1/2)

$$P(0 < X < 1/2) = \int_0^{1/2} 2e^{-2x}\, dx$$

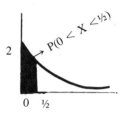

$$= -e^{-2x}\Big|_0^{1/2}$$

$$= 1 - e^{-1}$$

$$= .6321$$

## 2. *Normal Probability Distribution:*

The normal probability distribution plays a vital role in statistical theory and practice. The normal distribution is useful as an approximation to many other probability distributions, especially the binomial. In the area of statistical inference, the normal distribution is extremely useful as a model. Theoretical properties of the sample mean allows the application of the normal distribution in order to calculate probabilities of various sample results. This enables the decision maker to make inferences regarding the population mean when only the sample mean can be calculated directly.

The normal distribution is a bell-shaped curve where the mean divides it into two symmetrical halves. Because the curve is symmetrical, the median and the mode of the distribution fall at the center. A graphical representation of the normal distribution is shown in Figure 6-4.

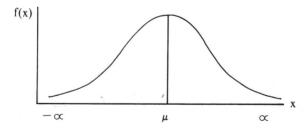

**Figure 6-4.** A normal probability distribution.

$$f(x) = \frac{1}{\sigma\sqrt{2\pi}} e^{-\frac{1}{2}\left(\frac{x-\mu}{\sigma}\right)^2}$$

**Probability density function:**

$$f(x) = \frac{1}{\sigma\sqrt{2\pi}} e^{-1/2\left(\frac{x-\mu}{\sigma}\right)^2} \qquad -\infty < x < \infty$$

*WRONG* $\left(\frac{x-\mu}{\sigma}\right)^2$ *is part of the exponent*

where $\mu$ and $\sigma$ are constants and $\sigma > 0$

The mean of the normal distribution, $\mu$, determines the location of the curve while the standard deviation, $\sigma$, shows the degree of dispersion or variation of the distribution around the mean $\mu$. The location of two normal curves with different means and having the same spread are shown in Figure 6-5.

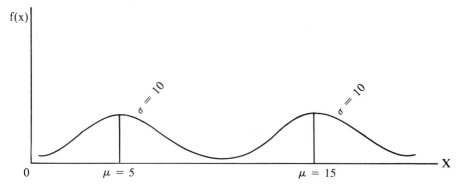

**Figure 6-5.** Two normal curves with different means and the same standard deviation.

The shape of the normal distribution is determined by the magnitude of its standard deviation. The smaller the value of the standard deviation, the smaller the spread and vice versa. This feature is shown for two normal curves with the same mean but having different standard deviations in Figure 6-6.

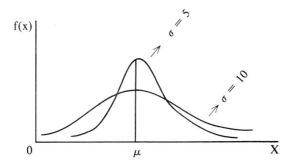

**Figure 6-6.** Two normal curves with different standard deviations.

If X is a random variable that follows a normal distribution, then one needs to know the mean and the standard deviation (or the variance) of X to draw the normal curve that represents the variable X and to find the probability of any interval or area of the function. The normal probability distribution of X is denoted as:

$$n(X; \mu, \sigma^2)$$

If X has a mean = 10, and variance = 4, then: $n(X;10,4)$ means that X is normally distributed or has a normal distribution with $\mu = 10$ and $\sigma^2 = 4$.

The total area under the normal curve which represents the probability of all the elements of the normal variable equals one, or the cumulative probability function:

$$F(X) = \int_{-\alpha}^{\alpha} f(x)\, dx = 1 = \Phi\left(\frac{X - \mu}{\sigma}\right)$$

where $\dfrac{X - \mu}{\sigma}$ is a standard normal distribution Z with $\mu = 0$ and $\sigma^2 = 1$, or $n(Z,0,1)$. A graphical representation of the standard normal distribution Z is shown in Figure 6-7:

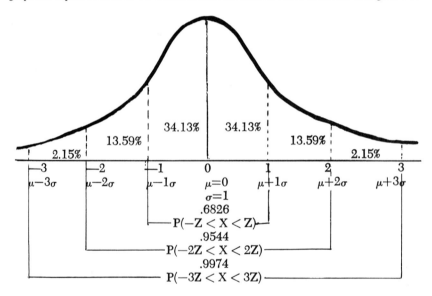

**Figure 6-7.** The standard normal distribution Z.

There are tables to provide the probability of any value of Z, where $Z = \dfrac{X - \mu}{\sigma}$.

If X is a normal distribution with mean = $\mu$ and variance = $\sigma^2$, then to find the probability of X for an interval ab or

$$P(a < x < b) = \int_{-\infty}^{b} f(x)dx - \int_{-\infty}^{a} f(x)dx$$

$$= \Phi\left(\frac{b - \mu}{\sigma}\right) - \Phi\left(\frac{a - \mu}{\sigma}\right)$$

$$= P\left(\frac{a - \mu}{\sigma} < Z < \frac{b - \mu}{\sigma}\right)$$

or find the $P(Z_b)$ and the $P(Z_a)$ and subtract $P(Z_a)$ from $P(Z_b)$ as follows:

$$Z_a = \frac{a - \mu}{\sigma}$$ then from the table find $P(Z_a)$

$$Z_b = \frac{b - \mu}{\sigma}$$ then from the table find $P(Z_b)$

$P(a < X < b) = P(Z_b) - P(Z_a)$ which represents the shaded area in the diagram.

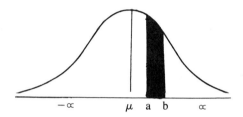

*Example 6.9:* The grades of 400 students in a statistics course are normally distributed with $\mu = 65$ and variance $= 100$, or if we let X represent these grades, then n(X;65,100). Find the probability that a student selected randomly from this group would score within any interval given below:

a. a grade between 60 and 65
b. a grade between 70 and 65
c. a grade between 52 and 68
d. a grade that is greater than 85
e. a grade that is less than 72
f. a grade between 70 and 78

**Solutions:** X is a normal distribution with $\mu = 65$ and $\sigma^2 = 100$. This information helps to draw the normal curve, which is essential to find the above mentioned probabilities, as shown by the shaded areas in the diagrams.

a. P(a grade between 60 and 65) = $P(60 \leqslant X \leqslant 65) = \int_{60}^{65} f(x)\, dx$

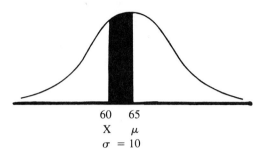

60    65
X     μ
σ = 10

P(60 ≤ X ≤ 65) is P(Z) between 65 and 60

$$Z = \frac{X - \mu}{\sigma}$$

$$= \frac{60 - 65}{10} = -0.5$$

The minus sign has no significance other than indicating that the shaded area is located within the left part of the normal curve.

P(60 ≤ X ≤ 65) = P(Z) = .1915

This result also means that 19.15% of the class scored a grade between 60 and 65.

b.  P(a grade between 70 and 65) = P(65 ≤ X ≤ 70) = $\int_{65}^{70} f(x)\ dx$

$$Z = \frac{X - \mu}{\sigma}$$

$$Z = \frac{70 - 65}{10} = 0.5$$

P(Z) = .1915

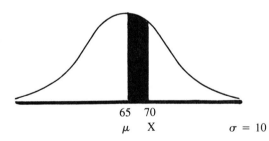

65    70
μ     X                              σ = 10

Then P(65 ≤ X ≤ 70) = .1915, or 19.15% of the class scored a grade between 65 and 70.

c.  P(a grade between 52 and 68) = P(52 ≤ X ≤ 68) = $\int_{52}^{68} f(x)\ dx$

$$P(52 \leqslant X \leqslant 68) = P(Z_1) + P(Z_2)$$

$$Z_1 = \frac{X_1 - \mu}{\sigma}$$

$$= \frac{52 - 65}{10} = -1.3$$

$$P(Z_1) = .0432$$

$$Z_2 = \frac{X - \mu}{\sigma}$$

$$= \frac{68 - 65}{10} = .3$$

$$P(Z_2) = .1179$$

$P(52 \leqslant X \leqslant 68) = .4032 + .1179 = .5211$, or 52.11% of the class scored a grade between 52 and 68.

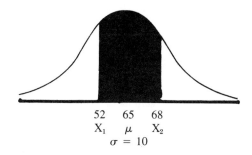

52    65    68
$X_1$    $\mu$    $X_2$
$\sigma = 10$

d.  P(a grade that is greater than 85) $= P(X \geqslant 85) = \int\limits_{85}^{\infty} f(x)\, dx$

$$P(65 \leqslant X \leqslant \infty) = .5$$

$$P(X \geqslant 85) = P(65 \leqslant X \leqslant \infty) - P(65 \leqslant X \leqslant 85)$$

$$P(65 \leqslant X \leqslant 85) = P(Z)$$

$$Z = \frac{X - \mu}{\sigma} = \frac{85 - 65}{10} = 2$$

$$P(Z) = .4772$$

$$P(X \geqslant 85) = .5 - .4772$$

$$= .0228$$

or 2.28% of the class scored a grade greater than 85.

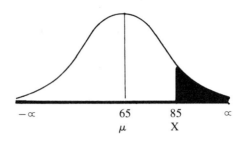

e. P(a grade less than 72) = P(X ⩽ 72) = $\int\limits_{-\infty}^{72}$ f(x) dx

P($-\infty$ ⩽ X ⩽ 65) = .5

P(X ⩽ 72) = P($-\infty$ ⩽ X ⩽ 65) + P(65 ⩽ X ⩽ 72)

P(65 ⩽ X ⩽ 72) = P(Z)

$Z = \dfrac{X - \mu}{\sigma}$

$= \dfrac{72 - 65}{10} = .7$

P(Z) = .2508

P(X ⩽ 72) = .5 + .2508 = .7508 or 75.08% of this class scored a grade less than 72.

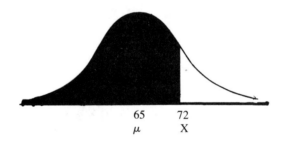

f. P(a grade between 70 and 78) = P(70 ⩽ X ⩽ 78) = $\int\limits_{70}^{78}$ f(x) dx

P(70 ⩽ X ⩽ 78) = P($-\infty$ ⩽ X ⩽ 78) − P($-\infty$ ⩽ X ⩽ 70)

or     = P($\mu$ ⩽ X ⩽ 78) − P($\mu$ ⩽ X ⩽ 70)

= P($Z_1$) − P($Z_2$)

$Z_2 = \dfrac{X_1 - \mu}{\sigma}$

$= \dfrac{78 - 65}{10} = 1.3$

$$P(Z_1) = .4032$$

$$Z_2 = \frac{X_2 - \mu}{\sigma}$$

$$= \frac{70 - 65}{10} = .5$$

$$P(Z_2) = .1915$$

$$P(70 \leqslant X \leqslant 78) = .4032 - .1915$$

$$= .2117 \text{ or } 21.17\% \text{ of the class scored a grade between 70}$$

and 78.

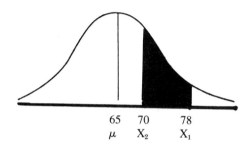

**Normal distribution to approximate binomial distribution:**

Most of the populations, including the binomial populations, can be described by a normal probability distribution. A binomial probability distribution can be approximated by a normal probability function. A binomial distribution with p = 0.50, p refers to the probability of success, and with small number of trials appears to be symmetric and therefore can be approximated by a normal distribution. An example is the binomial distribution presented in Table 6-2 with n = 5 and p = 0.50.

**TABLE 6-3**
Binomial Distribution with n = 5 and p = 0.50

| X | P(X) |
|---|------|
| 0 | 0.0312 |
| 1 | 0.1562 |
| 2 | 0.3125 |
| 3 | 0.3125 |
| 4 | 0.1562 |
| 5 | 0.0312 |

A graphical representation of the binomial distribution presented in Table 6-2 shown in Figure 6-8 appears to be symmetrical.

On the other hand, a normal distribution can be used to approximate a binomial distribution with large n(the number of trials) and small p.

**105**

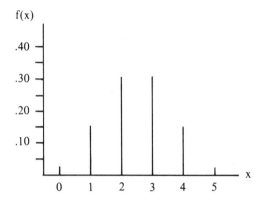

**Figure 6-8.** Binomial distribution with n = 5 and p = 0.50.

A binomial variable X approximated by a normal distribution is denoted as:

$$n(X; np, npq)$$

or X is normally distributed with:

$$\text{mean} = np \qquad \text{and} \qquad \text{variance} = npq$$

The Z value to be used to calculate the probabilities becomes:

$$Z = \frac{X - np}{\sqrt{npq}}$$

where $\sqrt{npq}$ is the standard deviation of the binomial variable.

*Example 6.10:* Let X be a binomial variable with n = 20 and p = 0.50. Find the probability of success of 15 to occur or P(x = 15), then find the approximation of such a probability using the normal distribution.

**Solution:**

1. Using the binomial distribution: (n = 20, p = 0.5, and x = 15)

$$P(x = 15) = 0.0148 \qquad \text{(binomial distribution table)}$$

2. Using the normal distribution:

$$P(x = 15) = \int_{15}^{15} f(x)\, dx = 0 \qquad \text{because the normal distribution is a continuous function.}$$

Therefore, one has to correct for moving from the discrete function (binomial) to the continuous function (normal). This can be done by moving a half unit to the left and a half unit to the right, so P(X = 15) becomes P(14.5 ≤ X ≤ 15.5). Such probability can be evaluated by a normal probability distribution with mean = np = (20) (1/2) = 10 and variance = npq = (20) (1/2) (1/2) = 5.

**106**

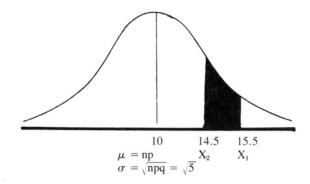

$$P(14.5 \leq X \leq 15.5) = \int_{14.5}^{15.5} f(x)\, dx$$

$$P(14.5 \leq X \leq 15.5) = P\left(\frac{15.5 - 10}{2.236}\right) - P\left(\frac{14.5 - 10}{2.236}\right)$$

$$= P(Z = 2.45) - P(Z = 2.01)$$

$$= 0.4929 - 0.4778$$

$$= 0.0151$$

Compare the results using the binomial (.0148) and using the normal to approximate the binomial (.0151), the difference is very small.

### 3. Sampling Probability Distribution

Characteristics of populations such as the mean and the proportion are of great interest to business decision makers. To find these parameters by using every element of the population is sometimes prohibitive (e.g., with infinite populations). For example, to find the mean annual income per family in the United States would be a time consuming and complicated project if data were collected from each household in the U.S. Not only would the project be costly but the data would be obsolete at the end of the research. Therefore, selection of samples (or a sample) from such populations enables the decision maker to infer the characteristics of a population parameter such as the mean $\mu$ or proportion $\pi$ by computing the corresponding sample statistics $\overline{X}$ and p.

Different samples of size n drawn randomly from a particular population will yield statistics $(\overline{X}$ or p) of different values. The possible samples of size n drawn from the N elements of a population is $\binom{N}{n}$. If a particular statistic, say $\overline{X}$, is computed from each of the possible samples, the value of the statistic is likely to vary from one sample to another. Theoretically, a frequency distribution constructed for the statistics of all the samples is called a sampling distribution. Some probability distributions such as binomial and hypergeometric probability distributions are appropriate models to describe the sampling distribution.

Under sampling probability distribution, one can differentiate between these two important distributions:

A. Sampling distribution of the Mean (or distribution of $\overline{X}$)
B. Proportion Sampling distribution (or distribution of p)

A. Sampling Distribution of the Mean

The means of selected random samples of size n from the N element of a population, such that $n \leq N$ and the probabilities assigned to these means, are called sampling probability distribution of the mean, or simply sampling distribution (or distribution of $\overline{X}$).

Theoretically, if samples of size n are selected randomly from the N elements of a population, then the distribution of $\overline{X}$ will have the following characteristics:

1. The mean of the sampling distribution, or the mean of the means, denoted by $\mu_x$, $\overline{X}_{\overline{x}}$, or $E(\overline{X})$ equals the mean of the population $E(\overline{X}) = \mu$.

This characteristic indicates that successive sample means computed from different samples of size n drawn from the same population will tend to cluster about the mean of the population so that, on the average, $\overline{X}_{\overline{x}}$ will equal $\mu$. The following example is presented to illustrate this feature.

*Example 6.11:* Consider a population of 5 elements: $X_1$, $X_2$, $X_3$, $X_4$, and $X_5$ that assume these values: 1, 2, 3, 4, and 5, respectively.

Select samples of size $n = 2$ from this population of $N = 5$. Calculate the mean of the population $\mu$, the mean of the sampling distribution $E(\overline{X})$, and compare the two values.

**Solution:**

1. $\mu = \dfrac{1 + 2 + 3 + 4 + 5}{5} = 3$

2. The possible samples of size $n = 2$ selected from the elements of the population $N = 5$ can be determined by $_5C_2$ which equals to 10 samples. A list of these samples, a sampling frequency distribution constructed from the means of these samples, and the calculation of the mean of the sampling distribution are presented in Table 6-3:

**TABLE 6-4**
Listing of Samples, Sampling Frequency Distribution,
and Calculation of $E(\overline{X})$

| Samples | | | Sampling Frequency Distribution | | |
|---|---|---|---|---|---|
| | | | $\overline{X}$ | f | $\overline{X}f$ |
| 1,2 | | | 1.5 | 1 | 1.5 |
| 1,3 | | | 2.0 | 1 | 2.0 |
| 1,4 | 2,3 | | 2.5 | 2 | 5.0 |
| 1,5 | 2,4 | | 3.0 | 2 | 6.0 |
| | 2,5 | 3,4 | 3.5 | 2 | 7.0 |
| | | 3,5 | 4.0 | 1 | 4.0 |
| | | 4,5 | 4.5 | 1 | 4.5 |
| | | | | 10 | 30.0 |

$$E(\overline{X}) = \frac{30}{10} = 3$$

3. $\mu = 3$ and $E(\overline{X}) = 3$, hence $E(\overline{X}) = \mu$ regardless of the shape of the underlying population distribution.

2. The variance of the sampling distribution of $\overline{X}$ denoted by $\sigma_{\overline{X}}^2$ or Var $(\overline{X})$ equals the variance of the population $\sigma^2$ divided by the size of the sample n:

$$\text{Var } (\overline{X}) = \frac{\sigma^2}{n}$$

This indicates the larger the size of the sample n, the smaller the variance of the sampling distribution of $\overline{X}$. The standard deviation of $\overline{X}$ which is the square root of Var $(\overline{X})$ or:

$$\sigma_{\overline{x}} = \frac{\sigma}{\sqrt{n}}$$

measures the dispersion of the sampling distribution of the mean around the mean of the population. As the standard deviation becomes smaller, the possible values of $\overline{X}$ will cluster more closely to the population mean $\mu$ and the shape of the normal curve will be more peaked. The standard deviation of $\overline{X}$ is known as the standard error of $\overline{X}$ because it measures the chance of error inherent in the sampling process. The standard error of $\overline{X}$ depends on the standard deviation of the population and the size of the sample. The larger the sample size, the smaller the chance of error inherent in the sampling process.

The sampling process can occur with or without replacement. Where the population is infinite, sample observations will always be independent when sampling is done with replacements. However, in many practical situations sampling is done without replacement. When populations are large in comparison to the sample size, then sampling without replacement will generally yield the same conclusion as in the case of infinite populations. If the population is small in comparison to the sample size, this fact must be reflected in computing the variance and the standard error of the sampling distribution.

For *small populations,* the variance and the standard error of X will be multiplied by a finite population correction factor $\dfrac{N - n}{N - 1}$, so the variance and the standard error becomes:

$$\text{Var } (\overline{X}) = \frac{\sigma^2}{n} \quad \frac{N - n}{N - 1}$$

and $\quad \sigma_{\overline{x}} \quad = \dfrac{\sigma}{\sqrt{n}} \sqrt{\dfrac{N - n}{N - 1}}$

When n is small in comparison to N, the limit of the finite population correction factor approaches unity and can be ignored. In practice, it is usually ignored when n is less than 10% of N.

To illustrate, the variance of the sampling distribution is computed from the data in *Example* 6.11 as follows:

1. The variance of the population $\sigma^2$.

| $X$ | $X^2$ |
|---|---|
| 1 | 1 |
| 2 | 4 |
| 3 | 9 |
| 4 | 16 |
| 5 | 25 |
| 15 | 55 |

$$\sigma^2 = \frac{55}{5} - \left(\frac{15}{5}\right)^2 = 11 - 9 = 2$$

2. The variance of the sampling distribution constructed from the population of $N = 5$ with samples size $n = 2$.

| $\overline{X}$ | $f$ | $\overline{X}f$ | $\overline{X}^2 f$ |
|---|---|---|---|
| 1.5 | 1 | 1.5 | 2.25 |
| 2.0 | 1 | 2.0 | 4.00 |
| 2.5 | 2 | 5.0 | 12.25 |
| 3.0 | 2 | 6.0 | 18.00 |
| 3.5 | 2 | 7.0 | 24.50 |
| 4.0 | 1 | 4.0 | 16.00 |
| 4.5 | 1 | 4.5 | 20.25 |
| | 10 | 30 | 97.5 |

$$\text{Var}(\overline{X}) = \frac{97.5}{10} - \left(\frac{30}{10}\right)^2$$

$$= 0.75$$

3. Var $(\overline{X})$ can be calculated as follows:

$$\text{Var}(\overline{X}) = \frac{\sigma^2}{n} \frac{N - n}{N - 1}.$$

$$= \frac{2}{2} \frac{5 - 2}{5 - 1}$$

$$= 0.75$$

## The Normal Distribution as a Model for the Distribution of $\overline{X}$

The distribution of $\overline{X}$ for samples selected randomly from a population provides the values of $E(\overline{X})$ and $\sigma_{\overline{x}}$. Having these two values, the appropriate normal curve can be determined to find the probability values of $\overline{X}$. The shape of the normal curve is determined by its standard error $\sigma_{\overline{x}}$ while $E(\overline{X})$ is located at the center of the curve.

Samples are drawn randomly from populations that are normal or not normal. These two situations are presented below.

### 1. *The Population is Normal*

Regardless of the sample size, the distribution of $\overline{X}$ for samples selected randomly from a normal population is considered to be normally distributed with mean $= \mu$ and variance $= \dfrac{\sigma^2}{n}$. Probabilities of $\overline{X}$ represented by any area under the normal curve can be evaluated by the standard normal variable:

$$Z = \frac{\overline{X} - \mu}{\sigma_{\overline{x}}}$$

*Example 6.12:* A production process is considered to be under control if the diameter of the parts produced may be looked upon as a normal population with mean $= 6''$ and variance $= .0036''$.

Thirty-six units are selected randomly from each shift production, and the process is considered under control if the mean diameter falls between $5.99''$ and $6.01''$.

What is the probability that a sample will fail to meet this criterion?

**Solution:**

a. A sample selected randomly from a normal population is also normally distributed with mean $= \mu$, and variance $= \dfrac{\sigma^2}{n}$ or:

$$n\left(\overline{X};\ 6,\ \frac{.0036}{36}\right)$$

b. The variance of the population $\sigma^2$ is known; then we can use the Z-distribution to evaluate probabilities:

$$Z = \frac{\overline{X} - \mu}{\sigma_{\overline{x}}} \quad ; \quad \sigma_{\overline{x}} = \frac{\sigma}{\sqrt{n}} = \frac{\sqrt{.0036}}{\sqrt{36}} = \frac{.06}{6} = .01$$

c. Draw a normal curve to determine the nonshaded area that indicates the process is under control.

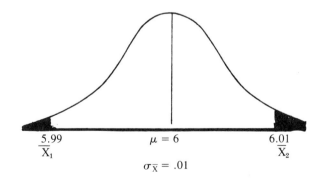

$$
\begin{array}{ccc}
5.99 & \mu = 6 & 6.01 \\
\overline{X}_1 & & \overline{X}_2 \\
 & \sigma_{\overline{x}} = .01 &
\end{array}
$$

d. P(a sample mean fails to meet this criterion) $= P(\overline{X}_2 < \overline{X} < \overline{X}_1)$

$$P(6.01 < \overline{X} < 5.99) = \int\limits_{6.01}^{\infty} f(x)\,dx + \int\limits_{-\infty}^{5.99} f(x)\,dx$$

$$= 1 - \left[ P\left(\frac{6.01 - 6.00}{.01}\right) + P\left(\frac{5.99 - 6.00}{.01}\right) \right]$$

$$= 1 - [P(Z = 1) + P(Z = -1)]$$

$$= 1 - (.3414 + .3414)$$

$$= 1 - .6828$$

$$= .3172 \quad \text{the probability of the shaded area.}$$

2. *The Population is Not Normal* (Central Limit Theorem)

The distribution of $\overline{X}$ for samples drawn randomly from a population whose frequency distribution is not normal can be approximated by a normal distribution, if the sample size n is large. This is known as the Central Limit Theorem. It allows the decision maker to develop procedures to make an inference about the mean of a population based on the sample results. The Central Limit Theorem applies to discrete as well as to continuous populations, regardless of their shape. The Central Limit Theorem is applied only when the population has a finite variance.

For most populations, samples do not have to be very large for the sampling distribution of $\overline{X}$ to be approximated by a normal distribution. The approach to a normal distribution is quite rapid as n increases.

3. *The Student or T-Distribution as a Model for the Distribution of* $\overline{X}$

As stated before, regardless of the sample size, the distribution of $\overline{X}$ for samples selected randomly from a normal population with mean $= \mu$ and variance $= \dfrac{\sigma^2}{n}$ is normally distributed and transformed into the standard normal distribution Z where $n\left(Z;\ \mu, \dfrac{\sigma^2}{n}\right)$ or

$$Z = \frac{\overline{X} - \mu}{\sigma_{\overline{x}}}$$

If the distribution of the population is not normal, a large sample selected randomly from this population can be viewed as normally distributed according to the Central Limit Theorem and $\overline{X}$ can be transformed into the standard normal distribution Z as stated above.

To calculate Z and the P(Z) one needs to know the variance $\sigma^2$ or the standard deviation of the population. However, there may be no information about the population and the only data that is available is the sample selected randomly from this population. Statistics such as the standard deviation s or the variance $s^2$ can be calculated for selected samples and may be used instead of the standard deviation $\sigma$ or the variance of the population $\sigma^2$. In this case, t-distribution (student's distribution)[5] may be used instead of the Z-distribution to evalulate the probabilities of sample means:

$$t = \frac{\overline{X} - \mu}{s_{\overline{x}}} \quad ; \quad s_{\overline{x}} = \frac{s}{\sqrt{n-1}}$$

The probability density function of t-distribution is:

$$f(X) = \frac{1}{\sqrt{r\pi}} \frac{\Gamma[(r+1)/2]}{\Gamma(r/2)} \left(1 + \frac{x^2}{r}\right)^{-(r+1)/2}$$

where r refers to the degrees of freedom.[6]

The t-distribution is flatter than the standard normal distribution for small samples of n < 30. Figure 6-9 shows a comparison of the two distributions.

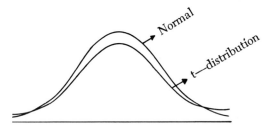

**Figure 6-9.** Comparison of standard normal distribution and the t-distribution.

There are tables to find the probability of the t-distribution. It is necessary to determine the degree of freedom for the sample or samples under investigation. For a single sample, the degree of freedom (denoted by r, d.f., or v.) is equal to n − 1. For example, if we have a sample size of 20, then v = 20 − 1 = 19. For two or more samples, find the degree of freedom for each sample ($n_1$ − 1) and simply add them together to find the degree of freedom. For example, if we have two samples, the first sample size of 10, and the second size 15, then:

$$v = n_1 + n_2 - 2 = 10 + 15 - 2 = 23$$

or  $v_1 = 10 - 1$  and  $v_2 = 15 - 1$  and  $v = v_1 + v_2$
   $= 9$  $= 14$  $= 9 + 14 = 23$

If n is large (n ⩾ 30), then t-distribution approaches the Z-distribution and one can use either to evaluate the probabilities of the sample means.[7]

---

(5) t-distribution has been derived by W.S. Gosset under "student" because his employer forbade him to publish under his name.

(6) Degrees of freedom refers to the free choices for the values of n observations used to calculate $\overline{X}$ and s. There are only n − 1 free choices in this case.

(7) The choice of n⩾ 30 is agreed upon but it is quite arbitrary.

*Example 6.13:* A normal population yielded an average of 50. What is the probability that a sample size of 101, selected randomly from this population and yielding a variance of 225, will have a mean less than 47?

**Solution:**

a. The variance of the population is unknown. Therefore, use the t-distribution.

b. Calculate $s_{\bar{x}}$ as follows:

$$s_{\bar{x}} = \frac{\sqrt{225}}{\sqrt{100 - 1}}$$

$$= 1.50$$

c. Determine the area where the sample mean is less than 47 (the shaded area in the diagram).

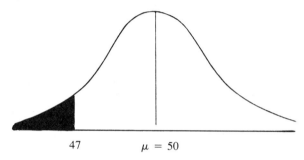

47     $\mu = 50$

d. P(a sample mean less than 47) $= \int_{-\infty}^{47} f(x)\, dx$

$$= P\left(\frac{47 - 50}{1.5}\right)$$

$$= P(t = -2)$$

$$= .0225 \quad \text{(from t-distribution table with degree of freedom} = 100)$$

e. Since n is large, one may use the Z-distribution instead of the t-distribution as follows:

$$P(\text{a sample mean less than } 47) = \int_{-\infty}^{47} f(x)\, dx$$

$$= \int_{-\infty}^{\mu} f(x)\, dx - \int_{47}^{\mu} f(x)\, dx$$

$$= .5 - P\left(\frac{47 - 50}{1.5}\right)$$

$$= .5 - P(Z = -2)$$

$$= .5 - .4772$$

$$= .0228$$

**114**

## B. Sampling Distribution of the Proportion

This probability distribution is similar to the probability distribution of the mean except that the population of the proportion sampling distribution is binomial. Many populations that describe business problems are binomial. For example, the production process can be viewed as a binomial population with two, and only two, outcomes: defectives and non-defectives. Proportioning of defectives of samples selected randomly from these populations either by simulation (Monte Carlo method) or by other techniques can be used to construct the proportion sampling distribution with the following characteristics: the mean of P is $E(p) = \pi$ or the expected value of the sampling proportion equals the population proportion. The variance of the sampling proportion is:

$$\text{Var }(p) = \frac{\pi(1 - \pi)}{n}$$

and the standard error of the sampling distribution is:

$$\sigma_p = \sqrt{\frac{\pi(1 - \pi)}{n}}$$

As mentioned before, for the sampling distribution of $\overline{X}$, the sampling distribution of P can be approximated by a normal curve where P may be transformed into the standard normal distribution Z as follows:

$$Z = \frac{P - \pi}{\sigma_p}$$

For small populations where sampling is done without replacement the standard error of P becomes:

$$\sigma_p = \sqrt{\frac{\pi(1 - \pi)}{n}} \sqrt{\frac{N - n}{N - 1}}$$

where $\dfrac{N - n}{N - 1}$ is the finite population correction factor introduced before with the sampling distribution of $\overline{X}$.

If $\pi$ is unknown then the variance and the standard error become:

$$S_p^2 = \frac{p(1 - p)}{n - 1}$$

$$\text{and} \quad S_p = \sqrt{\frac{p(1 - p)}{n}} \quad \text{(or)} \quad S_p = \sqrt{\frac{p(1 - p)}{n - 1}}$$

## EXERCISES

6.1  a.  A young man, while waiting for a young lady who is late, decided to amuse himself by walking either to the north or to the south 5 yards according to the following scheme. He tosses a fair die. If an odd number occurs he walks 5 yards to the south; if an even number occurs, he walks 5 yards to the north. His young lady was so late that he walked 50 yards. What is the probability of these:

1. he will be back at his starting point?
2. he will be exactly 10 yards either way from his starting point?

b. Given that a binomial variable X has mean $= 6$ and standard deviation $= 2$, find:

$$P(X = 0), \ P(X \geqslant 10), \ P(X \leqslant 1), \ P(2 \leqslant X \leqslant 4)$$

6.2 a. An airline company discovered that an average of 3% of the reservations for a particular flight have been cancelled for the last 5 years. Therefore, the company set a policy of selling three more tickets than the 85-seat capacity of the flight. What is the probability that for every passenger who shows up for the flight there will be a seat available?

b. It is known from past experience that machine A produces 2.5% defectives. A sample of size 10 is selected randomly from the output of the machine. Find the probability that the sample will contain:
1. no defectives.
2. at least one defective.
3. not more than three defectives.

6.3 Mr. X produces 100 units per day of which 10 are defective. A sample of size 5 is selected from the production of Mr. X without replacement. Find:
1. P(no defectives in the sample)
2. P(exactly two defectives)
3. P(at least one defective)
4. P(not more than two defectives)
5. P(less than three defectives)

6.4 a. Consider a lottery that sells 50 tickets and offers 3 prizes to be selected without replacement from the 50 tickets. If a person buys 4 tickets, what is the probability of winning two prizes, given that he has won at least one prize?

b. Solve part a. assuming that the prizes are awarded by drawing with replacement.

6.5 a. Consider 3 urns: urn I contains 2 white and 4 green balls, urn II contains 8 white and 4 green balls, and urn III contains one white and 3 green balls. Two fair coins are tossed. If two tails occur urn I is selected and 2 balls are drawn; if one head and one tail occur urn II is selected and 2 balls are drawn; if two heads occur, urn III is selected and 2 balls are drawn. Compute the conditional probability of selecting urn III given that 2 green balls are drawn with replacement.

b. Solve part a. assuming that the balls have been drawn without replacement.

6.6 a. Assume that the number of telephone calls made to a medical clinic during an hour can be viewed as a random phenomenon that follows a Poisson distribution with $\lambda = 10$. What is the probability of:
1. no phone calls received during an hour.
2. at least one phone call during an hour.
3. more than 5 phone calls during an hour.
4. not less than 8 phone calls during an hour.

b. Assume that surgical cases arriving in the accident ward of a hospital can be described by a Poisson probability function with $\lambda = 6$ per day. Find the probability of:

1. no surgical cases in a particular day.
2. exactly five surgical cases.
3. at least one surgical case.
4. more than 8 surgical cases.

6.7 a. The records show that machine A produces 2% defectives, and the number of defectives produced by the machine appears to follow a Poisson distribution. For a sample of 100 units selected randomly from the output of the machine find the probability of:
1. no defectives.
2. at least one defective.
3. exactly 5 defectives.
4. not less than 2 defectives.

b. For a liquor store, it has been determined that the number of times a customer gets drunk after 5 drinks is well approximated by a Poisson probability distribution with $\lambda = 3$. Calculate: $P(X = 2)$, $P(2 \leqslant X \leqslant 4)$, $P(X \leqslant 1)$.

6.8 a. The specifications to produce a certain part are as follows: the length should be $15'' \pm .2''$. A machine has been set up to produce this part. The production of the machine is normally distributed with $\mu = 15.01''$ and $\sigma = .10''$. What is the percentage of defectives produced by this machine?

b. Assume that the machine in part a. is adjusted so that the production is normally distributed with $\mu = 14.98''$ and $\sigma = .15''$. What is the percentage of defectives produced in this case?

6.9 Assume that the weight of football players is a random variable obeying a normal distribution with mean $= 210$ pounds and a variance $= 625$ pounds.
1. What is the percentage of football players whose weight is between 180 and 200 pounds?
2. What is the percentage of those who weigh 195 or less?
3. What is the percentage of those who weigh 240 or more?
4. Find the probability that the weight of one player is 235 pounds or more given that he weighs more than 220 pounds.

6.10 a. The life in hours of a radio tube is normally distributed with mean equals 1000 hours. Find the variance that makes the life of the tube between 750 and 1250 hours, and has a probability of 95%.

b. In part a., what is the probability that the tube will have a life of 1300 hours or more?

6.11 An automatic machine fills cans with coffee. The machine is set for 16 oz. net weight. The process is viewed as normally distributed with $\mu = 16$ and $\sigma = 0.5$. A sample of 20 cans is selected randomly. What is the probability that the mean of this sample:
1. falls between 15.9 and 16.2.
2. is less than 16.
3. is more than 16.

6.12 Past experience showed that the fiber strength of cotton purchased from a certain company appears to follow a normal probability distribution with mean $= 80$ and standard deviation of 10. What is the probability that the mean fiber strength of fifty samples selected randomly from a lot of cotton fibers differs from the population mean by 13?

6.13 a. The proportion of defectives in the total production of tires in factory A is found to be 20%. A sample of 100 tires has been selected randomly from the output of one week. Calculate the standard error of the sample proportion.

b. In part a. assume that the proportion of defectives in total production is unknown, and out of the 100 tires selected randomly 18 are found to be defective. Calculate the estimated standard error of proportion.

# CHAPTER

# 7

# Statistical Decision Making: Statistical Inference

The main objective of statistical theory is to be applied in decision making. In the area of decision making one has to distinguish between classical and non-classical approaches used to arrive at conclusions and decisions using statistical analysis.

The process of making decisions concerning the parameters of populations based on information contained in a sample or samples selected randomly from these populations is called "statistical inference," which is considered the classical approach of statistical decision making.

The non-classical approach for statistical decision making includes Decision theory and Econometrics. Decision makers sometimes face a situation where many alternative courses of action exist and they have to choose the one that optimizes their goal. This can be achieved through the application of decision theory. On the other hand, whenever historical data (time series) are readily available, an appropriate mathematical model can be constructed and the parameters of the model can be estimated by statistical techniques to provide the decision maker with a reliable predictive model. This is the econometric approach to statistical decision making.

In this chapter, the classical approach of statistical decision making, namely "Statistical Inference," will be covered, while subsequent chapters will provide the reader with the elements of the non-classical approach.

## Statistical Inference:

Statistical inference or analytical statistics refers to the process of arriving at decisions about the parameters of the populations by examining a sample or samples drawn from these populations. For example, to find the average diameter of a part produced by 10 machines, the diameter of a sample of n parts selected randomly from the production of the machines to be measured, and the average diameter of the seleted sample ($\overline{X}$), is used as an estimate of the average diameter of the total production or of the population ($\mu$). This statistical technique is called "Estimation." In other cases, the mean of the population is known (or predetermined) and the mean of a sample selected from this population may differ from the mean of the population. This difference can be viewed as significant or as due to sampling. Tests designed to arrive at such decisions are called "Test of Hypotheses." Statistical inference can be classified into two categories:

1. Estimation
2. Test of Hypotheses

## 1. *Estimation*

Estimation is a common statistical technique that plays an important role in applied statistics. For example, one may use the mean of the lifetime of 12 batteries to estimate the true average time of this brand of batteries. Estimation is classified into two types:

A. Point estimation
B. Interval estimation

### A. Point Estimation

A point estimation, a single-valued statistic calculated from a sample, may be used to make an inference or a decision about a parameter of the population from which the sample has been drawn. Single-valued statistics such as $\overline{X}$, $s^2$, s, and p are point estimators for the parent population parameters $\mu$, $\sigma^2$, $\sigma$, and $\pi$, respectively. Examples of point estimates are a single value assigned to the average hourly wage of workers in an industry (e.g. $5.50), or to the percentage of defective units produced by a machine, or to the variation in performance of a new car. Decisions based on point estimators will be absolutely correct only if the statistics hit the parameters they are supposed to estimate "on the nose." For instance, in the example where the sample average hourly wage of workers in an industry is $5.50 per hour, decisions based on this information are absolutely correct only if the population average which represents the average of the hourly wage of all workers in this industry is exactly $5.50. If the point estimator is not exactly equal to the population parameter which it is supposed to estimate, then a decision to accept it depends on the desirable properties the estimator possesses. Error is found in the case of a point estimator which does not equal its parent population parameter. There is no way to measure such error and hence there is no way to assess the reliability of the estimator. The situation is different if the interval estimation is used.

### B. Interval Estimation

The estimation of a population parameter $\theta$ using a random interval is called interval estimation or confidence interval $\hat{\theta}_L < \theta < \hat{\theta}_U$, where $\hat{\theta}_L$ and $\hat{\theta}_U$ refer to the lower limit and the upper limit of the confidence interval, respectively. A graphical representation of the confidence interval $\hat{\theta}_L < \theta < \hat{\theta}_U$ is shown in Figure 7-1:

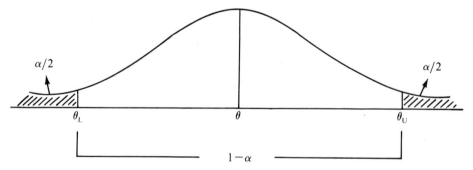

**Figure 7-1.** Confidence interval.

To determine a confidence interval for a population parameter one has to know two items: the degree of confidence, and information concerning the sample.

1. The degree of confidence: $(1 - \alpha)$

To determine the degree of confidence $(1 - \alpha)$, a value for $\alpha$ has to be assigned. The value of $\alpha$ determines the size of the confidence interval. There is an inverse relationship between the value of $\alpha$ and the size of the confidence interval. The smaller the magnitude of $\alpha$, the wider the size of the confidence interval and vice versa. Decreasing $\alpha$ to zero will produce an interval of 100% degree of confidence, but this outcome is meaningless. On the other hand, a narrow interval, say, 20% would be very costly to the decision maker because the probability of making an error is very high, and this shortcoming can only be corrected by selecting a very large sample. Therefore, it is desirable to set $\alpha$ at an acceptable level in order to balance the cost of sampling and the cost (loss) of a wrong decision. Conventionally, $\alpha$ assumes a value of 0.05 or 0.01 which yields a confidence interval of 0.95 or 0.99, respectively. It is common to divide $\alpha$ evenly between the two tails.

2. Sample information

The population parameter to be estimated is constant. To construct a confidence interval for this constant population parameter, samples of size n selected randomly from the parent population yield different intervals for the statistic known as probability intervals. These probability intervals are viewed as a random variable because they are the outcome of selecting different samples of a given size randomly from the population. For example, a random sample of size n selected from the population yields a probability interval for $\bar{x}$: $x_L < \bar{x} < x_U$. Repeated random samples of the same size will produce random probability intervals of $\bar{x}$. The confidence interval of the population mean $\mu$ consists of a whole range of $\overline{X}$ values estimated by probability intervals. Therefore, this confidence interval depends on the distribution of $\overline{X}$, which in turn depends on the distribution of the parent population. If the population is normal, then a random sample selected from this population is also normal. If the parent population is non-normal, then according to the Central Limit Theorem, a large sample $(n > 30)$ selected at random from this non-normal population is normally distributed.

The size of the sample (n) is an important factor in the construction of any confidence interval. As a matter of fact, the decision maker uses the information of the sample size n to construct an appropriate confidence interval for the estimation of the population parameter he needs as an input in his decisions. The larger the size of the sample, the more confident the decision maker will be about the estimation of $\theta$ the parameter of the population and hence he can use a smaller confidence interval. For example, as the sample size n is increased, the standard error of sampling or $\sigma/\sqrt{n}$ decreases and the distribution of $\overline{X}$ becomes more concentrated around the population parameter $\mu$ and the confidence interval of $\mu$ becomes narrow or more precise. However, using a large sample will increase the cost of sampling, and the decision maker has to decide whether or not such increase of cost is justifiable.

**Interpretation of Interval Estimation**

Given information about the distribution of the sample and its size n, confidence intervals of different degrees can be constructed by selecting different values of $\alpha$. For example, if $\alpha = 0.05$, then a 95% confidence interval can be constructed for the population parameter $\theta$, while a 99% confidence interval indicates that the value of $\alpha = .01$. In general, with different values of $\alpha$, one can construct $(1 - \alpha)\%$ confidence intervals. The following interpretation of a $(1 - \alpha)\%$ confidence interval will be very useful: a 95% confidence interval of $\mu$ shows that, on the average, the population parameter $\mu$ will be contained in the interval in 95 out of 100 calculations. In other words, a random sample of size n selected from the parent population can be used to calculate a probability interval of $\overline{X}$, where in 95 out of 100 cases, the calculated $\overline{X}$ will fall within the limits of the confidence interval of $\mu$ and, therefore can be used as an estimator of $\mu$.

## Confidence Intervals for the Population Mean

a. *With $\sigma$ Known*

Sampling distribution can be approximated by a normal distribution if the sample is large enough or if the sample selected randomly is from a normal population. In both cases the mean of the sample size n is normally distributed with mean $= \mu$, and variance $= \dfrac{\sigma^2}{n}$ or n $\left( \overline{X}; \mu, \dfrac{\sigma^2}{n} \right)$

To construct an interval estimation or confidence interval for $\mu$ the mean of the population, one has to determine the degree of the confidence interval $(1 - \alpha)$ as shown in the diagram.

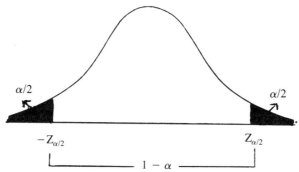

If $\alpha = .05$ then the degree of confidence is 95% or we have a 95% confidence interval. This means that the probability that the mean of the population $\mu$ falls between $-Z_{\alpha/2}$ and $Z_{\alpha/2}$ is 0.95, or we are confident 95% that $\mu$ falls between the lower limit at $-Z_{\alpha/2}$ and the upper limit at $Z_{\alpha/2}$. The confidence interval $1 - \alpha$ is:

$$\int_{-Z_{\alpha/2}}^{Z_{\alpha/2}} f(x)\, dx = P\left(-Z_{\alpha/2} < \frac{\overline{X} - \mu}{\sigma/\sqrt{n}} < Z_{\alpha/2}\right) = 1 - \alpha$$

The confidence interval $1 - \alpha$ for the population mean $\mu$, where the variance or the standard deviation of the population is known, can be derived from the double inequality:

$$-Z_{\alpha/2} < \frac{\overline{X} - \mu}{\sigma/\sqrt{n}} < Z_{\alpha/2}$$

$$-Z_{\alpha/2}\frac{\sigma}{\sqrt{n}} < \overline{X} - \mu < Z_{\alpha/2}\frac{\sigma}{\sqrt{n}}$$

**122**

$$-\overline{X} - Z_{\alpha/2}\frac{\sigma}{\sqrt{n}} < -\mu < -\overline{X} + Z_{\alpha/2}\frac{\sigma}{\sqrt{n}}$$

$$\overline{X} + Z_{\alpha/2}\frac{\sigma}{\sqrt{n}} > \mu > \overline{X} - Z_{\alpha/2}\frac{\sigma}{\sqrt{n}}$$

$$\overline{X} - Z_{\alpha/2}\frac{\sigma}{\sqrt{n}} < \mu < \overline{X} + Z_{\alpha/2}\frac{\sigma}{\sqrt{n}}$$

The lower limit of the confidence interval $(1 - \alpha)$ is: $\overline{X} - Z_{\alpha/2}\dfrac{\sigma}{\sqrt{n}}$ where $\overline{X}$ is the

sample mean and $\dfrac{\sigma}{\sqrt{n}}$ is the standard error of the sample mean. The upper limit of the

confidence interval $(1 - \alpha)$ is: $\overline{X} + Z_{\alpha/2}\dfrac{\sigma}{\sqrt{n}}$. This means the value $Z_{\alpha/2}\dfrac{\sigma}{\sqrt{n}}$ must be

subtracted from the sample mean to find the lower limit of the confidence interval, and to

add the same value $Z_{\alpha/2}\dfrac{\sigma}{\sqrt{n}}$ to the sample mean to find the upper limit of the confidence

interval $(1 - \alpha)$. In other words, the mean of the population $\mu = \overline{X} \pm Z_{\alpha/2}\dfrac{\sigma}{\sqrt{n}}$

*Example 7.1:* Past records showed that the length of time customers take to shop in a supermarket is normally distributed with variance of 100 minutes. A sample of 25 customers selected randomly yielded an average of 30 minutes to shop. Construct a 99% confidence interval for the mean length of time spent in shopping in this supermarket.

**Solution:**

a. The variance of the population $\sigma^2$ is known ($\sigma^2 = 100$)

b. $\overline{X} = 30$, $\sigma = \sqrt{100} = 10$, $\dfrac{\sigma}{\sqrt{n}} = \dfrac{10}{\sqrt{25}} = \dfrac{10}{5} = 2$

c. $1 - \alpha = .99$, $P(Z) = \dfrac{.99}{2} = .4950$, then $Z_{\alpha/2} = 2.58$

$$\overline{X} - Z_{\alpha/2}\frac{\sigma}{\sqrt{n}} < \mu < \overline{X} + Z_{\alpha/2}\frac{\sigma}{\sqrt{n}}$$

$$30 - (2.58)(2) < \mu < 30 + (2.58)(2)$$

$$24.84 < \mu < 35.16$$

We are confident 99% that the mean length of time spent in shopping in this supermarket is between 24.84 and 35.16 minutes. Or the probability that a customer will spend time between 24.84 and 35.16 minutes to shop at this store is 0.99.

b. *With σ Unknown*

If the population variance $\sigma^2$ (or standard deviation $\sigma$) is unknown, then the sample variance $s^2$ (or standard deviation s) will be used as a point estimate and t-distribution is applied instead of the Z-distribution. The confidence interval $(1 - \alpha)$ for the population mean $\mu$, when the population variance is unknown is:

$$\overline{X} - t\frac{s}{\sqrt{n-1}} < \mu < \overline{X} + t\frac{s}{\sqrt{n-1}}$$

where $\overline{X}$ = the sample mean, s = sample standard deviation, n − 1 = degree of freedom.
As mentioned before, to evaluate t one needs to know $\alpha$ and v (degree of freedom).

**123**

*Example 7.2:* A random sample of 101 students showed an average of $800 as summer earnings for a full-time job, with standard deviation of $100.

Construct a 95% confidence interval for the mean summer earnings.

**Solution:**

a. The variance of the population ($\sigma^2$) is unknown; however, the standard deviation of the sample s is available (s = 100).

b. $\overline{X} = 800, \dfrac{s}{\sqrt{n-1}} = \dfrac{100}{\sqrt{101-1}} = 10$

c. $1 - \alpha = .95$, find $t_{.05, \ 100}$ $\qquad (v = n - 1 = 101 - 1 = 100)$

$$\overline{X} - t\dfrac{s}{\sqrt{n-1}} < \mu < \overline{X} + t\dfrac{s}{\sqrt{n-1}}$$

$$800 - (1.97)(10) < \mu < 800 + (1.97)(10)$$

$$780.3 < \mu < 819.7$$

We are confident 95% that the average summer earnings of the students in this group is between $780.30 and $819.70.

**Confidence Interval for the Population Proportion:**

Many business phenomena follow a binomial distribution. Therefore the estimation of proportions or rates is desirable. For example, any production process can be viewed to follow a binomial population with $\pi$ as the population proportion of defectives. It is of great interest for the production management as well as for the quality control department to evaluate the confidence interval for the population proportion $\pi$ by selecting a random sample of this population.

A binomial population can be approximated by a normal distribution with mean = np and variance = npq or: n(X; np, npq) and the confidence interval $(1 - \alpha)$ for the population $\pi$ becomes:

$$p - Z_{\alpha/2}\,\sigma_p < \pi < p + Z_{\alpha/2}\,\sigma_p$$

$$\sigma_p = \sqrt{\dfrac{\pi(1-\pi)}{n}}$$

where p = sample proportion and $\sigma_p$ = standard error of the sampling proportion.

This formula assumes that $\pi$ is known, which is being estimated.

The realistic situation is when $\pi$ is unknown and the confidence interval of $\pi$ becomes:

$$p - Z_{\alpha/2}\,s_p < \pi < p + Z_{\alpha/2}\,s_p$$

$$s_p = \sqrt{\dfrac{P(1-P)}{n}}$$

The sample proportion p is a ratio of the number of successes X and the number of trials n as n $\rightarrow \infty$. This means that $p = \dfrac{X}{n}$ as the sample size increases. Using this feature the confidence interval for $\pi$ becomes:

$$\dfrac{X}{n} - Z_{\alpha/2}\,s_p < \pi < \dfrac{X}{n} + Z_{\alpha/2}\,s_p$$

$$s_p = \sqrt{\dfrac{\dfrac{x}{n}\left(1 - \dfrac{x}{n}\right)}{n}}$$

*Example 7.3:* A sample survey showed that 120 of 800 families interviewed in the midwest would like to move to the west coast. Find a 0.90 confidence interval for the actual proportion of families in the midwest willing to move to the west coast.

**Solution:**

a.  $x = 120$          $n = 800$          $\dfrac{x}{n} = .15$

b.  $s_p = \sqrt{\dfrac{(.15)(.85)}{800}} = .0126$

c.  $P(Z_{\alpha/2}) = .45$          $Z_{\alpha/2} = 1.65$

$\dfrac{x}{n} - Z_{\alpha/2}\, s_p < \pi < \dfrac{x}{n} + Z_{\alpha/2}\, s_p$

$15 - (1.65)(0.126) < \pi < .15 + (1.65)(0.126)$

$.12921 < \pi < .17079$

We are confident 99% that the actual proportion of families in the midwest willing to move to the west coast is between 12.921% and 17.079%.

**Determination of the Sample Size:**

A question frequently raised is the appropriate size of the sample to be selected from the population at random. The sample size n depends on the degree of confidence or the interval estimation $(1 - \alpha)$, standard deviation $(\sigma$ or $s)$, the deviation of the sample mean or proportion from the mean or proportion of the population $(\overline{X} - \mu$ or $p - \pi)$.

The sample size n can be derived from:

$$n = \left[\dfrac{Z\sigma}{\overline{X} - \mu}\right]^2 \qquad\qquad n = \dfrac{Z^2\pi(1 - \pi)}{(p - \pi)^2}$$

*Example 7.4:* A tire manufacturing company wishes to know the average useful life of Brand A tire. What is the proper sample size to estimate this average in order that the probability will be 95% that the true mean does not differ from the sample mean by more than 100 miles? The standard deviation for the useful life of all tires produced is 500 miles.

**Solution:**

$$n = \left[\dfrac{Z\sigma}{\overline{X} - \mu}\right]^2$$

a.  $P(Z_{\alpha/2}) = \dfrac{.95}{2} = .475$

$Z_{\alpha/2} = 1.96$

b. $\overline{X} - \mu = 100$ miles

c. $\sigma = 500$ miles

$$n = \left[ \frac{(1.96)(500)}{100} \right]^2 \cong 96$$

*Example 7.5:* A production manager wishes to know the proper size of a sample to estimate the rate of defectives produced by a new machine such as the probability is 99% that the true proportion or rate of defectives does not differ from the sample proportion by more than 3%. The rate of defectives produced by all machines is 5%.

**Solution:**

$$n = \frac{Z^2 \pi (1 - \pi)}{(p - \pi)^2}$$

a. $P(Z_{\alpha/2}) = \frac{0.99}{2} = 0.495$

$Z = 2.58$

b. $p - \pi = .03$

c. $\pi = .05$

$$n = \frac{(2.58)^2(.05)(.95)}{(.03)^2} \cong 351$$

### 2. Hypothesis Testing

It has already been pointed out that inference about population parameters are generally made in one of two ways: Estimation, which is covered previously, and Hypothesis testing, which is the subject under consideration.

Hypothesis testing is an important tool being widely used in business and industry for managerial decisions. Hypothesis testing differs from interval estimation. In interval estimation the statistical objective is to estimate the unknown value of a population parameter, while in hypothesis testing a value is assumed or hypothesized for the unknown population parameter. Such hypothesized value may represent some type of standard set by the decision maker. The collection of sample data will provide a statistic, called a test statistic, to determine whether the hypothesized value of the population parameter should be accepted or rejected.

A hypothesis is defined as an assumption. A statistical hypothesis is an assumption about a population parameter. The hypothesis to be tested is known as the null hypothesis and usually denoted by $H_0$.

Samples drawn randomly from a population are random variables having a sampling distribution that is viewed as normally distributed. Such samples usually provide different values of test statistics. Therefore, even with a correct assumption about a population parameter some difference due to chance or the sampling process is likely to occur. If a correct assumption is made about a population parameter ($\mu_0$, $\pi_0$, $\sigma_0$), then the corresponding computed test statistics ($\overline{X}$, p, s), will have a value close to the hypothesized value of the population parameter and the existing difference is considered to be insignificant or ($\overline{X} - \mu_0 = 0$). If such difference is insignificant, then the null hypothesis $H_0$ is accepted and actions based on that result will be taken by decision makers. However, if the difference is significant or $(\overline{X} - \mu_0 \neq 0)$ this indicates that the statistical hypothesis $H_0$ is incorrect and hence reject it. With a correct assumption about the population parameter there is a slim chance of rejecting the null hypothesis. However, if $H_0$ is rejected then an alternative hypothesis denoted as $H_1$ may be accepted.

In statistical testing, significance is specified in terms of the probability that a particular difference occurs by chance given that the hypothesis being tested ($H_0$) is correct. A decision maker must specify exactly what constitutes a significant difference through the application of a decision rule to determine when to accept or to reject a statistical hypothesis.

A decision maker usually runs a test of statistical hypothesis to find out what decisions should be made and what actions should be taken. A test of statistical hypothesis is merely a set of procedures to decide whether to accept or to reject a null hypothesis. The procedures to be followed to run a statistical test of hypothesis are summarized in six steps:

1. Formulate the null hypothesis and the alternative hypothesis.
2. Determine a decision rule (or level of significance) to establish critical values of Z or t.
3. Collect sample data and calculate the test statistic.
4. Convert the sample statistic into either $Z_c$ or $t_c$ values.
5. Make the decision: Accept $H_0$, or reject $H_0$ and accept $H_1$ by comparing the critical values of Z or t with those of $Z_c$ or $t_c$.
6. Use the outcome of the test to take the appropriate managerial action.

*Step 1:* Formulation of the test:

There are two types to formulate a statistical test of hypothesis: (a) Two-tail test; (b) One-tail test.

(a) Two-tail test:

If the decision maker wishes to test whether the sample statistic is different from the hypothesized population parameter, then the test is a two-tail test. In this situation the sample test statistic could be either greater or less than the population parameter and the level of significance $\alpha$ will be divided evenly to determine the shaded area (rejection region) for each tail as shown in the diagram below:

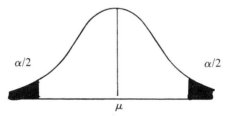

The null hypothesis and its alternative are set as follows:

$$H_0: \overline{X} - \mu_0 = 0$$
$$H_1: \overline{X} - \mu_0 \neq 0$$

(b) One-tail test:

If the decision maker wishes to test whether the sample statistic is greater than the population parameter, then he runs a one-tail test. In this case, the value of $\alpha$ will determine the shaded area on the right tail as shown in the following diagram:

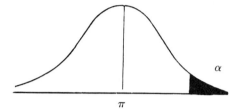

The test for this case is set as follows:

$$H_0: p \leqslant \pi_0$$
$$H_1: p > \pi_0$$

On the other hand, if the decision maker is concerned about whether the sample statistic is less than the population parameter, then he is faced with a one-tail test and $\alpha$ is located at the left tail as shown in the diagram:

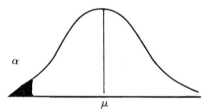

The null hypothesis and its alternative are set as follows:

$$H_0: \overline{X} \geqslant \mu_0$$
$$H_1: \overline{X} < \mu_0$$

*Step 2:* Determination of a decision rule:

As stated before, some difference between a correct hypothesized population parameter ($\mu_0$) and a test statistic ($\overline{X}$) is likely to occur due to chance. Such difference is either insignificant ($\overline{X} - \mu_0 = 0$) or significant ($X - \mu_0 \neq 0$). A decision maker, however, must specify exactly what constitutes a significant difference to determine when to accept or reject a null hypothesis by setting a decision rule. For example, if we are testing the hypothesis that a coin is fair against its alternative, then we may set the following decision rule: "Toss the coin 10 times; if the number of heads is between 4 and 6, accept the null hypothesis, otherwise reject it." This decision rule will help us to determine the acceptance and the rejection regions as shown in the following diagram.

**128**

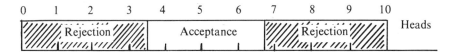

A decision rule is a statement which indicates the action to be taken in deciding whether to accept or to reject $H_0$, for all possible values of the test statistic. Test statistics, however, are random variables having a sampling distribution that is normally distributed. Accordingly, a decision rule divides the area under the normal curve into two mutually exclusive regions: the acceptance region and the rejection or critical region. The values of $\pm$ Z falling at the line of demarcation of the acceptance and rejection regions are referred to as the critical values of Z. Such values can be determined from the normal distribution table (in case of t-values, then use the t-distribution table). For example, if $\alpha = 0.05$ then the P(Z) for each shaded area in Figure 7-2 equals 0.025 and the critical values of Z = $\pm 1.96$ as shown in Figure 7-2.

**Figure 7-2.** The acceptance and rejection regions and the critical values of Z.

When the decision maker sets an assumption about the parameter of the population to be the null hypothesis, he is usually sure that his assumption is correct. For example, a marketing research department executive making an assumption or hypothesis about the potential average sales of a new product, his assumption is based on many factors as his experience, local and national data, surveys or any type of research concerning this commodity. In short, the executive or the decision maker will not make any hypothesis unless he is sure to a great extent that it is correct. Even with a correct $H_0$, some test statistics may differ significantly from the assumed population parameter and lead to the rejection of the true $H_0$. The probability of such rejection to occur is represented by the shaded area in Figure 7-2. The probability of the rejection region is denoted by $\alpha$ which is known as the level of significance. In other words, the difference is statistically significant to reject a true $H_0$. Rejecting a true $H_0$ is an error, referred to as Type I error. Accepting a false $H_0$ is known as Type II error. These two types of errors are presented in Table 7-1:

**TABLE 7-1**
Decision Table in Hypothesis Testing

|  | $H_0$ is True | $H_1$ is True |
|---|---|---|
| reject $H_0$ | Type I error | correct decision |
| accept $H_0$ | correct decision | Type II error |

**129**

Two types of errors are shown in the above table: Type I error and Type II error. The probability of committing these errors are denoted by $\alpha$ (alpha) or $\beta$ (beta) respectively, and can be expressed in the following form:

$$\alpha = P(\text{Type I error}) = P(\text{reject } H_0/H_0 \text{ is true})$$
$$\beta = P(\text{Type II error}) = P(\text{accept } H_0/H_0 \text{ is false})$$

Probability of Type I error ($\alpha$) is referred to as the level of significance, as mentioned before, and it represents the portion of the sample statistics that would occur in the rejection region. If $\alpha = 0.05$, this means that the probability of rejecting a true $H_0$ is 0.05, or 5% of the time a true null hypothesis will be rejected. The size of Type II error (or $\beta$) refers to the probability of accepting a false $H_0$.

Both types of errors are undesirable. Therefore, in selecting a decision rule for hypothesis testing, the decision maker should select the one that minimizes both types of errors. At a given sample size, the probability of reducing one error can be done at the expense of increasing the probability of the other. However, with the same decision rule, increasing the sample size will reduce the probabilities of the two types of errors to occur. Increasing the sample size has its cost and hence it becomes a matter of trade off between the cost of sampling and the reduction in the probability of committing those errors. The decision maker being faced with situation, usually makes the choice on the basis of subjective evaluations. However, a decision rule must be selected to provide a lower probability of the more serious error to occur ($\alpha$ or $\beta$). In common practice, the decision maker may designate $H_0$ as the one if rejected when it is true, then the more serious error to occur is Type I error, and hence a lower value of $\alpha$ is assigned, usually 0.05 or 0.01.

Another way to select a decision rule to determine the size of $\alpha$ is to construct a power curve which is the outcome of the power function: $1 - \beta = f(\mu)$. The power test $1 - \beta$ is the probability of rejecting $H_0$ when it is false. The power test $1 - \beta$ is the complement of Type II error $\beta$. The discussion of the relationship between $\beta$ and $\mu$ is essential to derive the power function.

For a given hypothesis testing, there is only one value for $\alpha$. However, there are many values of $\beta$, one for each possible value of the true parameter (e.g., $\mu$). Consequently, the probability of Type II error ($\beta$) is a function of the true parameter when the null hypothesis $H_0$ is false. If the true parameter ($\mu$) is close to its hypothesized value ($\mu_0$), the probability of accepting a false hypothesis is high, while if the true parameter is much different from the hypothesized value, the probability of Type II error ($\beta$) is low. The functional relationship between $\mu$ and $\beta$ is shown in Figure 7-3.

The relationship between $\beta$ and $\mu$: $\beta = f(\mu)$ can be described by an operating characteristic curve (OC). The power of a test is the probability of rejecting the null hypothesis when it is false. The power of the test is the complement of the probability of Type II error of accepting the null hypothesis when it is false. The power function: $1 - \beta = f(\mu)$ will produce a power curve having the opposite shape of the OC curve as shown in Figure 7-4.

A power curve can be constructed for each hypothesis testing and the appropriate value of $\alpha$ is determined. However, in practice, as previously mentioned, the decision maker selects a decision rule that provides a lower probability of $\alpha$ while designating $H_0$ as the true hypothesis.

**130**

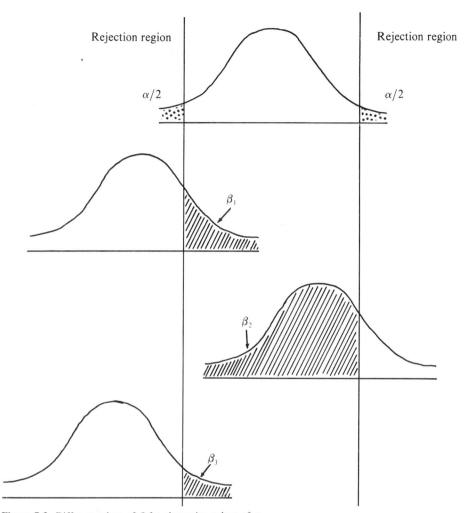

**Figure 7-3.** Different values of $\beta$ for alternative values of $\mu$.

P (Reject $H_0$)

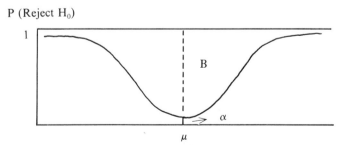

**Figure 7-4.** A power curve to determine the size of $\alpha$.

131

*Step 3:* Collection of sample data and calculation of the test statistic:

After the formulation of the null and alternative hypotheses and the determination of a decision rule, sample data are to be selected randomly from the population to provide sample test statistics ($\overline{X}$, p) to run the test. Sample statistics ($\overline{X}$) are random variable having a sampling distribution which is normally distributed.

*Step 4:* Conversion of the sample statistic into either $Z_c$ or $t_c$ value:

The decision rule divides the area under the normal curve into two regions, namely, the acceptance region and the rejection region. As presented in Step 2, the critical values of Z or t (values derived from the tables) fall at the line of demarcation of the two regions.

To be able to make the decision of accepting or rejecting the null hypothesis $H_0$ one needs to express the difference between the sample test statistic and the hypothesized value of the population parameter in terms of the standard error. Given the information about the sample statistic ($\overline{X}$), the hypothesized value of the population parameter ($\mu_0$), the sample size (n), and the variance or the standard deviation of the population ($\sigma^2$ or $\sigma$), one can calculate the value of Z referred as $Z_c$ which determines the difference between the sample test statistic and the hypothesized value of the population parameter ($\overline{X} - \mu_0$) in terms of the standard error as follows:

$$Z_c = \frac{\overline{X} - \mu_0}{\dfrac{\sigma}{\sqrt{n}}}$$

However, in many situations the variance or the standard deviation of the population will not be available. The standard deviation of the sample statistic (s) can be used as a point estimate of the standard deviation of the population ($\sigma$) and the sampling distribution will follow a Student t-distribution. The difference between the sample test statistic ($\overline{X}$) and the hypothesized value of the population parameter ($\mu_0$) will be expressed in terms of the standard error by $t_c$:

$$t_c = \frac{\overline{X} - \mu_0}{\dfrac{s}{\sqrt{n - 1}}}$$

where n − 1 refers to the degrees of freedom.

*Step 5:* Make the decision: Accept $H_0$, or reject $H_0$ and accept $H_1$:

To make the decision whether to accept or reject $H_0$, the calculated value of $Z_c$ is compared with the critical value (or values) of Z (determined in Step 2). If the value of $Z_c$ falls in the acceptance region, accept $H_0$. On the other hand, if the value of $Z_c$ falls in the rejection region, reject the null hypothesis $H_0$ and accept its alternative $H_1$. In managerial decision analysis one should recognize that there is a chance or risk that the accepted hypothesis may be incorrect.

The decision to accept the null hypothesis $H_0$ for a two-tail test and a one-tail test is portrayed in Figure 7-5.

If the Student t-distribution is applied, the decision will be similar to that of the Z-distribution.

*Step 6:* Use the outcome of the test to take the appropriate managerial action:

Regardless of the decision reached in the last step, to accept or to reject the null hypothesis, there is always a risk of making the incorrect decision. As previously mentioned, increasing the sample size will lower such risk. However, an increase of a sample size means an additional cost to be incurred. Now, the decision maker will be faced with another decision to make based on whether the additional cost can be justified. In general, hypothesis testing as well as any quantitative technique are merely tools in the hands of any decision maker in business and economics to use in taking the appropriate managerial action. Such tools provide the decision makers with more information to be used with what is available of other types of information to select the action that optimizes their goals.

Two-tail test:

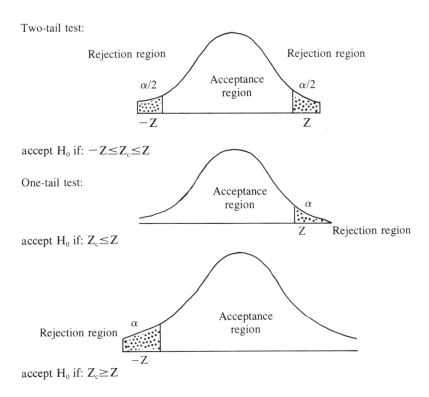

accept $H_0$ if: $-Z \leq Z_c \leq Z$

One-tail test:

accept $H_0$ if: $Z_c \leq Z$

accept $H_0$ if: $Z_c \geq Z$

**Figure 7-5.** Decision to accept $H_0$ for two-tail and one-tail tests.

**133**

## Methods of Testing Hypotheses

There are different methods to test the hypotheses concerning the mean, the proportion, and the variance. The following methods are considered:

1. Tests concerning the mean:
   a. Test for the difference between $\mu_0$ and $\overline{X}$; $\sigma$ is known.
   b. Test for the difference between $\mu_0$ and $\overline{X}$; $\sigma$ is unknown.
   c. Test for the difference between two means.
2. Tests concerning the proportion:
   a. Test for the difference between $\pi_0$ and p.
   b. Test for the difference between two proportions.
3. Tests concerning the variance:
   a. Chi-square test.
   b. F-distribution test.

### 1. Tests concerning the mean

These tests are conducted to show that the difference between the sample mean and the population mean is due to sampling and no significant difference exists. For example, a part produced by a factory according to specifications such that the diameter is 8" with standard deviation of .15". The machine is set to produce the part to meet these specifications. Nonetheless, a quality control or a production management man has to check whether the average length of the diameter of samples selected from the total production is within the acceptable range. If not, the machine needs to be adjusted.

These tests also show the validity of the average of two samples, one taken before introducing a new technique and the other after the application of the new technique. These two samples means may be tested to show whether the new technique has or has not caused any improvement or effect.

#### 1.a. Test for the hypothesized population mean; $\sigma$ is known

If the standard deviation of the population $\sigma$ is known, then the test for the difference between the population mean $\mu_0$ and the sample mean $\overline{X}$ requires the calculation of $Z_c$:

$$Z_c = \frac{\overline{X} - \mu_0}{\frac{\sigma}{\sqrt{n}}}$$

*Example 7.6:* An automobile company claimed that the six-cylinder cars produced by the company deliver an average of 20 miles per gallon in town with standard deviation of 4 miles. Fifty of these cars are tested for mileage and they yielded an average of 18.5 miles per gallon. At a .05 level of significance, would you agree with the company's claim?

**Solution:**

a. $\sigma$ is known $\rightarrow$ use $Z_c = \frac{\overline{X} - \mu_0}{\frac{\sigma}{\sqrt{n}}}$

b. $\mu_0 = 20$        $\sigma = 4$
   $\overline{X} = 18.5$       $n = 50$

**134**

c. Set up the test and determine the type of the test: Two-tail or One-tail test. Obtain the critical value of Z from the normal table to determine the acceptance and the rejection regions. Shade the rejection region in the diagram.

$H_0$: $\overline{X} \geq \mu_0$

$H_1$: $\overline{X} < \mu_0$ → One-tail test.

$\alpha = 0.05$

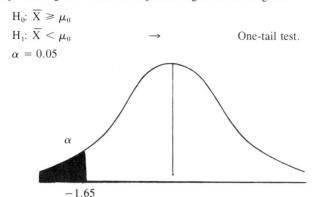

$\alpha$

$-1.65$

d. $Z_c = \dfrac{\overline{X} - \mu_0}{\dfrac{\sigma}{\sqrt{n}}} = \dfrac{18.5 - 20}{\dfrac{4}{\sqrt{50}}} = -2.65$

e. $Z_c$ falls in the shaded area (rejection region)

$H_0$ is rejected.

We are confident 95% that the average mileage per gallon delivered by these six-cylinder cars is less than the average mileage per gallon claimed by the company.

**1.b. *Test for the hypothesized population mean; $\sigma$ is unknown***

In many cases, the standard deviation of the universe $\sigma$ is unknown, and the standard deviation of the sample s is used as a point estimate of $\sigma$. When the standard deviation of the population is unknown, then the t-distribution will be used instead of the Z-distribution:

$t_c = \dfrac{\overline{X} - \mu_0}{\dfrac{s}{\sqrt{n-1}}}$

where $n - 1$ is the degree of freedom (v).

*Example 7.7:* A supermarket, used to issuing trading stamps, showed an average sale of $20.13 per customer in the past. The new manager decided to reduce prices and stop issuing trading stamps. A sample of 200 customers yielded an average sale of $21.50 and a standard deviation of $6.90.

At a level of significance of .01, would you conclude that there is a significant difference between the two policies?

**Solution:**

a. $\sigma$ is unknown $\rightarrow$ use $t_c = \dfrac{\overline{X} - \mu}{\dfrac{s}{\sqrt{n-1}}}$

b. $\mu_0 = 20.15$       $s = 6.90$
   $\overline{X} = 21.50$       $n = 200$
   $v = n - 1 = 199$

c. Set up the test and determine the acceptance and rejection regions in the diagram.

   $H_0: \overline{X} - \mu_0 = 0$
   $H_1: \overline{X} - \mu_0 \neq 0$      $\rightarrow$      Two-tail test
   $\alpha = .01$
   $t_{\alpha/2}, v = 2.58$

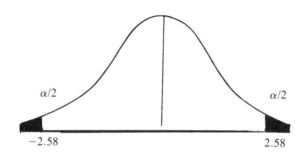

$\alpha/2$                                      $\alpha/2$

$-2.58$                                   $2.58$

d. $t_c = \dfrac{\overline{X} - \mu_0}{\dfrac{s}{\sqrt{n-1}}} = \dfrac{21.50 - 20.15}{\dfrac{6.90}{\sqrt{200-1}}} = 2.76$

e. $t_c$ falls in the shaded area (rejection region)
   $H_0$ is rejected.
   The probability that the two policies differ significantly is 0.99.

### 1.c. *Test for the difference between two means*

An important hypothesis testing is the one that involves testing for the difference between hypothesized means of two populations $(\mu_1 - \mu_2)_0$ to determine whether or not one population's values are significantly larger or smaller than the values of the other.

In testing for the difference between two population means, two sample means calculated from two samples selected independently, one from each population, and designated as $\overline{X}_1$ and $\overline{X}_2$. The sample test statistic is the difference between the two sample means $(\overline{X}_1 - \overline{X}_2)$. Samples selected from the population may differ in size: large samples (size of each is 30 or more observations) or small samples.

### For large samples:

With large samples, the sampling distribution of $\overline{X}_1$ and $\overline{X}_2$ are viewed as normally distributed. The sample test statistic representing the difference between the two samples $(\overline{X}_1 - \overline{X}_2)$ is also normally distributed with mean $= \mu_1 - \mu_2$ and standard error of:

**136**

$$\sigma_{(\overline{x}_1 - \overline{x}_2)} = \sqrt{\frac{\sigma_1^2}{n_1} + \frac{\sigma_2^2}{n_2}}$$

where $\sigma_1^2$ and $\sigma_2^2$ are the variance of the two populations while $n_1$ and $n_2$ refer to the size of each sample.

The only information known about the two populations is their hypothesized values of their means. The actual values of the means ($\mu_1$ and $\mu_2$) are unknown. Ordinarily, their variances $\sigma_1^2$ and $\sigma_2^2$ will also be unknown. Sample variances $s_1^2$ and $s_2^2$ may be used as estimators and the standard error of the test statistic ($\overline{X}_1 - \overline{X}_2$) becomes:

$$S_{\overline{x}1 - \overline{x}2} = \sqrt{\frac{s_1^2}{n_1} + \frac{s_2^2}{n_2}}$$

The standard normal $Z_c$ value for the test statistic is:

$$Z_c = \frac{(\overline{X}_1 - \overline{X}_2) - (\mu_1 - \mu_2)_0}{S_{(\overline{x}_1 - \overline{x}_2)}}$$

or

$$Z_c = \frac{\overline{X}_1 - \overline{X}_2}{S_{(\overline{x}_1 - \overline{x}_2)}}$$

The last formula to calculate $Z_c$ is based on the general assumption that the usual null hypothesis when testing for the difference between two means is set such that there is no difference between the means of the two populations.

*Example 7.8:* The manager of a gas station observed the amount of time required for his two attendants to serve the customers. For a sample of 50 customers, it takes Bill an average of 12 minutes with a standard deviation of 4 minutes, while for 52 customers it takes Dick an average of 10 minutes with a standard deviation of 5 minutes.

At a .01 level of significance, is there a significant difference between the performance of Bill and Dick?

**Solution:**

a.  Samples are large $\rightarrow s_{\overline{x}1 - \overline{x}2} = \sqrt{\frac{s_1^2}{n_1} + \frac{s_2^2}{n_2}}$

b.  $n_1 = 50 \qquad \overline{X}_1 = 12 \qquad s_1 = 4$

$\phantom{b.}\ n_2 = 52 \qquad \overline{X}_2 = 10 \qquad s_2 = 5$

$\qquad s_{\overline{x}1 - \overline{x}2} = \sqrt{\frac{16}{50} + \frac{25}{52}} = 0.895$

c.  Set up the test:

$H_0: \mu_1 = \mu_2$

$H_1: \mu_1 \neq \mu_2 \qquad\qquad \rightarrow \qquad$ Two-tail test

$\alpha = 0.01$

The values for $Z = \pm 2.58$ (from the normal distribution table) determines the rejection region (the shaded area) and the acceptance region as shown in the diagram.

**137**

d. $Z_c = \dfrac{12 - 10}{0.895} = 2.235$

e. $Z_c$ falls in the acceptance region, accept $H_0$.

With probability of 0.99, there is no significant difference between the performance of Bill and Dick.

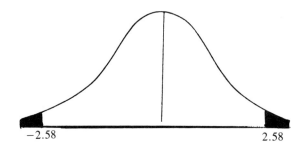

$-2.58$          $2.58$

**For small samples:**

When small samples are used (the size of each is less than 30) the Student t-distribution will be applied. The use of the t-distribution requires that both populations have the same variance (or $\sigma_1^2 = \sigma_2^2 = \sigma^2$). The standard error $\sigma_{(\bar{x}_1 - \bar{x}_2)}$ becomes:

$$\sigma_{(\bar{x}_1 - \bar{x}_2)} = \sqrt{\frac{\sigma^2}{n_1} + \frac{\sigma^2}{n_2}} = \sigma \sqrt{\frac{1}{n_1} + \frac{1}{n_2}}$$

A better estimate can be obtained by pooling the sample variances ($s_1^2$ and $s_2^2$) as follows:

$$s_{\bar{x}1 - \bar{x}2} = \sqrt{\frac{(n_1 - 1)s_1^2 + (n_2 - 1)s_2^2}{n_1 + n_2 - 2}} \sqrt{\frac{1}{n_1} + \frac{1}{n_2}}$$

The sample test statistic $(\bar{X}_1 - \bar{X}_2)$ can be approximated by a Student t-distribution with $n_1 + n_2 - 2$ degrees of freedom, having a mean $= \mu_1 - \mu_2$ and standard error of $s_{\bar{x}1 - \bar{x}2}$. The value of the test statistic expressed in $t_c$ terms is:

$$t_c = \frac{\bar{X}_1 - \bar{X}_2}{s_{\bar{x}1 - \bar{x}2}}$$

*Example 7.9:*  Two lots of cotton are tested for fiber strength. The following sample information is made available:

| I | II |
|---|---|
| (Pounds) | (Pounds) |
| 88.5 | 79.9 |
| 90.0 | 89.4 |
| 87.5 | 90.5 |
| 85.9 | 89.2 |
| 79.1 | 91.0 |
|  | 88.0 |

At a .05 level of significance, determine whether or not there is a significant difference in the fiber strength of the two lots.

**138**

**Solution:**

a. Small samples

$$s_{\bar{x}_1 - \bar{x}_2} = \sqrt{\frac{(n_1 - 1)s_1^2 + (n_2 - 1)s_2^2}{n_1 + n_2 - 2}} \sqrt{\frac{1}{n_1} + \frac{1}{n_2}}$$

b. $n_1 = 5 \qquad n_2 = 6 \qquad v = n_1 + n_2 - 2 = 9$ (degree of freedom)

Calculate: $\bar{X}_1$, $\bar{X}_2$, $s_1^2$, $s_2^2$:

| $X_1$ | $X_1^2$ | $X_2$ | $X_2^2$ |
|---|---|---|---|
| 88.5 | 7832.25 | 79.9 | 6384.01 |
| 90.0 | 8100.0 | 89.4 | 7992.36 |
| 87.5 | 7656.25 | 90.5 | 8190.25 |
| 85.9 | 7378.81 | 89.2 | 7956.64 |
| 79.1 | 6256.81 | 91.0 | 8281.00 |
| | | 88.0 | 7744.00 |
| 431.0 | 37224.12 | 528.0 | 46548.26 |

$$\bar{X}_1 = 86.20 \qquad\qquad \bar{X}_2 = 88.0$$

$$s_1^2 = \frac{\Sigma X_1^2}{n} - (\bar{X}_1)^2 \qquad\qquad s_2^2 = \frac{\Sigma X_2^2}{n} - (\bar{X}_2)^2$$

$$= \frac{37224.12}{5} - (86.2)^2 \qquad\qquad = \frac{46548.26}{6} - (88.0)^2$$

$$= 14.38 \qquad\qquad\qquad\qquad = 14.04$$

$$s_{\bar{x}_1 - \bar{x}_2} = \sqrt{\frac{(14.38)(5 - 1) + (14.04)(6 - 1)}{5 + 6 - 2}} \sqrt{\frac{1}{5} + \frac{1}{6}}$$

$$= 2.30$$

c. Set up the test:

$H_0: \mu_1 = \mu_2$

$H_1: \mu_1 \neq \mu_2 \qquad\qquad \rightarrow \qquad\qquad$ Two-tail test

$\alpha = 0.05$

The values for $t = \pm 2.262$ ($\alpha = 0.05$, $v = 9$) from the Student t-distribution table determines the rejection and the acceptance regions as shown in the diagram:

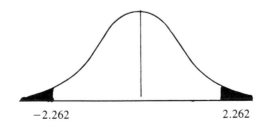

$-2.262 \qquad\qquad\qquad\qquad 2.262$

d. $t_c = \dfrac{86.20 - 88.0}{2.30} = -0.78$

e. $t_c$ falls in the acceptance region, accept $H_0$.

We are confident 95% that there is no significant difference in the fiber strength of the two lots of cotton.

### 2. Tests concerning the proportion

Testing proportions is commonly used in making business and economic decisions especially in the area of marketing and quality control.

#### 2.a. Test for the difference between $\pi_0$ and $p$

The binomial probability distribution may be used as a model for many random variables in business and economics. A binomial probability distribution with large n ($n \geqslant 30$) and a relatively large p (neither p or q is too close to zero)[1] can be approximated by a normal distribution. The sampling distribution of the proportions is viewed as normally distributed and the sample test statistic p will be used to test the hypothesized population proportion $\pi_0$. The hypothesized value $\pi_0$ is accepted as correct until or unless it is rejected. Therefore, the standard error of the proportion used in hypothesis testing is:

$$\sigma_p = \sqrt{\frac{\pi_0(1 - \pi_0)}{n}}$$

The value of the test statistic p can be converted into a $Z_c$ value as follows:

$$Z_c = \frac{P - \pi_0}{\sigma_p}$$

*Example 7.10:* A manager of a liquor store claimed that 85% of those whose ages range between 21 and 25 years prefer to drink beer. A market research analyst observed 200 customers (whose ages fall between 21 and 25 years) and found that 163 of them prefer to drink beer.

At a level of significance of .05, would the market research analyst agree with the manager's claim?

**Solution:**

a. $\pi_0 = .85$

$p = \dfrac{163}{200} = .815$

b. Set up the test:

$H_0: p \geqslant \pi_0$

$H_1: p < \pi_0$ $\rightarrow$ one-tail test

$\alpha = 0.05$

The value of $Z = -1.65$ (from the normal distribution table) determines the rejection and acceptance regions as shown in the diagram:

---

(1) Some statisticians use the following criterion:

$np \geqslant 5$ and $nq \geqslant 5$

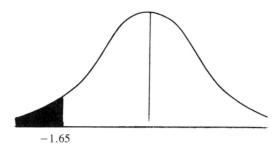

$-1.65$

c. $Z_c = \dfrac{p - \pi_0}{\sqrt{\dfrac{\pi_0(1 - \pi_0)}{n}}}$

$= \dfrac{8.15 - .85}{\sqrt{\dfrac{(.85)\,(.15)}{200}}}$

$= -1.38$

d. $Z_c$ falls in the acceptance region.
   $H_0$ is accepted.

e. The probability that the market research analyst agrees with the manager's claim is 0.95.

### 2.b. Test for the difference between two populations proportions

In many practical cases, decisions concerning the introduction of new techniques, new machines, or new operators can be made by testing the difference between hypothesized populations proportions ($\pi_1 - \pi_2$). Although it is possible to set the null hypothesis that the proportions in two populations are different, the most frequent test encountered in practice is that the proportions of the two populations are not different.

This test is similar to that for the difference between two population means. The value of the test statistic ($p_1 - p_2$) can be converted to $Z_c$ value as follows:

$$Z_c = \dfrac{(p_1 - p_2) - (\pi_1 - \pi_2)}{s_{p1 - p2}}$$

where $s_{p1 - p2} = \sqrt{\dfrac{p_t(1 - p_t)}{n_1} + \dfrac{p_t(1 - p_t)}{n_2}}$

and $p_t = $ a proportion constructed for the two samples combined.

*Example 7.11:*   Two new machines, one from Company X and the other from Company Y are tested. Machine X turns out 15 defectives in 250 units produced, and Machine Y turns out 25 defectives in 350 units produced.

At a .05 level in significance, is there a significant difference in the proportion of defectives turned out by the two machines?

**141**

**Solution:**

a. $P_1 = \dfrac{15}{250} = .06$    $P_2 = \dfrac{25}{350} = .071$

$P_t = \dfrac{15 + 25}{250 + 350} = .067$

b. Set up the test:

$H_0 : \pi_1 - \pi_2 = 0$

$H_1 : \pi_1 - \pi_2 \neq 0$

$\alpha = 0.05$

The value of $Z = \pm 1.96$ (from the normal distribution table) determines the rejection and acceptance regions as shown in the diagram:

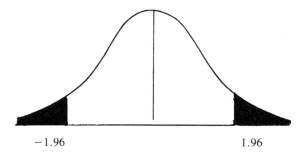

$-1.96$                    $1.96$

c. Calculate $Z_c$:

$$S_{p1-p2} = \sqrt{\dfrac{P_t(1-P_t)}{n_1} + \dfrac{P_t(1-P_t)}{n_2}}$$

$$= \sqrt{\dfrac{(.067)(.933)}{250} + \dfrac{(.067)(.933)}{350}}$$

$$= .0207$$

$$Z_c = \dfrac{P_1 - P_2}{S_{p1-p2}} = \dfrac{.06 - .071}{.0207} = -.5314$$

d. $Z_c$ falls in the acceptance region.
   $H_0$ is accepted.

e. We are confident 95% that there is no significant differences between the proportion of defectives turned out by the two machines.

### 3. *Tests concerning the variance*

In many cases it is important to test hypotheses concerning variances of populations. For example, a manufacturer of high-precision tools has to test the variability of his product in order to meet the rigid specifications set for these tools. Chi-square ($\chi^2$) distribution can be used for tests concerning variances of populations. Also, $\chi^2$ distribution can be used in other tests such as: test for goodness-of-fit, test of independency, and test of homogeneity.

Another distribution, the F-distribution, can be used for testing hypotheses concerning the equality of two estimated population variances. F-distribution is called the variance ratio and is named in the honor of R. Fisher, the great statistician who developed this distribution. F-distribution is also used in the analysis of variance where a decision should be made whether the difference among sample means are attributed to sampling error or that the difference is statistically significant.

In this section, the $\chi^2$ distribution and F-distribution are applied to different hypotheses.

### 3.a. The chi-square ($\chi^2$) distribution

#### 1. Hypothesis testing for the population variance

Knowledge of a population variability is important to the decision maker when he takes the appropriate managerial action to optimize his goal. For a trucking company, the variability in wear is an important factor in making a decision to buy tires.

The population variance $\sigma^2$ is usually unknown, therefore, the variance of a sample drawn from that population $s^2$ is used as a point estimate of $\sigma^2$. The role that $s^2$ plays as an estimator of $\sigma^2$ is similar to that role of $\overline{X}$ as an estimator of $\mu$. If $\sigma$ is known, then $\overline{X}$ is viewed as a random variable that is normally distributed, while if $\sigma$ is unknown, $\overline{X}$ follows the Student t-distribution. In the same manner, $s^2$ is viewed as a random variable follows a Chi-Square distribution with $n - 1$ degrees of freedom:

$$\chi^2 = \frac{s^2(n - 1)}{\sigma^2}$$

In testing for the population variance $\sigma^2$ which is unknown, a hypothesized value of the population variance $\sigma_0^2$ is being used instead of $\sigma^2$ and the test statistic $\chi_c^2$ becomes:

$$\chi_c^2 = \frac{s^2(n - 1)}{\sigma_0^2}$$

*Example 7.12:*  The diameter of a cylinder should meet rigid specifications and is allowed a variability of $\sigma = .005$. A sample of size 25 selected yielded variance $= .000028$.

At a level of significance of .01, what are the chances that the production of these cylinders will meet the specifications?

**Solution:**

a.  This is a test concerning the variance of the population, $\chi_c^2 = \frac{s^2(n - 1)}{\sigma_0^2}$ to be used.

b.  Set up the test:

$$H_0 : s^2 \leq \sigma_0^2$$
$$H_1 : s^2 > \sigma_0^2$$
$$\alpha = .01$$
$$\chi^2_{\alpha,v} = \chi^2_{.01,24} = 42.98$$

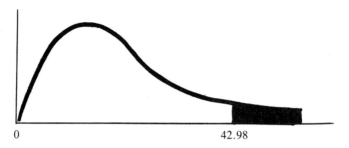

0                                         42.98

c. Calculate $\chi_c^2 = \dfrac{s^2(n-1)}{\sigma_0^2} = \dfrac{.000028(24)}{(.005)^2} = 26.88$

d. $\chi_c^2$ falls in the acceptance region
   $H_0$ is accepted.

e. The probability that the production of the cylinders meets its specification is 0.99.

2. *Goodness-of-fit test*

A frequent problem that faces those who apply statistics to business and economics is whether a set of observed data may be looked upon as values assumed by a random variable that follows a specific probability distribution. A test for that can be run by applying the following Chi-Square distribution $\chi^2$, with degrees of freedom $v = n - t - 1$ ($n$ = number of classes, and $t$ = the number of estimated parameters):

$$\chi_c^2 = \sum_{i=1}^{n} \frac{(O_i - E_i)^2}{E_i}$$

where $O_i$ = observed frequencies and $E_i$ = expected frequencies.

*Example 7.13:* A quality control supervisor claimed that the distribution of defectives contained in samples of size 50 taken at equal intervals from the weekly production can be looked upon as a random variable following a Poisson probability distribution.

At a .01 level of significance, what is the probability that his claim is right? The following frequency distribution of the defectives for 100 samples each of size 50 is provided to run the test.

| Number of Defectives | Observed frequency of samples |
|:---:|:---:|
| 0 | 5 |
| 1 | 10 |
| 2 | 14 |
| 3 | 17 |
| 4 | 16 |
| 5 | 13 |
| 6 | 10 |
| 7 | 6 |
| 8 | 5 |
| 9 | 2 |
| 10 | 1 |
| 11 | 1 |
| 12 | 0 |
| | 100 |

**144**

**Solution:**

a. To test whether this distribution fits a Poisson probability distribution,

$$\chi_c^2 = \sum_{i=1}^{n} \frac{(O_i - E_i)^2}{E_i} \text{ is used.}$$

b. Set up the test:

$H_0$: The distribution of defectives follows a Poisson distribution with $\lambda$ = the average of defectives (or $\overline{X}$).

$H_1$: The distribution of defectives does not follow a Poisson distribution.

$\alpha = .01$

$\chi^2_{.01, v=10} = 23.209$

$v = 12-1-1$, where $t = 1$ for the estimated $\lambda$

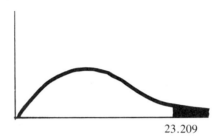

23.209

c. Calculate $\chi_c^2$:

$$\chi_c^2 = \sum_{i=1}^{n} \frac{(O_i - E_i)^2}{E_i}$$

Number of defectives = X

Observed frequency of samples = f

$\lambda = \Sigma Xf / \Sigma f = \overline{X}$

| X | f | Xf | Poisson Probabilities | $E_i$ | $O_i$ | $(O_i - E_i)^2$ | $(O_i - E_i)^2/E_i$ |
|---|---|---|---|---|---|---|---|
| 0 | 5 | 0 | .0183 | 1.83 | 5 | 10.05 | 5.49 |
| 1 | 10 | 10 | .0733 | 7.33 | 10 | 7.13 | .97 |
| 2 | 14 | 28 | .1465 | 14.65 | 14 | .42 | .03 |
| 3 | 17 | 51 | .1954 | 19.54 | 17 | 6.45 | .33 |
| 4 | 16 | 64 | .1954 | 19.54 | 16 | 12.53 | .64 |
| 5 | 13 | 65 | .1563 | 15.63 | 13 | 6.92 | .44 |
| 6 | 10 | 60 | .1042 | 10.42 | 10 | .18 | .02 |
| 7 | 6 | 42 | .0595 | 5.95 | 6 | .00 | .00 |
| 8 | 5 | 40 | .0298 | 2.98 | 5 | 4.08 | 1.37 |
| 9 | 2 | 18 | .0132 | 1.32 | 2 | .46 | .35 |
| 10 | 1 | 10 | .0053 | .53 | 1 | .22 | .42 |
| 11 | 1 | 11 | .0019 | .19 | 1 | .66 | 3.47 |
| 12 | 0 | 0 | .0006 | .06 | 0 | .00 | .00 |
| | 100 | 399 | | | | | $\chi_c^2 = 13.53$ |

$\overline{X} = \lambda = 3.99 \cong 4$

**145**

d. $\chi_c^2$ falls in the acceptance region.

   $H_0$ is accepted.

e. We are confident 99% that the distribution of defectives follows a Poisson distribution.

3. *Test of independence*

Chi-square can be used to test the hypothesis that two variables with different classifications are independent or there is no relationship between the two variables. The two variables are classified in a two-way table called contingency table. The formula of $\chi^2$ distribution used for the test of independence is:

$$\chi_c^2 = \sum_{i=j}^{r} \sum_{j=1}^{c} \frac{(O_{ij} - E_{ij})^2}{E_{ij}}$$

with $v = (r - 1)(c - 1)$, $r$ = number of rows, and $c$ = number of columns.

In this situation, the use of the Chi-square distribution as an approximation of the true sampling distribution of Chi-square is based on the assumption that the sample size is sufficiently large. In practice, the sample will be considered large enough when each cell in the contingency table has five or more expected frequencies. If not, this can be achieved by combining columns or rows.

*Example 7.14:* A sample of 500 wives with children was interviewed to test if the education level has anything to do with the number of children in the family. The following data is the result of the research:

| Education | Number of Children | | | |
| --- | --- | --- | --- | --- |
| | More than 8 | 6,7,8 | 3,4,5 | 1,2 |
| High School | 50 | 60 | 70 | 20 |
| College | 20 | 15 | 45 | 50 |
| Graduate | 10 | 20 | 80 | 60 |

At a .05 level of significance, test if these two variables are independent.

**Solution:**

a. Set up the test:

   $H_0$: The number of children in a family is not related to the mother's level of education.

   $H_1$: The number of children in a family is related (not independent) to education.

   $\alpha = .05$

b. $\chi^2_{\alpha,v}$

   $v = (r - 1)(c - 1)$

   $\quad = (2)(3) = 6$

   $\chi^2_{.05,6} = 12.592$

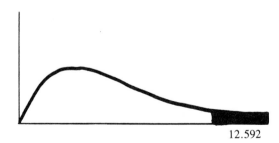

12.592

c. Calculate:

$$\chi_c^2 = \sum_{i=j}^{r} \sum_{j=1}^{c} \frac{(O_{ij} - E_{ij})^2}{E_{ij}}$$

1. **Observed data or $O_{ij}$:**

| Education | Number of Children | | | | | |
|---|---|---|---|---|---|---|
| | More than 8 | 6,7,8 | 3,4,5 | 1,2 | Total | |
| High School | 50 | 60 | 70 | 20 | 200 | .40 |
| College | 20 | 15 | 45 | 50 | 130 | .26 |
| Graduate | 10 | 20 | 80 | 60 | 170 | .34 |
| Total | 80 | 95 | 195 | 130 | 500 | 1.00 |

$$\text{High School} \qquad \frac{200}{500} = .40$$

$$\text{College} \qquad \frac{130}{500} = .26$$

$$\text{Graduate} \qquad \frac{170}{500} = .34$$

2. **Expected data or $E_{ij}$:**
For the first column, multiply .4, .26, .34, by 80; for the second column, multiply .4, .26, .34 by 95 and so on and $E_{ij}$ becomes:

| | More than 8 | 6,7,8 | 3,4,5 | 1,2 |
|---|---|---|---|---|
| High School | 32 | 38 | 78 | 52 |
| College | 20.8 | 24.7 | 50.7 | 33.8 |
| Graduate | 27.2 | 32.3 | 66.3 | 44.2 |

3. **Càlculate $\chi_c^2$:**

$$\chi_c^2 = \frac{(50 - 32)^2}{32} + \frac{(60 - 38)^2}{38} + \frac{(70 - 78)^2}{78} + \frac{(20 - 52)^2}{52}$$

$$+ \frac{(20 - 20.8)^2}{20.8} + \frac{(15 - 24.7)^2}{24.7} + \frac{(45 - 50.7)^2}{50.7} + \frac{(50 - 33.8)^2}{33.8}$$

$$+ \frac{(10 - 27.2)^2}{27.2} + \frac{(20 - 32.3)^2}{32.3} + \frac{(80 - 66.3)^2}{66.3} + \frac{(60 - 44.2)^2}{44.2}$$

$$= 10.13 + 12.74 + .82 + 19.69 + .03 + 3.81 + .64 + 7.76 + 10.88 +$$
$$4.68 + 2.83 + 5.65 = 79.66$$

d. $\chi_c^2$ falls in the shaded or rejection region.
   $H_0$ is rejected.
e. There is a relationship between the number of children in a family and the level of education of the mother, and the probability of that is 0.95.

### 4. Test of homogeneity

Chi-square distribution can be used to test for the homogeneity of two samples. The test can determine whether the two samples are drawn from the same population.

*Example 7.15:* Two statistics sections, one in the morning and the other in the evening, scored the following grades on the Final:

| Grade | Morning Class | Evening Class |
|-------|---------------|---------------|
| A | 6 | 7 |
| B | 10 | 15 |
| C | 15 | 10 |
| D | 14 | 12 |
| F | 5 | 6 |
| | 50 | 50 |

At a .05 level of significance, test the hypothesis that the two sections are homogeneous.

**Solution:**

a. Set up the test:

$H_0$: The two samples are homogeneous or drawn from the same population.

$H_1$: The two samples are not homogeneous.

$\alpha = .05$

b. $\chi^2_{\alpha,v}$

$\chi^2_{.05,4} = 9.488$ $\qquad v = (r - 1)(c - 1) = 4$

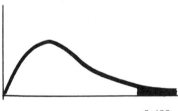

9.488

c. Calculate $\chi_c^2 = \Sigma \dfrac{(O - E)^2}{E}$

| Morning Class (O) | Evening Class (O) | Total | |
|-------------------|-------------------|-------|------|
| 6 | 7 | 13 | .13 |
| 10 | 15 | 25 | .25 |
| 15 | 10 | 25 | .25 |
| 14 | 12 | 26 | .26 |
| 5 | 6 | 11 | .11 |
| 50 | 50 | 100 | |

The expected values (E) are derived as follows:

For the Morning class: Multiply the percentages .13, .25, .25, .26, .11 all by 50, the total number of students in the class.

For the Evening class: Follow the same steps.

| Morning class (E) | Evening class (E) |
|---|---|
| 6.5 | 6.5 |
| 12.5 | 12.5 |
| 12.5 | 12.5 |
| 13 | 13 |
| 5.5 | 5.5 |

$$\chi_c^2 = \frac{(6 - 6.5)^2}{6.5} + \frac{(10 - 12.5)^2}{12.5} + \frac{(15 - 12.5)^2}{12.5} + \frac{(14 - 13)^2}{13}$$

$$+ \frac{(5 - 5.5)^2}{5.5} + \frac{(7 - 6.5)^2}{6.5} + \frac{(15 - 12.5)^2}{12.5} + \frac{(10 - 12.5)^2}{12.5}$$

$$+ \frac{(12 - 13)^2}{13} + \frac{(6 - 5.5)^2}{5.5}$$

$$= 2.34$$

d.  $\chi_c^2$ falls in the acceptance region

   $H_0$ is accepted.

e.  The probability that the two samples are drawn from the same population is 0.95.

## 3.b. *F-distribution*

F-distribution or the variance ratio can be used to test for the difference between two sample variances to determine whether they are from populations with the same variances. Also, F-distribution is being used in the analysis of variance to test for the difference among population means.

1. *Test for the difference between two variances*

F-distribution of the following form can be used to run the test for the difference between two variances:

$$F_{v1,v2} = \frac{\chi_1^2/v_1}{\chi_2^2/v_2} \; ; \qquad v_1 = n_1 - 1 \quad \text{and} \quad v_2 = n_2 - 1$$

or $\qquad F_{v1,v2} = \frac{\hat{s}_1^2}{\hat{s}_2^2}$

where $\hat{s}_1^2 = s_1^2 \left( \dfrac{n_1}{n_1 - 1} \right)$

$\qquad \hat{s}_2^2 = s_2^2 \left( \dfrac{n_2}{n_2 - 1} \right)$

and $v_1$, $v_2$ refer to the degrees of freedom.

*Example 7.16:* Two different types of machines are used to produce 12-inch cylinders in two separate factories. A sample of size 50 from the first factory production yielded a variance of 0.18" while a sample of size 42 of the second factory production yielded a variance of 0.12".

At 0.05 level of significance, test the hypothesis that the variance of the production in both factories is the same.

**Solution:**

a. Set the test:

$$H_0: \sigma_1^2 = \sigma_2^2$$
$$H_1: \sigma_1^2 \neq \sigma_2^2$$
$$\alpha = .05$$

$$v_1 = n_1 - 1 = 50 - 1 = 49$$
$$v_2 = n_2 - 1 = 42 - 1 = 41$$
$$F_{49,41} = 1.80$$

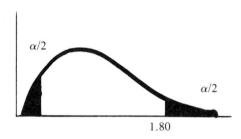

1.80

b. Calculate

$$F_{v1,v2} = s_1^2 \frac{n_1}{n_1 - 1} \div s_2^2 \frac{n_2}{n_2 - 1}$$

$$= .18 \frac{50}{49} \div .12 \frac{42}{41}$$

$$= 1.496$$

c. $F_{v1,v2}$ falls in the acceptance region
   $H_0$ is accepted.

d. The probability is .95 that the variance of the production in both factories is the same.

## 2. *Analysis of variance*

Analysis of variance consists of a set of models constructed to describe different cases for testing the hypothesis that the groups under investigation came from populations of equal means or $\mu_1 = \mu_2 = \mu_3 = \ldots$.

There is no intention to present the different models of analysis of variance in this section. However, a simple model is introduced to show the application of F-distribution in making decisions about the equality of estimated population means. This simple model is:

$$Y_{ij} = \mu + T_i + e_{ij}$$

where $Y_{ij}$ = observations

$\mu$ = population mean

T = Treatment or subdivision of the experiment observed associated with the $i^{th}$ group. It is assumed that $\Sigma T_i = 0$.

e = random error with mean = 0, and a constant variance. (The random error is assumed to be independent and normally distributed.)

F-distribution to be used for this test is as follows:

$$F_{v1,v2} = \frac{\text{variance among column means}}{\text{variance within columns}}$$

$v_1 = c - 1$      c = number of columns

$v_2 = n - c$      n = number of total observations

variance among column means = sum of squares among column means/$v_1$

variance within columns = sum of squares within colums/$v_2$

The data needed to calculate $F_{v1,v2}$ is arranged in a table called Analysis of Variance (ANOV) table where the term Mean Square (MS) is used to denote the variance. The form of ANOV table is as follows:

**ANOV TABLE**

| Due to | SS | v | MS |
|--------|-----|-----|-----|
| Among columns | $SS_a$ | $v_1 = c - 1$ | $SS_a/v_1 = MS_a$ |
| Within columns | $SS_w$ | $v_2 = n - c$ | $SS_w/v_2 = MS_w$ |
| Total | $SS_t$ | $v_1 + v_2 = n - 1$ | |

$$F_{v1,v2} = \frac{MS_a}{MS_w}$$

*Example 7.17:* The following is data collected concerning the time, in minutes, spent by 4 workers in producing one part during each day of the week:

| Day | A | B | C | D |
|-----|-----|-----|-----|-----|
| Monday | 20 | 19 | 16 | 22 |
| Tuesday | 18 | 18 | 19 | 19 |
| Wednesday | 17 | 19 | 20 | 18 |
| Thursday | 21 | 20 | 22 | 16 |
| Friday | 16 | 21 | 20 | 17 |

At a .05 level of significance, test the hypothesis that the average time spent by every worker in producing one unit is equal, or $\mu_1 = \mu_2 = \mu_3 = \mu_4$.

**151**

**Solution:**

a. Set the test:

$$H_0: \mu_1 = \mu_2 = \mu_3 = \mu_4$$

$H_1$: The means are not equal

$$\alpha = 0.05$$

$$F_{v1,v2} = F_{3,16} = 4.08$$

$$v_1 = c - 1 = 4 - 1 = 3$$

$$v_2 = n - c = 20 - 4 = 16$$

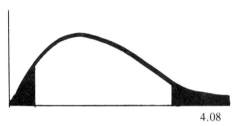

4.08

b. Calculate $F_{v1,v2}$:

|   | A | B | C | D | Total |
|---|---|---|---|---|---|
|   | 20 | 19 | 16 | 22 |   |
|   | 18 | 18 | 19 | 19 |   |
|   | 17 | 19 | 20 | 18 |   |
|   | 21 | 20 | 22 | 16 |   |
|   | 16 | 21 | 20 | 17 |   |
| $\Sigma$ | 92 | 97 | 97 | 92 | 378 |
| n | 5 | 5 | 5 | 5 | 20 |
|   | 18.4 | 19.4 | 19.4 | 18.4 | 18.9  Average |

Total sum of squares $= (20)^2 + (18)^2 + (17)^2 + (21)^2 + (16)^2 + (19)^2 + (18)^2 +$

$$\ldots + (16)^2 + (17)^2 - \frac{(378)^2}{20}$$

$$= 67.8$$

Sum of squares among column means $= \dfrac{(92)^2}{5} + \dfrac{(97)^2}{5} + \dfrac{(97)^2}{5}$

$$+ \frac{(92)^2}{5} - \frac{(378)^2}{20} = 5.0$$

With this information the Analysis of Variance table can be constructed as follows:

| Due to | SS | v | MS |
|---|---|---|---|
| Workers (columns) | 5.0 | 3 | 1.67 |
| Error (within columns) | 62.8 | 16 | 3.93 |
| Total | 67.8 | 19 |   |

**152**

$$F_{3,16} = \frac{1.67}{3.93} = .42$$

c. F calculated falls in the acceptance region;
$H_0$ is accepted.
d. We are confident 95% that there is no significant difference among the average times spent by each worker in producing one unit.

## EXERCISES

7.1    a. Differentiate between point and interval estimation.
      b. What are the reasons for running a test of hypothesis?
      c. What is the difference between type I and type II error?
      d. When do we use a one-tail test?

7.2    a. The variance for the life of incandescent lamps is known to be 1225 hours. A life test on 108 lamps yielded a mean of 1681 hours. Construct a 90% confidence interval for the actual population mean.
      b. The variance of usage of brand X tires is found to be 2500 miles. A random sample of 10 tires yielded an average usage of 3850 miles. At a level of confidence of .90, find the two limits between which the true average usage falls.

7.3    A car dealer wishes to determine the actual average miles per gallon (mpg) a medium sized car can deliver. Ten tests were run and the following results are made available:

| mpg |
|-----|
| 22 |
| 21 |
| 18 |
| 21 |
| 20 |
| 19 |
| 20 |
| 19 |
| 18 |
| 23 |

Construct a 99% confidence interval for the actual average miles per gallon.

7.4    a. A random sample of 25 test pieces of cotton yielded a sample mean of 84 pounds breaking strength with a standard deviation of 12 pounds. Construct an 80% confidence interval.
      b. If we want to estimate the average mechanical aptitude of a large group of people, how large a sample should we take to be 95% confident that our estimate will not differ from the true mean by more than 2.5 points? ($\sigma^2 = 25$)

7.5    a. A random sample of 200 units selected randomly from the total production of a machine contains 10 defectives. Construct a 99% confidence interval for the actual proportion of defectives.

**153**

b. A medical survey showed that 10 of 200 patients failed to recover completely from a given disease. Find .95 confidence interval for the proportion of those not completely recovered from the disease.

7.6 a. A sample survey indicated that 600 of 1100 families in the midwest would like to move to the west coast.

Construct a 99% confidence interval for the actual proportion of the families in the midwest that desire to move to the west coast.

b. A market research survey showed that 400 of 1000 families owning television sets watch a particular program regularly.

Construct a .99 confidence interval for the universe proportion.

7.7 In a big company a level of erroneous entries of 3% is considered acceptable. An internal auditor selected a sample of 250 files randomly and found 10 erroneous vouchers. What inference should you draw at .05 level of significance?

7.8 A machine, when in adjustment, produces parts that have a mean length of 12″ with standard deviation of 1.5″. A random sample of size 50 parts yielded a mean length of 12.5″.

At a .05 level of significance, would you conclude that the machine still is in adjustment?

7.9 The standard time for a certain assembly operation is 3.5 minutes with variance of .81 minutes. Mr. Smith, the new operator, has been observed and timed in this job 25 times and it is found that the average time he spent is 4.2 minutes.

At a level of significance of .01, would you recommend Mr. Smith be retrained?

7.10 The manager of a grocery chain store claimed that the average sales per customer has increased from $18.89 as a result of abolishing trading stamps and decreasing prices.

A random sample of 600 customers selected at random yielded an average sale of $19.06. At a level of significance of .01, would you agree with the manager's claim? (s = 4)

7.11 A survey of 50 families in a large city showed that their average net income is $12,000.00 with a standard deviation of $1,250.00. The state average income is $12,900.00. At a level of significance of .05, test the claim that there is no significant difference between the state and the survey average income.

7.12 A physical education teacher arranged a group of 10 students to test the effectiveness of a newly designed program to lose weight. The following are the weights before and after the program:

| Weights before (pounds) | Weights after (pounds) |
|:---:|:---:|
| 210 | 195 |
| 176 | 169 |
| 197 | 189 |
| 188 | 175 |
| 156 | 148 |
| 200 | 169 |
| 190 | 175 |
| 220 | 200 |
| 199 | 189 |
| 206 | 200 |

At a .01 level of significance, would you agree with the teacher's claim that the new program is effective?

7.13 An electric company claimed that in a certain area 90% of the families use electricity for heating. A sample survey showed that 825 of 1200 families interviewed use electricity for heating.

At a .05 level of significance, test the validity of the company's claim.

7.14 Random samples of the height of basketball players in two universities produced the following results:

$$n_1 = 50 \qquad \overline{X}_1 = 70.9'' \qquad s_1 = 2.4''$$
$$n_2 = 100 \qquad \overline{X}_2 = 69.8'' \qquad s_2 = 2.9''$$

At a .05 level of significance, test the difference between the average height of the two samples.

7.15 The percentage of defectives produced by a certain manufacturing process is .18. A supplier claimed that a special treatment to the new raw material will reduce the percentage of producing defectives. A trial run with the new raw material showed that an output of 700 yielded 90 defective units. Is the claim of the supplier valid? (Use $\alpha = .05$)

7.16 A production manager claimed that the absence of production workers during a particular month appears to follow a Poisson distribution.

To investigate his claim the following data have been collected:

| Absence (in days) | Frequency |
|:---:|:---:|
| 0 | 35 |
| 1 | 45 |
| 2 | 25 |
| 3 | 18 |
| 4 | 19 |
| 5 | 10 |
| 6 | 5 |

Would you agree with the production manager? Use $\alpha = .05$.

7.17 A sample of 750 persons of different ages were interviewed to test the relationship between age and smoking. The results of the study are presented in the following table:

| | Age | | | |
|---|---|---|---|---|
| | less than 20 | 20–40 | 40–60 | above 60 |
| Smoker | 40 | 160 | 100 | 50 |
| Nonsmoker | 50 | 100 | 150 | 100 |

At a .05 level of significance, would you conclude that a relationship exists between age and smoking?

# 8

# Statistical Decision Making: Bayesian Decision Theory Without Sampling

Estimation and Test of Hypotheses, the two components of the classical approach to statistical decision making presented in the previous chapter, provide the decision maker with more information concerning the population parameters. Such information can help the decision maker refine subjective decisions, those based on practical experience and personal intuition. As the economy becomes more complex, alternative courses of actions become available and the decision maker relies to a greater extent on scientific method to achieve the goals of maximization the profit or minimization the loss. The body of this scientific method is the modern approach to statistical decision making called ''Bayesian Decision Theory.'' This method is being applied to both business and government to enable the decision maker to select the course of action that optimizes his goal.

Managerial decision makers are faced with two or more alternative courses of action in every aspect of management. Inventory, production, marketing, and investment are a few of the fields where the decision maker has to select one of the many courses of action available for optimization of goals. In the area of inventory, the decision may be made to maintain an optimum level of stock to minimize the cost of storage as well as the loss of sales due to shortage. However, the optimum level of stock depends on the expected sale of the commodity. The potential level of sales of any commodity is a strong determining factor in any decision concerning the production of the commodity. A decision concerning the market share for the commodity can effect the production of this commodity. On the other hand, decisions concerning the production of a commodity may lead to the replacement of an obsolete machine with new and more advanced equipment. This investment decision involves many financial factors such as alternative rate of returns, etc. In addition, production depends on the marketability of the product which in turn depends on the probabilities of sale.

A decision maker in today's complex economy faces many problems that require solutions, therefore, he should be equipped with the scientific methods that help him select the course of action that maximizes profit or minimizes loss.

## Structure of a Decision Problem

The decision maker has to find the optimal course of action in any problem situation where two or more alternative courses of action are available. The decision should be based on scientific method to obtain the appropriate course of action in a problem situation.

To make a decision, the decision maker must determine the outcome or outcomes of each alternative course of action. Each outcome is called an "event" or "state of nature." The occurrence of any state of nature is unpredictable because it depends on uncontrollable outside forces.

If the state of nature for a decision problem is known with absolute certainty, then decisions made under this condition are said to be "decisions under certainty." For example, if the decision maker wants to produce a different product mix, where each mix can be viewed as a course of action and the decision maker knows for sure the production as well as storage capacity, then the decision concerning production can be made under certainty. In this case, linear programming techniques may be used to reach an optimum decision. On the other hand, if the state of nature and its occurrence are not known to the decision maker, then decisions made under this circumstance are "decisions under ignorance." The game theory would play an important role in arriving at a rational decision in this case.

Most of the decision making problems in business are made under "uncertainty" where the occurrence of the state of nature is not known with certainty to the decision maker. Therefore, the decision maker should assign probabilities to the states of nature. These probabilities are called "prior probabilities" and they may be subjective in nature or they may be based on the outcome of relative frequencies constructed from past data. Prior probability can be improved when additional information on the states of nature becomes available. The availability of more information can be incorporated with the prior probability to provide the decision maker with a new set of probabilities for the states of nature called "posterior probabilities." The Bayesian Theorem is based on the revision of the prior probability with additional information to arrive to the posterior probability.

For each course of action there exists many consequences or payoffs. A measure is needed to evaluate the consequences or payoff of each action in order to select the one which optimizes the decision. Payoffs can be expressed in either monetary or utility units. Payoffs expressed in monetary units may also be transformed into utils or units of utility. The calculation of payoffs for each outcome provides the decision maker with a matrix called "Payoff Matrix" or "Decision Matrix." The general form of a decision table is shown below:

| Events \\ Actions | $E_1$ | $E_2$ | $E_3 \ldots \ldots E_n$ |
|---|---|---|---|
| $A_1$ | $X_{11}$ | $X_{12}$ | $X_{13} \ldots \ldots X_{1n}$ |
| $A_2$ | $X_{21}$ | $X_{22}$ | $X_{23} \ldots \ldots X_{2n}$ |
| . | | | |
| . | | | |
| . | | | |
| . | | | |
| $A_m$ | $X_{m1}$ | $X_{m2}$ | $X_{m3} \ldots \ldots X_{mn}$ |

The body of this decision table is the decision matrix $X_{ij}$ whose elements are:

$$X_{ij} = \begin{bmatrix} X_{11} & X_{12} & X_{13} \ldots \ldots X_{1n} \\ X_{21} & X_{22} & X_{23} \ldots \ldots X_{2n} \\ . \\ . \\ . \\ X_{m1} & X_{m2} & X_{m3} \ldots \ldots X_{mn} \end{bmatrix}$$

A numerical example to illustrate the construction of a payoff table is presented below:

*Example 8.1:* A company is planning to produce a specific part. A set of machines used at the full capacity level during any month can produce 10,000 units of this part. The sale price is set at $5.00 per unit, the total cost is estimated at $3.00 per unit, and the storage cost is estimated at $0.20 per unit per month. The total cost per unit includes an average storage cost for a month. The decision is to be made about the total parts to be produced during any month to maximize the profit.

Payoffs which are the elements of the decision matrix can be expressed as a function $X_{ij} (A_i, E_j)$ called "payoff function" as follows:

$$X_{ij} (A_i, E_j) = \alpha A_i^0 \qquad \text{for } E_j \geqslant A_i^0$$
$$= \alpha E_j - \beta (A_i^0 - E_j) \qquad \text{for } E_j < A_i^0$$

where $\alpha$ refers to profit per unit, $\beta$ stands for loss per unit, and $A_i^0$ is a given optimal course of action.

**Solution:** The first step is to construct a payoff table as follows:

| Event \ Action | Sales | | |
|---|---|---|---|
| | $E_1 = 10,000$ | $E_2 = 20,000$ | $E_3 = 30,000$ |
| $A_1$: Produce 10,000 | $10,000 \times \$2.00$ $= \$20,000$ | $10,000 \times \$2.00$ $= \$20,000$ | $10,000 \times \$2.00$ $= \$20,000$ |
| $A_2$: Produce 20,000 | $10,000 \times \$2.00$ $- 10,000 \times \$0.20$ $= \$18,000$ | $20,000 \times \$2.00$ $= \$40,000$ | $20,000 \times \$2.00$ $= \$40,000$ |
| $A_3$: Produce 30,000 | $10,000 \times \$2.00$ $- 20,000 \times \$0.20$ $= \$16,000$ | $20,000 \times \$2.00$ $- 10,000 \times \$0.20$ $= \$38,000$ | $30,000 \times \$2.00$ $= \$60,000$ |

After the construction of the payoff table, the decision maker can apply a "decision strategy" or "criterion" for the selection of a course of action among the alternatives to maximize the profit or minimize the loss. There are many criterions available to the decision maker. Some of these criterions are:

1. The maximin criterion of payoffs.
2. The maximax criterion of payoffs.
3. The Hurwitz criterion of payoffs.

4. The minimax criterion of opportunity losses.

5. The expected value criterion or Bayesian criterion.

The following is a brief explanation of the first four criteria. "The Bayesian Criterias" will be discussed later in greater detail.

### 1. *The Maximin Criterion:*

This criterion is called the "pessimistic criterion" because the decison maker selects the action with the maximum payoff from all other actions with the minimum payoff. In other words, after the construction of a payoff table the decison maker selects the minimum payoffs of each action. From those minimums he selects the course of action that has the highest payoff. For example, in the payoff table constructed in example 8.1, the maximin criterion can be determined as follows:

| Action | Minimum payoff |
|--------|----------------|
| $A_1$ | 20,000 → Maximin |
| $A_2$ | 18,000 |
| $A_3$ | 16,000 |

$A_1$ will be selected by the decision maker according to the Maximin Criterion because $A_1$ has the highest of all minimum payoffs produced by the three actions available to the decision maker.

This criterion is considered to be very conservative. The extreme of Maximin is the Maximax Criterion.

### 2. *The Maximax Criterion:*

This criterion is considered optimistic decision making. According to this criterion the decision maker determines the maximum payoff for each action, then selects the action with the highest payoff of those maximums.

The Maximax Criterion can be applied to the payoff table constructed in example 8.1 as follows:

| Action | Maximum payoff |
|--------|----------------|
| $A_1$ | 20,000 |
| $A_2$ | 40,000 |
| $A_3$ | 60,000 → Maximax |

$A_3$ is the course of action to be selected according to this criterion.

### 3. *Hurwitz Criterion:*

Both the maximin and the maximax criterions are considered to be extreme strategies. One is completely pessimistic while the other is completely optimistic. Therefore, Hurwitz developed a new criterion based on the use of a coefficient "$\alpha$" called coefficient of optimism. His approach is a weighted average of the two extreme criterions: the maximin and the maximax. If $\alpha$ the coefficient of optimism equals zero, then Hurwitz criterion will be the same as the Maximin criterion, while if $\alpha = 1$ then Hurwitz criterion will be the same as the Maximax Criterion. For each course of action, the decision maker can express his degree of optimism as a value to be given to the coefficient of optimism $\alpha$ and then calculate an H value for each action using the following formula:

$$H(A_i) = \alpha(Max\ X_{ij}) + (1 - \alpha)(Min\ X_{ij})$$

The highest H value is the course of action selected.

*Example 8.2:* From the payoff table constructed in example 8.1, find the optimal course of action using Hurwitz Criterion with $\alpha = .60$.

**Solution:** From the payoff table of example 8.1, the following are the Maximum and the Minimum payoffs of each course of action:

| Action | Maximum | Minimum |
|--------|---------|---------|
| $A_1$ | 20,000 | 20,000 |
| $A_2$ | 40,000 | 18,000 |
| $A_3$ | 60,000 | 16,000 |

$\alpha = .60$ indicates that the decision maker is somewhat optimistic.

The optimal action to be selected according to the Hurwitz criterion can be determined as follows:

$$H(A_1) = (.60)\ (20,000) + (.40)\ (20,000) = 20,000$$
$$H(A_2) = (.60)\ (40,000) + (.40)\ (18,000) = 31,200$$
$$H(A_3) = (.60)\ (60,000) + (.40)\ (16,000) = 42,400$$

$H(A_3)$ has the maximum value of all $H(A_i)$, therefore, $A_3$ is the optimal course of action to be selected by the decision maker according to the Hurwitz criterion.

**Another Solution:** Matrix algebra can be used to calculate $H(A_i)$:

$$\begin{bmatrix} Max\ X_{ij} & Min\ X_{ij} \end{bmatrix} \cdot \begin{bmatrix} \alpha \\ 1 - \alpha \end{bmatrix} = \begin{bmatrix} H(A_1) \\ H(A_2) \\ \cdot \\ \cdot \\ \cdot \\ H(A_n) \end{bmatrix}$$

or

$$\begin{bmatrix} 20,000 & 20,000 \\ 40,000 & 18,000 \\ 60,000 & 16,000 \end{bmatrix} \cdot \begin{bmatrix} .60 \\ .40 \end{bmatrix} = \begin{bmatrix} 20,000 \\ 31,300 \\ 42,400 \end{bmatrix} \rightarrow A_3$$

$$3,2 \qquad\qquad 2,1 \qquad\qquad 3,1$$

4. *The Minimax of Opportunity Losses:*

The minimax of opportunity losses is the same as the maximin of payoffs, both of the two criterion are described to be pessimistic.

To apply this criterion, an opportunity loss or regret table should be constructed. An opportunity loss is the difference between realized profit and that profit which would have been realized if a specific action had been selected. For example, if $A_1$ is selected as the optimal course of action, the decision maker wil! produce 10,000 units per month as stated in example 8.1, then if $E_1$, event of selling 10,000 units per month occurs, the opportunity loss in this case is zero. However, if $E_2$ event of selling 20,000 units per month occurs, then there is an opportunity loss of $20,000 which equals 10,000 units demanded but not produced × $2.00 (profit/unit).

Opportunity losses or regrets, the elements of a regret matrix or an opportunity loss matrix, can be calculated from the following "loss or regret function":

$$X_{ij}(A_i, E_j) = \alpha\,(E_j - A_i^0) \qquad \text{for } E_j > A_i^0$$
$$= \beta\,(A_i^0 - E_j) \qquad \text{for } E_j \leq A_i^0$$

where $\alpha$ stands for profit per unit, $\beta$ refers to loss per unit, and $A_i^0$ is a given optimal course of action.

The following is an illustration of an opportunity loss table constructed for example 8.1:

| Events \ Action | Sales | | |
|---|---|---|---|
| | $E_1 = 10,000$ | $E_2 = 20,000$ | $E_3 = 30,000$ |
| $A_1$: Produce 10,000 | 0 | $20,000 | $40,000 |
| $A_2$: Produce 20,000 | $2,000 (storage cost) | 0 | $20,000 |
| $A_3$: Produce 30,000 | $4,000 (storage cost) | $2,000 (storage cost) | 0 |

The Minimax Criterion of opportunity losses requires the decision maker to determine the maximum opportunity loss for each action and then to select the action that has the minimum opportunity loss as shown below:

| Action | Maximum Opportunity Loss | |
|---|---|---|
| $A_1$ | 40,000 | |
| $A_2$ | 20,000 | |
| $A_3$ | 4,000 | Minimax |

The course of action $A_3$ is the optimal one to be selected according to the Minimax Criterion of regrets or opportunity losses.

All the criteria presented consider the extreme value of payoffs or opportunity losses and disregard the other elements of the payoff or opportunity loss matrix. This shortcoming has been compensated for in the Expected Value Criterion.

5. *The Expected Value Criterion:*

The decision maker is usually faced with alternative courses of action where each of those actions has one or more outcome. To make an optimal decision the decison maker has to calculate the "decision matrix" which consists of payoffs or regrets (opportunity losses) for the outcomes of the alternative courses of action available. The calculation of the decision matrix is the first step in the process of choosing an optimal course of action.

**162**

The calculation can be completed when the decision maker implements a decision strategy or criterion to his decison matrix. Four decision criteria have been presented and any of these four may be implemented in order to arrive at the optimal decision. However the four criteria depend only on extreme values and disregard the other values. This is a shortcoming that has been corrected in the fifth and the most popular decision strategy: The Expected Value Criterion.

The Expected Value Criterion utilizes all the elements of the decision matrix. The expected value of any course of action $E(A_i)$ is derived by multiplying the decision matrix $(X_{ij})$ by a column vector representing the probabilities assigned to each state of nature or event $P(E_i)$. In matrix notation, the expected values of alternative courses of action are expressed as follows:

$$E(A_i) = (X_{ij}) \cdot P(E_i)$$

In order to determine the probabilities of the states of nature, the decision maker has to make some assumption concerning the probability of the occurrence of those events. The assignment of those probabilities can be based on judgment, intuition, or experience. Or the probability of the states of nature occurring may follow a specific probability distribution. Also the probabilities assigned may be determined by a previous experiment where relative frequencies of the occurrence of each event have been calculated from the collected data. As a last resort, the decision maker may assign equal probabilities for all the states of nature. These probabilities assigned to the states of nature or events are called "Prior Probabilities." As explained later, prior probabilities can be improved as more information becomes available to produce a new set of probabilities called "Posterior Probabilities." The optimal decision, desired by the decision maker using the calculation of the expected values with prior or posterior probabilities, is the major emphasis of the Bayesian Decision Theory.

An illustration of arriving at an optimal course of action through the use of the Expected Value Criterion is shown in example 8.3. Example 8.3 is identical to example 8.1 except for the additional information concerning the probabilities of the states of nature.

*Example 8.3:* A company is planning to produce a specific part. A set of machines used at full capacity during any month can produce 10,000 units of this part. The sale price is set at $5.00 per unit, the total cost is estimated at $3.00 per unit, and the storage cost is estimated at $0.20 per unit per month. The total cost per unit includes an average storage cost for a month.

A market research analyst suggested the following probabilities of different sales levels:

| Event $(E_i)$ | $P(E_i)$ |
|---|---|
| $E_1$: Sales of 10,000 units | 0.50 |
| $E_2$: Sales of 20,000 units | 0.45 |
| $E_3$: Sales of 30,000 units | 0.05 |
| $E_4$: Sales of 40,000 units | 0.00 |
| | 1.00 |

The decision to be made concerns the total parts to be produced during any month for maximization of profit.

**Solution:** This is a decision making problem where the Expected Value Criterion is to be implemented. Therefore, a payoff table is constructed, the decision matrix is derived, and the expected payoff for each course of action is calculated. To maximize profits the action with the highest expected payoff is selected.

1. The payoff table can be constructed from the payoff function introduced previously:

$$X_{ij}(A_i, E_j) = \alpha A_i^0 \qquad\qquad \text{for } E_j \geq A_i^0$$
$$= \alpha E_j - \beta(A_i^0 - E_j) \qquad \text{for } E_j < A_i^0$$

| Events  Actions | $E_1 = 10{,}000$ | Sales  $E_2 = 20{,}000$ | $E_3 = 30{,}000$ |
|---|---|---|---|
| $A_1$: Produce  10,000 | $20,000 | $20,000 | $20,000 |
| $A_2$: Produce  20,000 | 18,000 | 40,000 | 40,000 |
| $A_3$: Produce  30,000 | 16,000 | 38,000 | 60,000 |

A sales level of 40,000 has been eliminated because of its zero probability. From the payoff table, the decision matrix $X_{ij}$ can be derived as follows: (in thousands of dollars)

$$X_{ij} = \begin{bmatrix} 20 & 20 & 20 \\ 18 & 40 & 40 \\ 16 & 38 & 60 \end{bmatrix}$$

2. Probabilities of the states of nature arranged in a column vector are:

$$P(E_i) = \begin{bmatrix} 0.50 \\ 0.45 \\ 0.05 \end{bmatrix}$$

3. Expected payoff for the alternative courses of action $A_1$, $A_2$, and $A_3$ can be calculated by multiplying the decision matrix ($X_{ij}$) by the probabilities of the states of nature $P(E_j)$ as follows:

$$X_{ij} \cdot P(E_i) = E(A_i)$$

or

$$\begin{bmatrix} 20 & 20 & 20 \\ 18 & 40 & 40 \\ 16 & 38 & 60 \end{bmatrix} \cdot \begin{bmatrix} 0.50 \\ 0.45 \\ 0.05 \end{bmatrix} = \begin{bmatrix} (20)(.50) + (20)(.45) + (20)(.05) \\ (18)(.50) + (40)(.45) + (40)(.05) \\ (16)(.50) + (38)(.45) + (60)(.05) \end{bmatrix}$$

or

$$\begin{bmatrix} E(A_1) \\ E(A_2) \\ E(A_3) \end{bmatrix} = \begin{bmatrix} 20 \\ 29 \\ 28.10 \end{bmatrix} \rightarrow \text{Optimal action}$$

**164**

The expected payoff of $A_2$ is the highest, therefore, the decision maker will select $A_2$ as the optimal course of action and the production will be set at 20,000 units per month in order to maximize profit.

**Another Solution:** Instead of constructing a Payoff Table, an Opportunity Loss Table or Regret Table may be constructed according to the following "loss or regret function":

$$X_{ij}(A_i, E_j) = \alpha (E_j - A_i^0) \qquad \text{for } E_j > A_i$$
$$= \beta (A_i^0 - E_j) \qquad \text{for } E_j \leq A_i^0$$

After deriving the opportunity loss table, the decision matrix, which constitutes the body of the opportunity loss table $(X_{ij})$, is multiplied by the probabilities of the states of nature $P(E_i)$. This will produce the expected opportunity loss of all alternative courses of action $E(A_i)$. To minimize the loss, the course of action with the lowest expected opportunity loss value will be chosen.

1. Opportunity Loss Table

| Events / Actions | $E_1 = 10,000$ | Sales $E_2 = 20,000$ | $E_3 = 30,000$ |
|---|---|---|---|
| $A_1$: Produce 10,000 | 0 | 20,000 | 40,000 |
| $A_2$: Produce 20,000 | 2,000 | 0 | 20,000 |
| $A_3$: Produce 30,000 | 4,000 | 2,000 | 0 |

From the opportunity loss table, the decision matrix $(X_{ij})$ is derived as follows: (in thousands of dollars).

$$X_{ij} = \begin{bmatrix} 0 & 20 & 40 \\ 2 & 0 & 20 \\ 4 & 2 & 0 \end{bmatrix}$$

2. Probabilities of the states of nature $P(E_i)$ are:

$$P(E_i) = \begin{bmatrix} 0.50 \\ 0.45 \\ 0.05 \end{bmatrix}$$

3. Expected opportunity loss for the three alternative course of actions, $A_1$, $A_2$, and $A_3$, can be calculated as follows:

$$E(A_i) = X_{ij} \cdot P(E_i)$$

$$\begin{bmatrix} 0 & 20 & 40 \\ 2 & 0 & 20 \\ 4 & 2 & 0 \end{bmatrix} \cdot \begin{bmatrix} 0.50 \\ 0.45 \\ 0.05 \end{bmatrix} = \begin{bmatrix} 11 \\ 2 \\ 2.9 \end{bmatrix} \rightarrow \text{optimal action}$$

The decision maker may select $A_2$ as the optimal action because its expected opportunity loss value is minimum. The decision to be followed is to produce 20,000 units per month.

The Expected Value Criterion is the Bayesian criterion and, in this case, probabilities of the states of nature (sales) are called the prior probabilities.

## Probabilities of the Events and the Break-Even Analysis

To apply the Expected Value Criterion the decision maker has to determine a set of probability values that describe the occurrence of the states of nature or events. As stated previously, those probabilities can be assigned either objectively or subjectively and they are called prior probabilities. As more information concerning the occurrence of the events becomes available, the assigned prior probabilities can be improved and another set of probabilities can be calculated using the Bayesian rule to produce the posterior probabilities. However, in some cases it is very difficult for the decision maker to assign probabilities for the states of nature, e.g., introduction of a new, expensive product. In this case, the probability of different levels of sales (or the states of nature) of this new, expensive product cannot be easily assigned. If the decision maker assigns equal probabilities for the states of nature this approach may not lead to the selection of the optimal course of action. Therefore, the decision maker needs guidelines to help him select the course of action that optimizes his goal and in this case, knowledge of the probability of the state of nature that produces a break-even point can help him to select the optimal action. Regarding states of nature and break-even analysis, a simple decision problem with two alternative courses of action where each action has two states of nature as well as multiaction decision problems, will be presented.

## Two-Action and Two-State of Nature with Linear Payoff Functions

The break-even point is reached when expected values of the alternative courses of action are equal. In a simple decision problem where the decision maker is faced with two actions, and every course of action has only two states of nature, the break-even point is that point where $E(A_1) = E(A_2)$; $A_1$ refers to the first course of action while $A_2$ is its alternative. At the break-even point the decision maker will be indifferent with respect to the selection of $A_1$ or $A_2$, because the expected value of the two actions is equal. If the decision maker calculates the probability of any of the two states of nature at the break-even point, then an optimal course of action will be selected (given a different probability value for the state of nature).

The calculation of the probability values of the states of nature at the break-even point is based on the fact that the sum of the probabilities of the first state of nature ($E_1$) and the second state of nature ($E_2$) equals unity and so the payoff function is linear.

Accordingly, if (p) is set to become the probability of $E_1$, then $(1 - p)$ is the probability of $E_2$. The expected value of the two actions $A_1$ and $A_2$ can be calculated as follows:

$$\begin{bmatrix} E(A_1) \\ E(A_2) \end{bmatrix} = \begin{bmatrix} X_{11} & X_{12} \\ X_{21} & X_{22} \end{bmatrix} \cdot \begin{bmatrix} P \\ 1 - p \end{bmatrix}$$

or

$$E(A_1) = X_{11}\, p + X_{12}\,(1 - p)$$
$$E(A_2) = X_{21}\, p + X_{22}\,(1 - p)$$

At the break-even point $E(A_1) = E(A_2)$

or $\quad X_{11} p + X_{12} (1 - p) = X_{21} p + X_{22} (1 - p)$

To calculate p or $1 - p$, probability of $E_1$ and $E_2$ respectively, solve for p from the above equation. The calculation of p will determine the $P(E_1)$ and the $P(E_2)$ that makes $E(A_1) = E(A_2)$ or simply the break-even point. The decision maker can then select the course of action $A_1$ or its alternative $A_2$ as the optimal action. An example of this simple decision problem is shown below.

*Example 8.4:* Let us assume that the company in Example 8.3 is being faced with two actions and two states of nature for each course of action. The payoff table for this simple decision problem is as follows:

**PAYOFF TABLE**

| Action \ Event | Sales | |
|---|---|---|
| | $E_1 = 10,000 \ P(E_1) = p$ | $E_2 = 20,000 \ P(E_2) = 1 - p$ |
| $A_1$: Produce 10,000 | \$20,000 | \$20,000 |
| $A_2$: Produce 20,000 | 18,000 | 40,000 |

Find the probability of $E_1$ or $E_2$ at the break-even point and then decide on the optimal action to be used.

**Solution:** The payoff table in the problem provides the decision maker with the decision matrix: (in thousands of dollars)

$$X_{ij} = \begin{bmatrix} 20 & 20 \\ 18 & 40 \end{bmatrix}$$

The probabilities of the two states of nature are set to equal p and $(1 - p)$ for $E_1$ and $E_2$ respectively and the probabilities can be expressed in a column vector of $P(E_i)$ as:

$$P(E_i) = \begin{bmatrix} p \\ 1 - p \end{bmatrix}$$

Now the expected value of the two alternative courses of action can be evaluated as follows:

$$\begin{bmatrix} E(A_1) \\ E(A_2) \end{bmatrix} = \begin{bmatrix} 20 & 20 \\ 18 & 40 \end{bmatrix} \cdot \begin{bmatrix} p \\ 1 - p \end{bmatrix} = \begin{bmatrix} 20p + 20(1 - p) \\ 18p + 40(1 - p) \end{bmatrix}$$

At the break-even point $E(A_1) = E(A_2)$, solve for p as follows:

$$20p + 20 (1 - p) = 18p + 40 (1 - p)$$
$$22p = 20$$
$$p = 20/22 = .91$$

This means if $P(E_1) = .91$ then the decision maker is indifferent to selecting either $A_1$ or $A_2$ since $E(A_1) = E(A_2)$.

If $P(E_1) < .91$ let say .90 then:

$$\begin{bmatrix} E(A_1) \\ E(A_2) \end{bmatrix} = \begin{bmatrix} 20 & 20 \\ 18 & 40 \end{bmatrix} \cdot \begin{bmatrix} .90 \\ .10 \end{bmatrix} = \begin{bmatrix} 20 \\ 22 \end{bmatrix} \rightarrow A_2 \text{ (optimal)}$$

then $A_2$ to be selected as the optimal course of action.

While if $P(E_1) > .91$ let say .96 then:

$$\begin{bmatrix} E(A_1) \\ E(A_2) \end{bmatrix} = \begin{bmatrix} 20 & 20 \\ 18 & 40 \end{bmatrix} \cdot \begin{bmatrix} .96 \\ .04 \end{bmatrix} = \begin{bmatrix} 20 \\ 18.88 \end{bmatrix} \rightarrow A_1 \text{ (optimal)}$$

then the optimal course of action becomes $A_1$.

These results can be shown graphically in Figure 8-1:

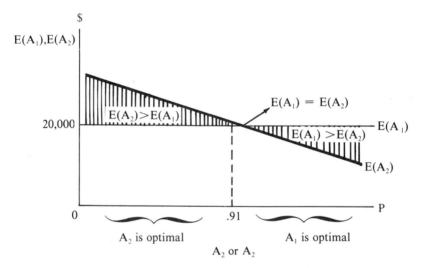

**Figure 8-1.** Break-even analysis.

The same results can be achieved by using the loss opportunity table and calculating the expected opportunity loss values as shown below:

### The Opportunity Loss Table

| Event _____ Action | Sales | |
|---|---|---|
| | $E_1 = 10,000$ | $E_2 = 20,000$ |
| $A_1$: Produce 10,000 | 0 | 20,000 |
| $A_2$: Produce 20,000 | 2,000 | 0 |

**168**

At the break-even point the $E(A_1) = E(A_2)$ or:

$$\begin{bmatrix} E(A_1) \\ E(A_2) \end{bmatrix} = \begin{bmatrix} 0 & 20 \\ 2 & 0 \end{bmatrix} \cdot \begin{bmatrix} p \\ 1 - p \end{bmatrix} = \begin{bmatrix} 0p + 20(1 - p) \\ 2p + 0 \ (1 - p) \end{bmatrix}$$

The probability of $E_1$ can be calculated from the following equation where $E(A_1) = E(A_2)$:

$$(0) p + 20 (1 - p) = (2) p + (0) (1 - p)$$
$$p = 10/11 = .91$$

## Multiaction Decision Problem with Linear Payoff Functions

In real-life situations the decision maker is faced with decision problems that include more than two actions and certainly more than two states of nature. Such problems include decisions involving order quantities, stock levels, production levels, etc.

*Example 8.5:* A decision has to be made to determine the optimal level of inventory for a specific part needed to repair frequently broken machines. There are six machines in the firm. The following is the number of machines and the probability of being broken down during any week:

| Number of Machines ($E_i$) | P (being broken down)/week $P(E_i)$ |
|:---:|:---:|
| 1 | 0.15 |
| 2 | 0.20 |
| 3 | 0.25 |
| 4 | 0.20 |
| 5 | 0.10 |
| 6 | 0.10 |

If the part is out of stock then the opportunity loss for leaving the machine idle is estimated at $20.00. On the other hand, any overstock part has an estimated cost of $5.00.

Find the optimal number of parts to be stocked.

**Solution:** To solve this same problem by applying the expected value criterion, construct an opportunity loss table as follows:

| Event / Action | Number of Broken Down Machines | | | | | |
|:---|:---:|:---:|:---:|:---:|:---:|:---:|
| | $E_1 = 1$ | $E_2 = 2$ | $E_3 = 3$ | $E_4 = 4$ | $E_5 = 5$ | $E_6 = 6$ |
| $A_1$: Stock 1 | 0 | 20 | 40 | 60 | 80 | 100 |
| $A_2$: Stock 2 | 5 | 0 | 20 | 40 | 60 | 80 |
| $A_7$: Stock 3 | 10 | 5 | 0 | 20 | 40 | 60 |
| $A_4$: Stock 4 | 15 | 10 | 5 | 0 | 20 | 40 |
| $A_5$: Stock 5 | 20 | 15 | 10 | 5 | 0 | 20 |
| $A_6$: Stock 6 | 25 | 20 | 15 | 10 | 5 | 0 |

The expected value of alternative courses of action is calculated by multiplying the decision matrix by the probabilities of the states of nature as shown below:

$$
\begin{bmatrix} E(A_1) \\ E(A_2) \\ E(A_3) \\ E(A_4) \\ E(A_5) \\ E(A_6) \end{bmatrix}
\begin{bmatrix}
0 & 20 & 40 & 60 & 80 & 100 \\
5 & 0 & 20 & 40 & 60 & 80 \\
10 & 5 & 0 & 20 & 40 & 60 \\
15 & 10 & 5 & 0 & 20 & 40 \\
20 & 15 & 10 & 5 & 0 & 20 \\
25 & 20 & 15 & 10 & 5 & 0
\end{bmatrix}
\begin{bmatrix} 0.15 \\ 0.20 \\ 0.25 \\ 0.20 \\ 0.10 \\ 0.10 \end{bmatrix}
=
\begin{bmatrix} 44.00 \\ 27.75 \\ 16.50 \\ 11.50 \\ 11.50 \\ 14.00 \end{bmatrix}
$$

The expected value criterion indicates that $A_4$ or $A_5$ may be selected as the optimal course of action because both actions have the lowest opportunity loss. The decision maker being indifferent as to $A_4$ (stock 4 units) or $A_5$ (stock 5 units), he may consider stocking 4 units ($A_4$). This method then is

## Utility and Utility Function

With any decision problem the decision maker has to evaluate payoffs or opportunity losses for each possible course of action under every state of nature. Those payoffs, or opportunity losses, are the elements of the decision matrix and are usually expressed in monetary units. Given a decision matrix and probabilities of the events or the states of nature, the decision maker will be able to select an optimal course of action; the one with the highest expected payoff or with the lowest expected opportunity loss. Usually, the expected payoffs or opportunity losses are measured in monetary units and the selection of an optimal course of action is based on the value in dollars. However, there are situations when rational decision makers do not follow the Bayesian Criterion. Instead of selecting an optimum course of action in terms of the expected value, other courses of action are selected which give a lower expected value. The question is would a decision maker ignore the Bayesian Criterion? Is the Bayesian Criterion for decision making invalid? Absolutely not. However, there are some cases where the payoffs, or the expected payoffs, measured in monetary units, do not necessarily reflect their true worth to the decision maker. The following examples are presented to illustrate this point:

*Example 8.6:* Mr. X, a president of a firm and a strong believer in the effective use of the Bayesian Criterion in making decisions, is leaving for a business trip from New York to Toronto on a commercial airline. Mr. X is a family man and his death would cause a financial disaster for his growing family. Mr. X paid $25.00 for flight insurance of $100,000 even though he knows that the probability of an air accident is as low as .003.

Is Mr. X, as a decision maker and a strong believer in the effective use of the Bayesian Criterion, making the correct decision in choosing to buy flight insurance?

**Solution:** To answer the question raised by example 8.6, set the opportunity loss table for the two possible actions available to Mr. X, namely: to buy or not to buy flight insurance. Then after the construction of the decision matrix for this problem use the probabilities of the two states of nature that may occur: having or not having an air accident. The next step will be to select the optimal action according to the Bayesian Criterion, which in this case, will be the course of action that has the lowest expected opportunity loss.

**170**

The Opportunity Loss Table:

| Event / Action | $E_1$: Air Accident $P(E_1) = .003$ | $E_2$: No Air Accident $P(E_2) = .997$ |
|---|---|---|
| $A_1$: Buy a Flight Insurance | $-25$ | $-25$ |
| $A_2$: Do Not Buy a Flight Insurance | $-100,000$ | $0$ |

From the opportunity loss table the following decision matrix is derived:

$$X_{ij} = \begin{bmatrix} -25 & -25 \\ -100,000 & 0 \end{bmatrix}$$

The probabilities of the events are arranged in a column vector $P(E_i)$ as follows:

$$P(E_i) = \begin{bmatrix} .003 \\ .997 \end{bmatrix}$$

The expected values of the alternative courses of action $E(A_i)$ are:

$$E(A_i) = \begin{bmatrix} -25 & -25 \\ -100,000 & 0 \end{bmatrix} \begin{bmatrix} .003 \\ .997 \end{bmatrix} = \begin{bmatrix} -25 \\ -300 \end{bmatrix} \rightarrow \text{optimal action}$$

According to the Bayesian Criterion, the second action, **Do Not Buy Flight Insurance,** is the optimal action because it has the lowest opportunity loss. The decision maker, Mr. X, did not follow the Bayesian Criterion. In fact, many people buy flight insurance although it is not the optimal action to follow. Are all these people irrational? They are not irrational because they are buying security for their families. Such security cannot be measured in monetary units.

A similar situation exists when homeowners buy fire insurance even though the probability of a home fire occurring is very low. Example 8.7 is an illustration of this case.

*Example 8.7:* Mr. Y, a homeowner, bought nondeductible fire insurance for $35,000 for his home and he paid $90.00 as a premium. Mr. Y knows that the probability of a home fire occurring is .004.

Is Mr. Y's decision consistent with the Bayesian Criterion?

**Solution:** To find out whether Mr. Y's decision adheres to the Bayesian Criterion, calculate the expected value of the two alternative courses of action available to him, to buy or not to buy home fire insurance. The first step is to calculate the opportunity loss table as follows:

| Action | $E_1 = Fire$ $P(E_1) = .004$ | $E_2 = No\ Fire$ $P(E_2) = .996$ |
|---|---|---|
| $A_1$ Buy a home fire insurance | $-90$ | $-90$ |
| $A_2$ Do Not Buy | $-35,000$ | $0$ |

The expected value of the two alternative actions is calculated as follows:

$$E(A_i) = \begin{bmatrix} -90 & -90 \\ -35,000 & 0 \end{bmatrix} \cdot \begin{bmatrix} .004 \\ .996 \end{bmatrix} = \begin{bmatrix} -90 \\ -140 \end{bmatrix} \rightarrow \text{optimal action}$$

Again, this example shows that people should not buy home fire insurance because the expected opportunity loss of this action is lower than its alternative. Nonetheless, most homeowners buy insurance to cover their homes in case of any fire. The real reason behind buying insurance is protection against losing a very valuable investment and "peace of mind". Neither can be measured in monetary units.

The two examples show that people do not always act in accordance with the Bayesian expected value criterion expressed in monetary units. People try to select the course of action that gives the maximum security and satisfaction. Such satisfaction is called "utility." Utility is a state of mind, it is subjective in nature. What satisfies one person may not satisfy another person and even the same person who derived satisfaction from a commodity at one time might later become dissatisfied.

The utility of money is very important to the decision maker because payoffs and expected value of alternative courses of action are usually expressed in monetary units. The utility of money, as well as other commodities, has been investigated by neoclassical theorists, and economists for many centuries. Total utility (TU) has been described by a curve that increases at a decreasing rate. This means that total utility increases as more units of money or commodity is added. However, the first unit adds more utility than the

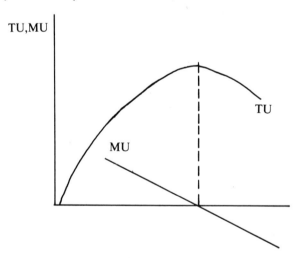

**Figure 8-2.** Total and marginal utility.

second unit, and the second unit adds more utility than the third unit, and so on. When the total utility curve reaches a stationary point, then the additional unit adds nothing to the total utility, this point is called maximum utility. Any additional unit after the maximum adds negative utility to total utility and hence the TU curve will decrease. Marginal utility (MU) represents a gain or a loss in the TU by the addition of one unit of the commodity. MU is the slope of TU. A graphic presentation of the TU and MU is shown in Figure 8-2.

The modern utility theory establishes a method of measuring satisfaction or utility under certain conditions. The pioneer of the modern utility theory is Daniel Bernoulli, a Swiss mathematician who lived during the 18th century. He came to the significant conclusion that rational decisions under circumstances of risk would not be made on the basis of expected monetary value, but rather on the basis of expected utility. This conclusion was the result of his investigation of people's behavior in the "St. Petersburg paradox" game. He explored the reason why a person is not willing to make a bet at better than 50-50 odds, when the expected value of winning in monetary units increases as the bet increases.

The "St. Petersburg paradox" game is fairly tossing a balanced coin until it falls heads up; then a reward equals $2^x$ is offered, where X is the number of tosses required to obtain a head. Since the coin is assumed to be balanced or fair, then the probability of a head showing is 1/2 and the expected monetary value can be calculated as follows:

$$E(\text{Monetary Value}) = \sum_{x=1}^{n}(1/2)^x (2)^x = \infty$$

where n = number of trials

or $\quad E(\text{Monetary Value}) = (1/2)(2) + (1/2)^2(2)^2 + (1/2)^3(2)^3 + \ldots$

$$= \quad 1 \quad + \quad 1 \quad + \quad 1 \quad + \ldots$$

$$= \infty$$

Even though the expected monetary value is infinite, no rational person would want to play the game. But why? Bernoulli's justification was that marginal utility of money diminishes as money increases. According to the law of diminishing marginal utility, any rational person would never engage in gambling because the gain in utility is always smaller than the loss for the same amount of money. In Figure 8-3, ab equals ac in terms

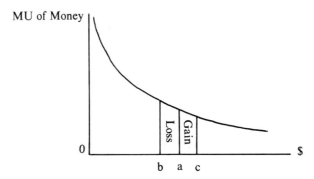

**Figure 8-3.** Gain and loss in total utility of money for an equal amount of money (ab = ac).

**173**

of dollars but $\int_a^b f(x)\,dx$ which represents the loss in total utility of money is larger than $\int_a^c f(x)\,dx$ the gain in total utility of money.

## Neumann-Morgenstern Method

More modern treatment of utility has been reported by John von Neumann and Oscar Morgenstern in their book: *Theory of Games and Economic Behavior* (Princeton University Press, 1947). Neumann and Morgenstern use a method known as "The Standard Gamble Method" where the risk element of Bernoulli's idea is used to measure utility. Constructing a tailored utility function for a decision maker proves that most decisions are based on the expected utility rather than the expected monetary payoff. Neumann and Morgenstern developed a "utility index", which describes the individual's preferences numerically. The preferences, however, must be fairly consistent. They listed the following five assumptions about the consistency of any individual's preferences:

### 1. *Transitivity*

If any individual is indifferent as to two prizes, say A and B, and he is also indifferent as to the two prizes A and C, then it can be deducted that the person is indifferent as to B and C.

### 2. *Continuity of preferences*

This assumption plays an important role in the standard gamble method which N-M used to construct any utility index. If a standard lottery ticket is preferred to another prize and the probability of winning for a standard lottery ticket is unity, or if another prize is preferred to the standard lottery ticket and the probability of winning the lottery ticket is zero, there exists a point that has different values of probability other than one or zero, where the person is indifferent between the standard lottery ticket and another prize. N-M use this point of indifference to calculate the value in utils to construct a utility index.

### 3. *Independence*

If a person is indifferent as to two prizes A and B, then he will be indifferent as to two identical lottery tickets. For example, if a person is indifferent as to a specific car and $6,000 in cash, then he will be indifferent as to two lottery tickets; one offers the specific car and the other offers $6,000 in cash.

### 4. *Desire of high probability of success*

Even though there are exceptions to this assumption, it is a valid assumption.

If a person is offered two identical lottery tickets with the same prize, then the ticket with the higher probability of winning will be preferred.

### 5. *Compound probabilities*

If the prize of a lottery ticket is another lottery ticket, then it is assumed that the person has calculated the probability of winning or losing the compound lottery ticket.

N-M utility index is based on the five assumptions mentioned above. The evaluation of a utility number is calculated according to the following general rule:

$$U(X) = U(X_1)\,p + U(X_2)\,(1 - p) \qquad (1)$$

where X is a lottery ticket that offers two prizes $X_1$ and $X_2$ and p is the probability of winning $X_1$ while $(1 - p)$ is the probability of winning $X_2$. The utility of $X_1$ is $U(X_1)$ and $U(X_2)$ stands for the utility of $X_2$. The general rule or equation (1) calculates the utility of

**174**

the lottery ticket or U(X) which is merely the expected utility value of the two prizes the lottery offers.

The first step in constructing any utility index or curve is to assign a utility number to the two values $U(X_1)$ and $U(X_2)$, which represent the extreme prices of an artificial lottery ticket for a standard comparison. $X_1$ and $X_2$ are the two extreme prizes of the standard lottery ticket. Say $X_1$ offers a prize of $5,000 and $X_2$ offers $10. The probability of winning $X_1$ is p and the probability of winning $X_2$ is $1 - p$. The utilities of $X_1$ and $X_2$ can be determined by observing the attitude of the individual and let us assume that $U(X_1)$ $= 100$ utils while $U(X_2) = 1$ util.

The next step is to consider another prize, say Y, where Y $=$ $2,000. To calculate a utility number for Y, the preference of the individual between the prizes of the standard lottery ticket and this prize (Y) is observed. If the probability of winning $X_1$, or p, equals unity then any individual will prefer $X_1$ to Y. As the probability of winning $X_1$ decreases, $X_1$ becomes less and less attractive than the prize Y. At a point, called the certainty monetary equivalent of the gamble, the individual or decision maker will be indifferent as to $X_1$ with a specific value of the probability of winning (say p $=$ .25) and the prize Y. Reaching this point, the U(Y) can be calculated by using the general rule expressed in equation (1) as follows:

$$U(Y) = U(X_1) p + U(X_2) (1 - p)$$
$$= (100)(.25) + (1)(.75)$$
$$= 25.75 \text{ utils}$$

So far, three values of the utility curve are known:

$$U(Y) \text{ or } U(\$2,000) = 25.75 \text{ utils}$$
$$U(X_1) \text{ or } U(\$5,000) = 100 \quad \text{utils}$$
$$U(X_2) \text{ or } U(\$10) = 1 \quad \text{util}$$

More utility values can be calculated by following the same procedure and a utility curve can be constructed for any individual.

A decision maker's attitude toward risk determines the shape of the utility function for money. In this respect, one can distinguish among three different utility functions for money: one for the risk seeker, another for the risk averter, and the last for the risk neutral individual. The risk seeker decision maker is one who is experiencing an increasing marginal utility while the risk averter decision maker faces a diminishing marginal utility function for money. The risk neutral decision maker has a constant marginal utility function for money. The three cases are shown graphically in Figure 8-4.

If a decision maker is observed or interviewed then a utility index can be constructed to fit his preferences. The payoff matrix or the decision matrix in monetary units can be changed in utility units or utils and an optimal action can be determined by maximizing the expected utility value rather than the expected monetary value of alternative actions.

The expected monetary value and the expected utility value have been discussed. If a decision maker elects to use the expected value criterion then he will be faced by the two approaches: monetary and utility expected value. Any decision maker can be interviewed or observed to determine whether he prefers to apply the expected monetary value or the expected utility value as a criterion to select the optimal action. A simple test described by

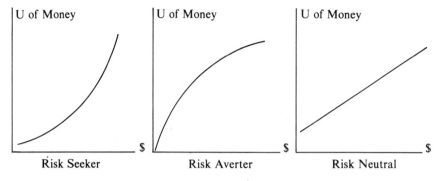

**Figure 8-4.** Utility function for money for the three different degrees of risk.

Robert Schlaiffer in his book, *Introduction to Statistics for Business Decision,* can be applied to select an optimal action. In this test any decision maker is given a choice of two courses of action:

A₁: a cash offer with certainty

A₂: a prize that has two outcomes: a very large amount of cash with a probability of p and a very small amount of cash with a probability of $1 - p$.

If the decision maker prefers $A_1$ over $A_2$ because the cash offer in $A_1$ is greater than the expected value of $A_2$ or if he prefers $A_2$ over $A_1$ because the cash offer in $A_1$ is less than the expected value of $A_2$ or if he becomes indifferent as to $A_1$ and $A_2$ because the cash offer in $A_1$ equals the expected value of $A_2$, then this decision maker prefers to apply the expected monetary value over the expected utility value in arriving at the optimal course of action.

EXERCISES

8.1 a. What is the difference between statistical inference and decision theory as tools for statistical decision making?

   b. What are the components of any decision problem structure?

8.2 a. What is the role of Bayesian Theorem in decision theory?

   b. How does the decision maker select the optimum solution?

8.3 An appliance dealer makes a net profit of $100 on each washer he sells. The monthly storage is $10 per washer. The dealer is facing a space problem and he has to decide on the optimum number of washers to stock to meet the monthly demand. Knowing that the highest number of washers that had been sold per month over the last five years is seven washers, and the probabilities of monthly sales are as follows:

| Sales (washers) | P(sales) |
|:---:|:---:|
| 1 | .40 |
| 2 | .22 |
| 3 | .18 |
| 4 | .10 |
| 5 | .05 |
| 6 | .03 |
| 7 | .02 |

find the optimum number of washers to be stocked.

8.4  In Exercise 8.3, if the monthly sales of washers appear to follow a binomial probability distribution with $p = .45$, find the optimal decision.

8.5  In Exercise 8.3, if the monthly sales of washers appear to follow a Poisson probability distribution with $\lambda = .80$, find the optimum number of washers the dealer should keep in stock monthly.

# CHAPTER

# 9

# Statistical Decision Making: Econometrics

Econometrics is widely used in decision making and policy formation by government and business. Econometrics is a branch of economics. Its main objective is the analysis of economic phenomena through the application of mathematics and statistical inference. The econometric task is to observe the behavior of economic variables; attempt to construct a mathematical model to describe the relationship among these variables; estimate the parameters of the model; and test the reliability of the model to predict such behavior in non-observed situations.

Econometric models play an important role in portraying economic relationships. They are built to make more systematic use of time series (or statistical data) in assessing their reliability. In general, econometric models are designed to describe the way in which a system actually operates. This system could be the whole economy, a sector or an activity of the economy, or a firm operating in the economy. Econometric models can be used to provide managerial decision makers with information concerning the behavior of national uncertain variables such as income, consumption, prices, interest rates, and others. On the other hand, they can be designed to predict the activity or activities of a firm such as sales. National and/or area econometric models are as important to management as those designed at the industry and the firm level.

A sales forecast econometric model is becoming a popular managerial tool to provide top management with information that can be used to make decisions concerning inventory control, production scheduling, and financial, marketing and hiring strategy for their organizations.

Application of econometrics as a tool of statistical decision making requires the construction of a model. But, what is a model? A model is a set of relationships among a group of variables. This relationship can be expressed algebraically by an equation. Therefore, a business or economic activity can be described by either a one-equation or a multi-equation model expressing the interrelationships among the measurable variables.

The relationship among the variables can be expressed in models of different types: linear and nonlinear, static and dynamic, stochastic and deterministic, and others. There are also many methods for the estimation of parameters of these models: ordinary least square, two-stage and three-stage least square, indirect least square, limited information and full information in maximum likelihood methods. Statistical methods are used to estimate parameters of the model, test their significance and the reliability of the overall model, and for prediction. This function constitutes what is usually known as regression and correlation analysis. In addition, econometricians test for the existence of problems such as identification, multicollinearity, and heteroscedesticity. These problems are beyond the scope of this book.

The ordinary least square method will be considered in this chapter to estimate parameters of linear and nonlinear model.

### 1. *Linear Models:*

If the relationship that exists among the variables is linear, then the model is called linear. Linear models are of two types: simple and multiple. A simple linear model represents a linear relationship that exists between two variables, while a multiple linear model has more than two variables.

A. Simple or Two-variable model:

The simplest relationship between two variables can be expressed in a linear form as:

$$Y = \alpha + \beta X$$

where Y and X are variables; Y is the dependent while X is the independent variable. $\alpha$ and $\beta$ are constants or parameters of the model, $\alpha$ representing the intercept, or the value Y assumes if $X = 0$, and $\beta$ is the slope or the rate of change of Y on X. The estimate equation is: $\hat{Y} = \hat{\alpha} + \hat{\beta}X$.

Usually a stochastic term e may be added to this linear model to become:

$$\hat{Y} = \hat{\alpha} + \hat{\beta}X + e$$

where $\hat{Y}$, $\hat{\alpha}$, and $\hat{\beta}$ are estimates of Y, $\alpha$, and $\beta$ respectively. The disturbance term e represents the residuals $(Y - \hat{Y})$, or the deviation of observed value Y from its estimate $\hat{Y}$. The function of $e_i$ stochastic terms have been included in the model to compensate for the inexact statistical formulation resulting from the following:

1. The omission of certain possible explanatory (or independent) variables.
2. The possible nonlinearity of the above model.
3. The basic and unpredictive element of randomness in human responses.
4. The imperfection of data and the errors of measurement.

The statistical theory upon which the estimating techniques are based requires the following assumption with regard to the distribution of $e_i$:

1. To be normally distributed random variables, with $\mu = 0$, and $\sigma^2 < \propto$ or $N(e_i; 0, \sigma^2 < \propto)$.
2. Not to be autocorrelated or $Cov(e_i e_j) = 0$ for all $i \neq j$ and $E(e_i e_{i-t}) = 0$ for all i and $t \neq 0$.

The least square method is used to estimate the parameters $\alpha$ and $\beta$ of the linear model. This method is based on the minimization of the residuals (or $\Sigma e_i$). If the estimated straight line could pass through every observed value $(Y_i)$, then there would be no residuals or errors, and $\Sigma e_i = 0$; but this is an unattainable situation. The closer the estimates or $\hat{Y}_i$ to the observations $Y_i$, the smaller the magnitude of the residuals $e_i$. To calculate the parameters, $\alpha$ and $\beta$ such that the estimated straight line $\hat{Y} = \hat{\alpha} + \hat{\beta}X$ is the best fit, one has to minimize the residuals $\Sigma e_i$.

The residuals are either positive, negative, or zero. Nonetheless, $\Sigma e_i = 0$. This means that the sum of all residuals around the regression or estimated line equals zero. There-

**180**

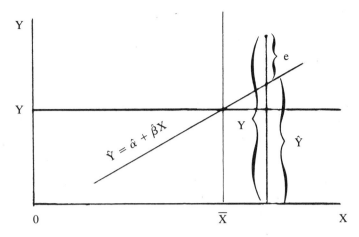

fore, to find the best straight line that fits the observations, one has to minimize $\Sigma e_i^2$ rather than $\Sigma e_i$. $\Sigma e_i^2$ is a non-negative value and it varies with the dispersion of the observations from the estimated line. This indicates that $\Sigma e_i^2$ is a function of the parameter estimates $\hat{\alpha}$ and $\hat{\beta}$. In other words, different pairs of $\hat{\alpha}$ and $\hat{\beta}$ yield different values of $\nu e_i^2$. Mathematically this function can be written as follows:

$$\Sigma e_i^2 = f(\hat{\alpha}, \hat{\beta})$$

Now in order to apply the least square method, simply minimize this function:

Min.  $\qquad \Sigma e_i^2 = f(\hat{\alpha}, \hat{\beta})$

The necessary condition of minimization is to calculate the first partial derivatives of the function and equate that to zero to locate the stationary points.

$$\Sigma e^2 = \Sigma(Y - \hat{Y})^2 \qquad \text{because } e = Y - \hat{Y}$$

but  $\quad \hat{Y} = \hat{\alpha} + \hat{\beta}X$

then  $\quad \Sigma e^2 = \Sigma(Y - \hat{\alpha} - \hat{\beta}X)^2$ to be minimized

$$\frac{\partial}{\partial\hat{\alpha}} \Sigma e^2 = (-1)(2)\, \Sigma(Y - \hat{\alpha} - \hat{\beta}X) = 0 \qquad\qquad (1)$$

$$\frac{\partial}{\partial\hat{\beta}} \Sigma e^2 = (-2)\, \Sigma X(Y - \hat{\alpha} - \hat{\beta}X) = 0 \qquad\qquad (2)$$

Equations (1) and (2) can be simplified to yield a system of two-normal equation to calculate the two unknown parameters $\alpha$ and $\beta$ of the linear model. These two equations are:

$$\Sigma Y = n\alpha + \beta\Sigma X$$
$$\Sigma XY = \alpha\Sigma X + \beta\Sigma X^2$$

The parameters $\alpha$ and $\beta$ can be calculated by solving for their estimated values $\hat{\alpha}$ and $\hat{\beta}$ in the above two normal equations. There are different methods of calculation: by elimination, substitution, graphs, or by using determinants (Cramer's Rule). The last method is considered to calculate $\hat{\alpha}$ and $\hat{\beta}$:

$$\hat{\alpha} = \frac{\begin{vmatrix} \Sigma Y & \Sigma X \\ \Sigma XY & \Sigma X^2 \end{vmatrix}}{\begin{vmatrix} n & \Sigma X \\ \Sigma X & \Sigma X^2 \end{vmatrix}} = \frac{(\Sigma Y)(\Sigma X^2) - (\Sigma XY)(\Sigma X)}{n(\Sigma X^2) - (\Sigma X)^2}$$

$$\hat{\beta} = \frac{\begin{vmatrix} n & \Sigma Y \\ \Sigma X & \Sigma XY \end{vmatrix}}{\begin{vmatrix} n & \Sigma X \\ \Sigma X & \Sigma X^2 \end{vmatrix}} = \frac{n(\Sigma XY) - (\Sigma X)(\Sigma Y)}{n(\Sigma X^2) - (\Sigma X)^2}$$

*Example 9.1:* Test the relationship that exists between Y and X, where Y represents personal consumption expenditures on total goods and services and X represents personal outlays (disposable income − personal savings) for the period 1957–1966 in the U.S.A. (The values are in hundred billions of dollars, seasonally adjusted.)

|      | Y    | X    |
|------|------|------|
| 1957 | 2.81 | 2.88 |
| 1958 | 2.90 | 2.97 |
| 1959 | 3.11 | 3.18 |
| 1960 | 3.25 | 3.33 |
| 1961 | 3.35 | 3.43 |
| 1962 | 3.55 | 3.64 |
| 1963 | 3.75 | 3.85 |
| 1964 | 4.01 | 4.12 |
| 1965 | 4.33 | 4.45 |
| 1966 | 4.66 | 4.79 |

*Source:* Business Statistics 1967, U.S. Department of Commerce, Office of Business Economics.

Estimate the parameters of the model that represents the relationship between personal consumption and personal outlays.

**Solution:**

a. For a visual test of the type of relationship that exists between Y and X, a scatter diagram is plotted:

From the scatter diagram, it appears that a linear relationship exists between personal consumption Y and personal outlays X. This relationship can be expressed by a linear model:

$$Y = \alpha + \beta X$$

b. The next step is to estimate the parameters $\alpha$ and $\beta$ of the above linear model:

$$\hat{\alpha} = \frac{(\Sigma Y)(\Sigma X^2) - (\Sigma XY)(\Sigma X)}{n(\Sigma X^2) - (\Sigma X)^2}$$

and  $$\hat{\beta} = \frac{n(\Sigma XY) - (\Sigma Y)(\Sigma X)}{n(\Sigma X^2) - (\Sigma X)^2}$$

To calculate $\alpha$ and $\beta$, one needs to find the value of:

n, $\Sigma Y$, $\Sigma X$, $\Sigma XY$, $\Sigma X^2$

This has been done in the following table:

| | Y | X | XY | X² |
|---|---|---|---|---|
| | 2.81 | 2.88 | 8.0928 | 8.2944 |
| | 2.90 | 2.97 | 8.6130 | 8.8209 |
| | 3.11 | 3.18 | 9.8898 | 10.1124 |
| | 3.25 | 3.33 | 10.8225 | 11.0889 |
| | 3.35 | 3.43 | 11.4905 | 11.7649 |
| | 3.55 | 3.64 | 12.9220 | 13.2496 |
| | 3.75 | 3.85 | 14.4375 | 14.8225 |
| | 4.01 | 4.12 | 16.5212 | 16.9744 |
| | 4.33 | 4.45 | 19.2685 | 19.8025 |
| | 4.66 | 4.79 | 22.3214 | 22.9441 |
| | 35.72 | 36.64 | 134.3792 | 137.8746 |
| n = 10 | $\Sigma Y$ | $\Sigma X$ | $\Sigma XY$ | $\Sigma X^2$ |

$$\hat{\alpha} = \frac{(35.72)(137.8746) - (134.3792)(36.64)}{10(137.8746) - (36.64)^2}$$

$$= \frac{1.2268}{36.2564}$$

$$= .03383 = .0338$$

$$\hat{\beta} = \frac{(10)(134.3792) - (35.72)(36.64)}{10(137.8746) - (36.64)^2}$$

$$= \frac{35.0112}{36.2564}$$

$$= .96565 = .9657$$

The estimated equation (or the regression line) is:

$$\hat{Y} = .0338 + .9657X$$

To plot the estimated linear equation, determine two points by using two different values for X and connect the two points to a straight line.

**183**

Another **method** to calculate $\hat{\alpha}$ and $\hat{\beta}$ follows:

The system of two normal equations used to derive $\alpha$ and $\beta$ is:

$$\Sigma Y = n\hat{\alpha} + \hat{\beta}\Sigma X \tag{1}$$

$$\Sigma XY = \hat{\alpha}\Sigma X + \hat{\beta}\Sigma X^2 \tag{2}$$

If we divide the elements of equation (1) by n:

$$\frac{\Sigma Y}{n} = \frac{n\hat{\alpha}}{n} + \frac{\hat{\beta}\Sigma X}{n}$$

or     $\overline{Y} = \hat{\alpha} + \hat{\beta}\overline{X}$

then    $\hat{\alpha} = \overline{Y} - \hat{\beta}\overline{X}$

Also, if we write the second equation in deviation form:

$$\Sigma xy = 0 + \Sigma \beta x^2$$

where    $x = X - \overline{X},$       $y = Y - \overline{Y}$

then   $\hat{\beta} = \dfrac{\Sigma xy}{\Sigma x^2}$

$\Sigma xy$ is the covariance that measures the joint variation of X and Y.
Let us apply these new formulas to the previous example:

| Y | X | y | x | yx | $x^2$ |
|---|---|---|---|----|-------|
| 2.81 | 2.88 | −.7620 | −.7840 | .5974 | .6147 |
| 2.90 | 2.97 | −.6720 | −.6940 | .4664 | .4816 |
| 3.11 | 3.18 | −.4620 | −.4840 | .2236 | .2343 |
| 3.25 | 3.33 | −.3220 | −.3340 | .1075 | .1116 |
| 3.35 | 3.43 | −.2220 | −.2340 | .0519 | .0548 |
| 3.55 | 3.64 | −.0220 | −.0240 | .0005 | .0006 |
| 3.75 | 3.85 | .1780 | .1860 | .0331 | .0346 |
| 4.01 | 4.12 | .4380 | .4560 | .1997 | .2079 |
| 4.33 | 4.45 | .7580 | .7860 | .5958 | .6178 |
| 4.36 | 4.79 | 1.0880 | 1.1260 | 1.2251 | 1.2679 |
| 35.72 | 36.64 | | | 3.5010 | 3.6258 |

$\overline{Y} = 3.572$       $\overline{X} = 3.664$

$\hat{\beta} = \dfrac{\Sigma xy}{\Sigma x^2}$

$= \dfrac{3.5010}{3.6258}$

$= \quad .9656$

$\hat{\alpha} = \overline{Y} - \beta\overline{X}$

$= (3.572) - (.9656)(3.664)$

$= .034$

$\hat{Y} = .034 + .9656X$

## Standard Error of Estimate

The standard deviation of estimate or the standard error of estimate $\hat{\sigma}_{yx}$ is a measure that shows the degree of dispersion of the observed values $Y_i$ around the estimated regression line: $\hat{Y} = \hat{\alpha} + \hat{\beta}X$. It is an absolute measure of goodness-of-fit. If it happened that the standard error $\hat{\sigma}_{yx}$ equals zero, then the regression line should pass through every actual observation of the scatter diagram. This is almost impossible, however, the smaller the magnitude of $\hat{\sigma}_{yx}$, the closer the fit, as shown in the following diagrams:

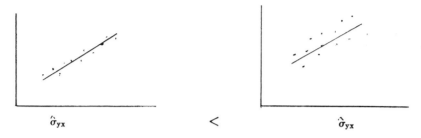

$\hat{\sigma}_{yx}$      $<$      $\hat{\sigma}_{yx}$

The standard error when regressing Y on X can be computed by the following formula:

$$\hat{\sigma}_{yx} = \sqrt{\frac{\Sigma(Y - \hat{Y})^2}{n - 2}}$$

where $n - 2$ refers to the degree of freedom.

The loss of two degrees of freedom results from using the estimated parameters $\hat{\alpha}$ and $\hat{\beta}$ in computing the standard error of estimate $\hat{\sigma}_{yx}$.

The standard error of estimate $\sigma_{yx}$ is computed for the estimated equation: $\hat{Y} = .0338 + .9657X$ as follows:

| Y | X | $\hat{Y}$ | $(Y - \hat{Y})^2$ |
|---|---|---|---|
| 2.81 | 2.88 | $.0338 + .9657(2.88) = 2.8150$ | .00002500 |
| 2.90 | 2.97 | $.0338 + .9657(2.97) = 2.9019$ | .00000361 |
| 3.11 | 3.18 | $= 3.1047$ | .00002809 |
| 3.25 | 3.33 | $= 3.2496$ | .00000016 |
| 3.35 | 3.43 | $= 3.3462$ | .00001444 |
| 3.55 | 3.64 | $= 3.5489$ | .00000121 |
| 3.75 | 3.85 | $= 3.7517$ | .00000289 |
| 4.01 | 4.12 | $= 4.0125$ | .00000625 |
| 4.33 | 4.45 | $= 4.3312$ | .00000144 |
| 4.66 | 4.79 | $= 4.6595$ | .00000025 |
| | | | .00008334 |
| | | | $\Sigma(Y - \hat{Y})^2$ |

$$\hat{\sigma}_{yx} = \sqrt{\frac{.00008334}{10 - 2}}$$

$$= .003249$$

**185**

The magnitude of the standard error of estimate .003249 is very small and indicates that the estimated regression line fits the observed data very well.

Another way to interpret $\hat{\sigma}_{yx} = .003249$ is that about 68% of the observed values $Y_i$ are expected to fall between $Y \pm \hat{\sigma}_{yx}$ as shown in the diagram:

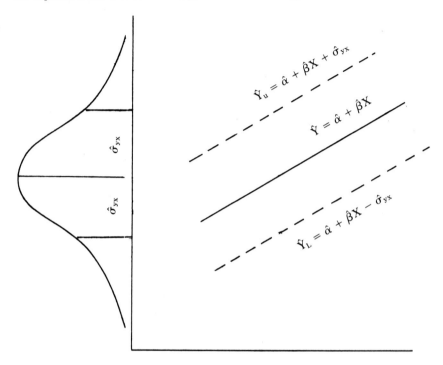

$\hat{Y}_u$ and $\hat{Y}_L$ computed for the previous example are:

$$\hat{Y}_u = .0338 + .9657X + .003249$$
and $$\hat{Y}_L = .0338 + .9657X - .003249$$

$\hat{Y}_u$ is the upper limit of the band while $\hat{Y}_L$ is the lower limit of the band.

Bands of different limits can be computed; e.g., a band that is expected to contain 95% of the observed data is calculated as follows:

$$\hat{Y}_u = \hat{\alpha} + \hat{\beta}X + 1.96\hat{\sigma}_{yx}$$
$$\hat{Y}_L = \hat{\alpha} + \hat{\beta}X - 1.96\hat{\sigma}_{yx}$$

### Standard Error of $\beta$

Errors are expected to occur in the process of estimating the parameters of the model. The standard error of $\beta$, $\hat{\sigma}_{\beta}$ measures the amount of error to determine the significance of the estimated parameter $\hat{\beta}$.

The standard error of $\beta$ can be calculated by the following formula:

$$\hat{\sigma}_\beta = \frac{\hat{\sigma}_{yx}}{\sqrt{\Sigma(X - \overline{X})^2}}$$

From this formula, one can conclude that the $\hat{\sigma}_\beta$ varies directly with $\hat{\sigma}_{yx}$, and inversely with the variation of the independent variable X.

The following is the standard error of $\beta$ calculated for the previous example:

| X | $(X - \overline{X})^2$ | |
|---|---|---|
| 2.88 | .6147 | |
| 2.97 | .4816 | |
| 3.18 | .2343 | $\hat{\sigma}_{yx} = .0032249$ |
| 3.33 | .1116 | |
| 3.43 | .0548 | $\hat{\sigma}_\beta = \dfrac{.0032249}{\sqrt{3.6258}}$ |
| 3.64 | .0006 | |
| 3.85 | .0346 | |
| 4.12 | .2079 | $= .0017$ |
| 4.45 | .6178 | |
| 4.79 | 1.2679 | |
| $\Sigma X = 36.64$ | 3.6258 | |
| n = 10 | | |
| $\overline{X} = 3.664$ | | |

## Statistical Significance of $\beta$

A non zero value computed for $\hat{\beta}$ as an estimate of $\beta$ does not mean in all cases that $\beta$ is significantly different than zero. To test for the statistical significance of $\beta$, one has to apply the following t-test:

$$t_c = \frac{\hat{\beta} - \beta}{\hat{\sigma}_\beta}$$

If $\beta = 0$, then the linear relationship between Y and X is insignificant and one can claim that it does not exist. On the other hand, if $\beta \neq 0$, then there is a significant linear relationship between Y and X. Accordingly, the set up of the test of hypothesis becomes:

$H_0: \beta = 0$

$H_1: \beta \neq 0$

With a level of significance $\alpha$, if $H_0$ is accepted, then $\beta$ is not significantly different from zero and there is no linear relationship between Y and X. If $H_0$ is rejected, then accepting $H_1$ indicates that $\beta$ is significantly different from zero and hence a linear relationship exists between the two variables Y and X.

*Example 9.2:* Test for the significance of $\beta$ in the previous example at .05 level of significance. Calculate the confidence limits.

**Solution:**

a. $\hat{\beta} = .9657$ and $\hat{\sigma}_\beta = .0017$

b. Set up the test of hypothesis:

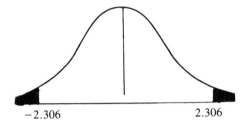

$H_0$: $\beta = 0$

$H_1$: $\beta \neq 0$

$\alpha = .05$

$v = n - 2 = 8$

$t_{.05,8} = \pm 2.306$

c. Calculate $t_c$:

$$t_c = \frac{.9657 - 0}{.0017}$$

$$= 568.0588$$

$-2.306 \qquad\qquad 2.306$

$t_c$ falls in the rejection region, $H_0$ is rejected.

$H_1$ is accepted. This means that $\beta$ is significantly different from zero or there is a linear relationship between the two variables Y and X.

d. Confidence limits ($\alpha = .05$):

The confidence interval $(1 - \alpha)$ of 95% has the two limits: upper limit of $\beta$ and lower limit of $\beta$. The two limits of this confidence interval can be calculated as follows:

$$t_{.05,8} = \pm 2.306$$

$$t_{.05,8} = \frac{\hat{\beta} - \beta}{\hat{\sigma}_\beta}$$

$$\pm 2.306 = \frac{.9657 - \beta}{.0017}$$

then solve for the two limits of $\beta$:

$$2.306 = \frac{.9657 - \beta}{.0017}$$

$$\beta = .9618$$

$$-2.306 = \frac{.9657 - \beta}{.0017}$$

$$\beta = .9696$$

We are confident 95% that the true value of $\beta$ falls between .9696 and .9618.

### Simple Correlation, or Correlation of Two Variables

Correlation is a measure of the association that exists between the variables. Simple correlation is a measure of the degree to which the two variables Y and X are related or associated. Correlation analysis does not imply the functional relationship that regression analysis requires between the two variables.

Correlations coefficient R is a measure of closeness of fit. Its magnitude reflects the significance of the overall equation. Instead of using the correlation coefficient R, statisticians are using $R^2$, the rate of determination, as a measure of association. If the magnitude of $R^2$ is one, then there is a perfect correlation between the variables. If the magnitude of $R^2$ is zero, this indicates no correlation between the variables. The closer the magnitude of $R^2$ to one, the stronger the correlation and the closer to zero, the weaker the association.

$R^2$, the rate or coefficient of determination, is a ratio of the variation explained by the regression line and the total variation of the observed value.

$$R^2 = \frac{\text{Explained variation}}{\text{Total variation}}$$

or $\quad R^2 = 1 - \dfrac{\text{Unexplained variation}}{\text{Total variation}}$

The concepts total, explained, and unexplained variation can be understood from the inspection of the following diagrams:

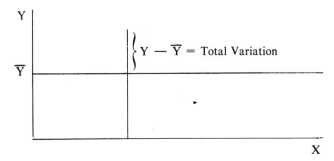

Total variation is the difference between an observation and the mean of all observations: $Y - \overline{Y}$.

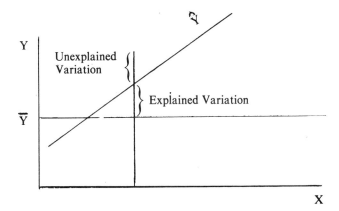

A portion of the total variation will be known or explained or absorbed by the estimated regression line: $\hat{Y} = \hat{\alpha} + \hat{\beta}X$, and is called the explained variation. The explained variation is equal to $\hat{Y} - \overline{Y}$ for each observation.

The portion of the total variation that has not been explained by the regression line is called the unexplained variation which is equal to $Y - \hat{Y}$ for each observation.

**189**

Total variation $Y - \overline{Y}$ is equal to the sum of explained and unexplained variation:

$$Y - \overline{Y} = \hat{Y} - \overline{Y} + Y - \hat{Y} \qquad \text{(for each observation)}$$
$$\Sigma(Y - \overline{Y}) = \Sigma(\hat{Y} - \overline{Y}) + \Sigma(Y - \hat{Y}) \qquad \text{(for all observations)}$$

Total variation = Explained + Unexplained.

The sum of the total variation $\Sigma(Y - \overline{Y})$ as well as the other sums on the right side of the equation equals zero. Therefore, the sum of squared deviations is used:

$$\Sigma(Y - \overline{Y})^2 = \Sigma(\hat{Y} - \overline{Y})^2 + \Sigma(Y - \hat{Y})^2$$

$R^2$ formula can be derived from the above equation:

$$R^2 = \frac{\Sigma(\hat{Y} - Y)^2}{\Sigma(Y - \overline{Y})^2}$$

or $\qquad R^2 = 1 - \dfrac{\Sigma(Y - \hat{Y})^2}{\Sigma(Y - \overline{Y})^2}$

*Example 9.3:* Calculate $R^2$ for the previous example and interpret the results.
**Solution:**

$$R^2 = 1 - \frac{\Sigma(Y - \hat{Y})^2}{\Sigma(Y - \overline{Y})^2}$$

$\Sigma(Y - \hat{Y})^2 = .00008334 \qquad\qquad$ calculated to find $\hat{\sigma}_{yx}$

So, we need $\Sigma(Y - \overline{Y})^2$.

| Y | $(Y - \overline{Y})^2$ | |
|---|---|---|
| 2.81 | .5806 | |
| 2.90 | .4516 | |
| 3.11 | .2134 | |
| 3.25 | .1037 | $R^2 = 1 - \dfrac{.00008334}{3.3809} = .9999$ |
| 3.35 | .0493 | |
| 3.55 | .0005 | |
| 3.75 | .0137 | |
| 4.01 | .1918 | |
| 4.33 | .5746 | |
| 4.66 | 1.1837 | |
| $\Sigma Y = \overline{35.72}$ | $\overline{3.3809}$ | |

$n = 10$

$\overline{Y} = 3.572$

The calculation of $R^2 = .9999$ means that 99.99% of the total variation has been explained by the estimated regression line. In other words, there is a very strong (almost perfect) correlation between Y and X.

Other **methods** of calculation of $R^2$:

There are many formulas developed to calculate R and $R^2$. However, the following two are considered because of their practical use:

(1) $\quad R^2 = \hat{\beta}_{yx} \cdot \hat{\beta}_{xy}$

$\hat{\beta}_{yx}$ is the slope of the regression line Y on X.
$\hat{\beta}_{xy}$ is the slope of the regression line X on Y, or the slope of the regression when rotating the axis.

*Example 9.4:* Calculate $R^2$ for the previous example by using the new formula:

$$R^2 = \hat{\beta}_{yx} \cdot \hat{\beta}_{xy}$$

**Solution:**

$$\hat{\beta}_{yx} = .96565$$

To calculate $\beta_{xy}$ we can use the following formula:

$$\hat{\beta}_{xy} = \frac{n(\Sigma XY) - (\Sigma X)(\Sigma Y)}{n(\Sigma Y^2) - (\Sigma Y)^2}$$

We know the value of:

$$n = 10 \qquad\qquad \Sigma XY = 134.3792$$
$$\Sigma X = 36.64 \qquad\qquad \Sigma Y = 35.72$$

then we need the value of: $\Sigma Y^2$

| Y | Y² |
|---|---|
| 2.81 | 7.8961 |
| 2.90 | 8.4100 |
| 3.11 | 9.6721 |
| 3.25 | 10.5625 |
| 3.35 | 11.2225 |
| 3.55 | 12.6025 |
| 3.75 | 14.0625 |
| 4.01 | 16.0801 |
| 4.33 | 18.7489 |
| 4.66 | 21.7156 |
| 35.72 | 130.9728 |
| $\Sigma Y$ | $\Sigma Y^2$ |

$$\hat{\beta}_{xy} = \frac{10(134.3792) - (36.64)(35.72)}{10(130.9728) - (35.72)^2}$$

$$= 1.03554$$

$$R^2 = \hat{\beta}_{yx} \cdot \hat{\beta}_{xy}$$

$$= (.96565)(1.03554)$$

$$= .9999$$

(2) $\quad R^2 = \dfrac{(\Sigma xy)^2}{\Sigma x^2 \, \Sigma y^2}$

From the previous Example 9.1 we have the value of:

$$\Sigma xy = 3.5010 \qquad\qquad \Sigma x^2 = 3.6258$$

We need the value of $\Sigma y^2$:

| $y$ | $y^2$ |
|---|---|
| −.7620 | .5806 |
| −.6720 | .4516 |
| −.4620 | .2134 |
| −.3220 | .1037 |
| −.2220 | .0493 |
| −.0220 | .0005 |
| .1780 | .0317 |
| .4380 | .1918 |
| .7580 | .5746 |
| 1.0880 | 1.1837 |
| | 3.3809 |
| | $\Sigma y^2$ |

$$R^2 = \frac{(3.5010)^2}{(3.6258)(3.3809)}$$

$$= \frac{12.2570}{12.2585}$$

$$= .9999$$

## B. Multiple Linear Model:

A simple linear model that represents the relationship between a dependent variable and an independent variable cannot yield satisfactory results if there are other independent variables which can affect that relationship. In other words, if the dependent variable is a function of two or more independent variables: $Y = f(X_1, X_2, \ldots, X_n)$ then a multiple linear model can be constructed to express this function provided that the relationship among the variables is linear. A multiple linear model for this function is:

$$Y = \alpha + \beta_1 X_1 + \beta_2 X_2 + \ldots + \beta_{n-1} X_{n-1} + \beta_n X_n$$

The least square method may be used to estimate the parameters of this multiple linear model: $\alpha, \beta_1, \beta_2 \ldots, \beta_n$. The basic analysis of deriving a system of $n + 1$ normal equation to solve for the $n + 1$ unknowns is the same as shown in the simple linear model. The minimization of the function:

$$\Sigma e_i^2 = f(\hat{\alpha}, \hat{\beta}_1, \hat{\beta}_2, \ldots, \hat{\beta}_n)$$

yields the partial derivatives:

$$\frac{\partial}{\partial \hat{\alpha}} \Sigma e_i^2, \frac{\partial}{\partial \hat{\beta}_1} \Sigma e_i^2, \ldots, \frac{\partial}{\partial \hat{\beta}_n} \Sigma e_i^2$$

when set to equal zero, the following system of n + 1 normal equation can be derived:

$$\Sigma Y \quad = n\hat{\alpha} \quad + \hat{\beta}_1 \Sigma X_1 \quad + \hat{\beta}_2 \Sigma X_2 \quad + \ldots + \hat{\beta}_n \Sigma X_n$$

$$\Sigma X_1 Y = \hat{\alpha}\Sigma X_1 + \hat{\beta}_1 \Sigma X_1^2 \quad + \hat{\beta}_2 \Sigma X_1 X_2 + \ldots + \hat{\beta}_n \Sigma X_1 X_n$$

$$\Sigma X_2 Y = \hat{\alpha}\Sigma X_2 + \hat{\beta}_1 \Sigma X_1 X_2 + \hat{\beta}_2 \Sigma X_2^2 \quad + \ldots + \hat{\beta}_n \Sigma X_2 X_y$$

$$\Sigma X_n Y = \hat{\alpha}\Sigma X_n + \hat{\beta}_1 \Sigma X_1 X_n + \hat{\beta}_2 \Sigma X_2 X_n + \ldots + \hat{\beta}_n \Sigma X_n^2$$

This system of normal equation can be expressed in matrix notation as follows:

$$
\begin{bmatrix}
\Sigma Y \\
\Sigma X_1 Y \\
\Sigma X_2 Y \\
\vdots \\
\Sigma X_n Y
\end{bmatrix}
=
\begin{bmatrix}
n & \Sigma X_1 & \Sigma X_2 & \ldots & \Sigma X_n \\
\Sigma X_1 & \Sigma X_1^2 & \Sigma X_1 X_2 & \ldots & \Sigma X_1 X_n \\
\Sigma X_2 & \Sigma X_1 X_2 & \Sigma X_2^2 & \ldots & \Sigma X_1 X_n \\
\vdots & & & & \vdots \\
\Sigma X_n & \Sigma X_1 X_n & \Sigma X_2 X_n & \ldots & \Sigma X_2 X_n
\end{bmatrix}
\cdot
\begin{bmatrix}
\hat{\alpha} \\
\hat{\beta}_1 \\
\hat{\beta}_2 \\
\vdots \\
\hat{\beta}_n
\end{bmatrix}
$$

or $\quad \mathbf{Y} = \mathbf{XA}$

To solve for the unknown estimated parameters: $\hat{\alpha}, \hat{\beta}_1, \hat{\beta}_2, \ldots, \hat{\beta}_n$, simply calculate the inverse of the matrix X and multiply both sides of the matrix equation ($\mathbf{Y} = \mathbf{XA}$) by $\mathbf{X}^{-1}$:

$$\mathbf{X}^{-1}\mathbf{Y} = \mathbf{X}^{-1}\mathbf{XA}$$

$$\mathbf{X}^{-1}\mathbf{Y} = \mathbf{IA}$$

$\mathbf{X}^{-1}\mathbf{Y}$ is a column vector of n + 1 rows or values representing the values of $\hat{\alpha}, \hat{\beta}_1, \hat{\beta}_2,$ $\ldots, \hat{\beta}_n$ respectively, the rows or elements of the column vector A.

This is the general method used to calculate the parameters of a multiple linear model. The computer has been utilized for these calculations and multiple linear computer programs of different types are readily available.

For illustration, a multiple linear model for the function: $Y = f(X_1, X_2)$ is considered. The dependent variable Y is affected by two independent variables $X_1$ and $X_2$. The relationships existing among the variables are found to be linear. The multiple linear model for this function is:

$$Y = \alpha + \beta_1 X_1 + \beta_2 X_2$$

The least square method is used to estimate the parameters $\alpha$, $\beta_1$, and $\beta_2$ of this model. As mentioned before, the least square method requires the minimization of $\Sigma e^2$ (the sum of squared residuals) which yields a system of three normal equations to solve for the three unknown parameters:

$$\Sigma Y \quad = n\hat{\alpha} \quad + \hat{\beta}\Sigma X_1 \quad + \hat{\beta}_2\Sigma X_2$$
$$\Sigma X_1 Y = \hat{\alpha}\Sigma X_1 + \hat{\beta}\Sigma X_1^2 \quad + \hat{\beta}_2\Sigma X_1 X_2$$
$$\Sigma X_2 Y = \hat{\alpha}\Sigma X_2 + \hat{\beta}\Sigma X_1 X_2 + \hat{\beta}\Sigma X_2^2$$

In matrix notation, the system becomes:

$$\begin{bmatrix} \Sigma Y \\ \Sigma X_1 Y \\ \Sigma X_2 Y \end{bmatrix} = \begin{bmatrix} n & \Sigma X_1 & \Sigma X_2 \\ \Sigma X_1 & \Sigma X_1^2 & \Sigma X_1 X_2 \\ \Sigma X_2 & \Sigma X_1 X_2 & \Sigma X_2^2 \end{bmatrix} \begin{bmatrix} \hat{\alpha} \\ \hat{\beta}_1 \\ \hat{\beta}_2 \end{bmatrix}$$

$$Y = XA$$

Calculate $X^{-1}$ and multiply both sides:

$$X^{-1}Y = X^{-1}XA$$
$$X^{-1}Y = IA$$

*Example 9.5:*  Assume the data collected for Y, $X_1$, and $X_2$ is as follows:

| Y | $X_1$ | $X_2$ |
|---|---|---|
| 3 | 2 | 4 |
| 8 | 1 | 2 |
| 11 | 4 | 1 |
| 15 | 7 | 3 |
| 9 | 13 | 1 |
| 4 | 7 | 3 |
| 17 | 15 | 4 |
| 13 | 19 | 2 |
| 11 | 12 | 5 |
| 9 | 19 | 4 |

The relationship among the three variables is linear. Estimate the parameters of the model that appropriately represents this relationship.

**Solution:**

a.  The model:

$$Y = \alpha + \hat{\beta}X_1 + \hat{\beta}_2X_2$$

b.  The system of normal equation:

$$\Sigma Y \quad = n\hat{\alpha} \quad + \hat{\beta}_1\Sigma X_1 \quad + \hat{\beta}_2\Sigma X_2$$
$$\Sigma X_1 Y = \hat{\alpha}\Sigma X_1 + \hat{\beta}_1\Sigma X_1^2 \quad + \hat{\beta}_2\Sigma X_1 X_2$$
$$\Sigma X_2 Y = \hat{\alpha}\Sigma X_2 + \hat{\beta}_1\Sigma X_1 X_2 + \hat{\beta}_2\Sigma X_2^2$$

We need the values of: $\Sigma Y$, $\Sigma X_1$, $\Sigma X^2$, $\Sigma X_1 Y$, $\Sigma X_2 Y$, $\Sigma X_1^2$, $\Sigma X_2^2$, $\Sigma X_1 X_2$, n.

| Y | $X_1$ | $X_2$ | $X_1 Y$ | $X_2 Y$ | $X_1^2$ | $X_2^2$ | $X_1 X_2$ |
|---|---|---|---|---|---|---|---|
| 3 | 2 | 4 | 6 | 12 | 4 | 16 | 8 |
| 8 | 1 | 2 | 8 | 16 | 1 | 4 | 2 |
| 11 | 4 | 1 | 44 | 11 | 16 | 1 | 4 |
| 15 | 7 | 3 | 105 | 45 | 49 | 9 | 21 |
| 9 | 13 | 1 | 117 | 9 | 169 | 1 | 13 |
| 4 | 7 | 3 | 28 | 12 | 49 | 9 | 21 |
| 17 | 15 | 4 | 255 | 68 | 225 | 16 | 60 |
| 13 | 19 | 2 | 247 | 26 | 361 | 4 | 38 |
| 11 | 12 | 5 | 132 | 55 | 144 | 25 | 60 |
| 9 | 19 | 4 | 171 | 36 | 361 | 16 | 76 |
| 100 | 99 | 29 | 1113 | 290 | 1379 | 101 | 303 |
| $\Sigma Y$ | $\Sigma X_1$ | $\Sigma X_2$ | $\Sigma X_1 Y$ | $\Sigma X_2 Y$ | $\Sigma X_1^2$ | $\Sigma X_2^2$ | $\Sigma X_1 X_2$ |

Substituting the values in the system of normal equation:

$$100 = 10\hat{\alpha} + 99\hat{\beta}_1 + 29\hat{\beta}_2$$
$$1113 = 99\hat{\alpha} + 1379\hat{\beta}_1 + 303\hat{\beta}_2$$
$$290 = 29\hat{\alpha} + 303\hat{\beta}_1 + 101\hat{\beta}_2$$

$$\begin{bmatrix} 100 \\ 1113 \\ 290 \end{bmatrix} = \begin{bmatrix} 10 & 99 & 29 \\ 99 & 1379 & 303 \\ 29 & 303 & 101 \end{bmatrix} \cdot \begin{bmatrix} \hat{\alpha} \\ \hat{\beta}_1 \\ \hat{\beta}_2 \end{bmatrix}$$

$$Y = XA$$

Calculate $X^{-1}$

1. Find $|X|$
2. Find Adj X
3. $X^{-1} = \dfrac{1}{|X|}$ Adj X

$$X^{-1} = \begin{bmatrix} .7316 & -.0187 & -.1540 \\ -.0187 & .0026 & -.0025 \\ -.1540 & -.0025 & .0615 \end{bmatrix}$$

To solve for $\hat{\alpha}$, $\hat{\beta}_1$, and $\hat{\beta}_2$ multiply $X^{-1}$ by Y:

$$\begin{bmatrix} .7316 & -.0187 & -.1540 \\ -.0187 & .0026 & -.0025 \\ -.1540 & -.0025 & .0615 \end{bmatrix} \cdot \begin{bmatrix} 100 \\ 1113 \\ 290 \end{bmatrix} = \begin{bmatrix} 7.6869 \\ .2988 \\ -.3475 \end{bmatrix} \begin{matrix} \hat{\alpha} \\ \hat{\beta}_1 \\ \hat{\beta}_2 \end{matrix}$$

The regression line becomes:

$$\hat{Y} = 7.6869 + .2988 X_1 - .3475 X_2$$

Another **method** for parameters estimation:

The system of three normal equations can be expressed in deviation form where $y = Y - \overline{Y}; x_1 = X_1 - \overline{X}_1; x_2 = X_2 - \overline{X}$, and so on.

$$\Sigma y = n\hat{\alpha} \quad + \hat{\beta}_1 \Sigma x_1 \quad + \hat{\beta}_2 \Sigma x_2$$
$$\Sigma x_1 y = \hat{\alpha} \Sigma x_1 + \hat{\beta}_1 \Sigma x_1^2 \quad + \hat{\beta}_2 \Sigma x_1 x_2$$
$$\Sigma x_2 y = \hat{\alpha} \Sigma x_2 + \hat{\beta}_1 \Sigma x_1 x_2 + \hat{\beta}_2 \Sigma x_2$$

$\Sigma y$, $\Sigma x_1$, $\Sigma x_2$ are equal to zero, therefore, the three equation system can be reduced to a two equation system:

$$\Sigma x_1 y = \hat{\beta}_1 \Sigma x_1^2 + \hat{\beta}_2 \Sigma x_1 x_2$$
$$\Sigma x_2 y = \hat{\beta}_1 \Sigma x_1 x_2 + \hat{\beta}_2 \Sigma x_2^2$$

Now we need $\Sigma x_1 y$, $\Sigma x_1^2$, $\Sigma x_1 x_2$, $\Sigma x_2 y$, $\Sigma x_2^2$ for this two equation system to solve for $\beta_1$ and $\beta_2$.

| $X_1 - \overline{X}_1$ | $X_2 - \overline{X}_2$ | $Y - \overline{Y}$ | $x_1^2$ | $x_2^2$ | $x_1 x_2$ | $x_1 y$ | $x_2 y$ |
|---|---|---|---|---|---|---|---|
| −7.9 | 1.1 | −7 | 62.41 | 1.21 | − 8.69 | 55.3 | −7.7 |
| −8.9 | − .9 | −2 | 79.21 | .81 | 8.01 | 17.8 | 1.8 |
| −5.9 | −1.9 | 1 | 34.81 | 3.61 | 11.21 | − 5.9 | −1.9 |
| −2.9 | .1 | 5 | 8.41 | .01 | − .29 | − 14.5 | .5 |
| 3.1 | −1.9 | −1 | 9.61 | 3.61 | − 5.89 | − 3.1 | 1.9 |
| −2.9 | .1 | −6 | 8.41 | .01 | − .29 | 17.4 | − .6 |
| 5.1 | 1.1 | 7 | 26.01 | 1.21 | 5.61 | 35.7 | 7.7 |
| 9.1 | − .9 | 3 | 82.81 | .81 | − 8.19 | 27.3 | −2.7 |
| 2.1 | 2.1 | 1 | 4.41 | 4.41 | 4.41 | 2.1 | 2.1 |
| 9.1 | 1.1 | −1 | 82.81 | 1.21 | 10.01 | − 9.1 | −1.1 |
| | | | 398.9 | 16.9 | 15.9 | 123 | 0 |

$$\overline{X}_1 = 9.9 \qquad \overline{X}_2 = 2.9 \qquad \overline{Y} = 10$$
$$123 = 398.9\,\hat{\beta}_1 + 15.9\,\hat{\beta}_2$$
$$0 = 15.9\,\hat{\beta}_1 + 16.9\,\hat{\beta}_2$$

$$\beta_1 = \frac{\begin{vmatrix} 123 & 15.9 \\ 0 & 16.9 \end{vmatrix}}{\begin{vmatrix} 398.9 & 15.9 \\ 15.9 & 16.9 \end{vmatrix}} = .3204 \qquad \beta_2 = \frac{\begin{vmatrix} 398.9 & 123 \\ 15.9 & 0 \end{vmatrix}}{\begin{vmatrix} 398.9 & 15.9 \\ 15.9 & 16.9 \end{vmatrix}} = -.3014$$

$\alpha$ is calculated from the following equation:

$$\hat{\alpha} = \overline{Y} - \hat{\beta}_1 \overline{X}_1 - \hat{\beta}_2 \overline{X}_2$$
$$= 10 - (.3204)(9.9) - (.3014)(2.9)$$
$$= 7.7021$$

The estimated multiple linear equation is:

$$\hat{Y} = 7.7021 + .3204\,X_1 - .3014\,X_2$$

The results achieved by this method of calculation differs insignificantly from those calculated by the application of matrix algebra.

There are many other methods available to calculate the parameters of multiple linear models, however, it will become a problem as the number of the independent variables increases. Therefore, it is easier and more accurate to use computer programs developed for these models.

The regression parameters $\hat{\beta}_1$, $\hat{\beta}_2$, . . . , $\hat{\beta}_n$ are sometimes called partial regression coefficients. They reflect the influence of each independent variable on the dependent variable in units of the original observed data. For example, $\hat{\beta}_1 = .3204$ as calculated above means that holding $X_2$ constant, a one unit change in $X_1$ will lead to an average change of .3204 units of Y.

## Standard Error of Estimate

The standard error of estimate, an absolute measure of closeness of fit, can be calculated by a formula similar to that of the simple linear model except for the degrees of freedom which become $n - k$; k refers to the number of parameters being estimated. The formula for the previous example is:

$$\hat{\sigma}_{yx_1x_2} = \sqrt{\frac{\Sigma(Y - \hat{Y})^2}{n - 3}}$$

To calculate the standard error of estimate for the previous example, we need to compute $\hat{Y}$ from the estimated equation:

$$\hat{Y} = 7.6869 + .2988X_1 - .3475X_2$$

$\hat{Y}$ for the 1st observation $= 7.6869 + .2988(2) - .3475(4)$

| $\hat{Y}$ | $(Y - \hat{Y})^2$ |
|---|---|
| 6.8945 | 15.1671 |
| 7.2907 | .5031 |
| 8.5346 | 6.0782 |
| 8.7360 | 39.2377 |
| 11.2238 | 4.9453 |
| 8.7360 | 22.4297 |
| 10.7789 | 38.7021 |
| 12.6691 | .1095 |
| 9.5350 | 2.1462 |
| 11.9741 | 8.8453 |
| | 138.1642 |

$$\hat{\sigma}_{yx_1x_2} = \sqrt{\frac{138.1642}{10 - 3}}$$

$$= 4.443$$

The magnitude of the standard error of estimation is quite high which may indicate that the regression plane does not fit the data well. To find more about the fit, we need to compute the multiple determination coefficient.

**Multiple Determination Coefficient: $R^2$**

Coefficients of multiple determination measures the percentage of variation in Y that has been explained by variations in the independent variables. It is a measure of closeness of the fit of the regression plane to the observed points relative to the fit of the plane going through the means: $\overline{Y}$, $\overline{X}_1$, and $\overline{X}_2$.

The formula for $R^2$ is the same as the simple linear model:

$$R^2 = \frac{\text{Explained variation (squared)}}{\text{Total variation (squared)}}$$

$$= \frac{\Sigma(\hat{Y} - \overline{Y})^2}{\Sigma(Y - \overline{Y})^2}$$

or $\quad R^2 = 1 - \dfrac{\text{Unexplained variation (squared)}}{\text{Total variation (squared)}}$

$$= 1 - \frac{\Sigma(Y - \hat{Y})^2}{\Sigma(Y - \overline{Y})^2}$$

We will use the last formula for $R^2$ because we have already computed $\Sigma(Y - \hat{Y})^2$, so we need to compute $\Sigma(Y - \overline{Y})^2$.

| $Y - \overline{Y}$ | $(Y - \overline{Y})^2$ |
|---|---|
| $-7$ | 49 |
| $-2$ | 4 |
| 1 | 1 |
| 5 | 25 |
| $-1$ | 1 |
| $-6$ | 36 |
| 7 | 49 |
| 3 | 9 |
| 1 | 1 |
| $-1$ | 1 |
| | 176 |

$$R^2 = 1 - \frac{138.1642}{176}$$

$$= 1 - .7850$$

$$= .2150$$

The magnitude of $R^2$ is very small indicating that the linear association among the variables Y, $X_1$, and $X_2$ is very weak. In addition, the small value of $R^2$ means that the regression equation is a poor fit of the plane.

$R^2$ indicates the degree of association of all variables combined. If the magnitude of $R^2$ is small, as is the case of the current example, then it might be of interest to determine the independent variable(s) with the lowest association to Y to eliminate. This can be done by examining the simple or two variable correlation coefficients as well as the partial correlation coefficients. Formulas to compute these two types of correlation are presented below.

## Two Variable Correlation Coefficients

For the function of: $Y = f(X_1, X_2)$, we can calculate three simple or two variable correlation coefficients: $R_{yx_1}$, $R_{yx_2}$, and $R_{x_1x_2}$ applying the following formulas:

$$R_{yx_1} = \frac{\Sigma yx_1}{\sqrt{\Sigma y^2 \Sigma x_1^2}} \qquad \text{correlation coefficient for Y and } X_1$$

$$R_{yx_2} = \frac{\Sigma yx_2}{\sqrt{\Sigma y^2 \Sigma x_2^2}} \qquad \text{correlation coefficient for Y and } X_2$$

$$R_{x_1x_2} = \frac{\Sigma x_1 x_2}{\sqrt{\Sigma x_1^2 \Sigma x_2^2}} \qquad \text{correlation coefficient for } X_1 \text{ and } X_2$$

The simple correlation coefficients for the previous example are calculated and arranged in a matrix. The following values have been previously computed:

$$\Sigma yx_1 = 123 \qquad \Sigma y^2 = 176 \qquad \Sigma x_1^2 = 398.9$$

$$\Sigma yx_2 = 0 \qquad \Sigma x_2^2 = 16.9 \qquad \Sigma x_1 x_2 = 15.9$$

$$R_{yx_1} = \frac{123}{\sqrt{(176)(398.9)}} = .4642$$

$$R_{yx_2} = \frac{0}{\sqrt{(176)(16.9)}} = 0$$

$$R_{x_1x_2} = \frac{15.9}{\sqrt{(398.9)(16.9)}} = .1937$$

### Correlation Matrix

|       | Y     | $X_1$  | $X_2$  |
|-------|-------|--------|--------|
| Y     | 1     | .4642  | 0      |
| $X_1$ | .4642 | 1      | .1937  |
| $X_2$ | 0     | .1937  | 1      |

## Partial Correlation Coefficients

A partial correlation coefficient measures the separate effect of each X on Y provided that the influence of all other X's has been removed. For example, $R_{yx_1 \cdot x_2}$ refers to the partial correlation between Y and $X_1$ holding $X_2$ constant, so that its influence on both Y and $X_1$ has been removed. The calculation of partial correlation coefficients can be done by using the following formulas:

$$R_{yx_1 \cdot x_2} = \frac{R_{yx_1} - R_{yx_2} \cdot R_{x_1x_2}}{\sqrt{1 - R^2_{yx_2}} \sqrt{1 - R^2_{x_1x_2}}}$$

$$R_{yx_2 \cdot x_1} = \frac{R_{yx_2} - R_{yx_1} \cdot R_{x_1x_2}}{\sqrt{1 - R^2_{yx_1}} \sqrt{1 - R^2_{x_1x_2}}}$$

$$R_{x_1x_2 \cdot y} = \frac{R_{x_1x_2} - R_{yx_1} \cdot R_{yx_2}}{\sqrt{1 - R^2_{yx_1}} \sqrt{1 - R^2_{yx_2}}}$$

For the previous example:

$$R_{yx_1 \cdot x_2} = \frac{.4642 - (0)(.1937)}{\sqrt{1 - (0)^2}\ \sqrt{1 - (.1937)^2}} = .4731$$

$$R_{yx_2 \cdot x_1} = \frac{0 - (.4642)(.1937)}{\sqrt{1 - (.4642)^2}\ \sqrt{1 - (.1937)^2}} = -.1035$$

$$R_{x_1 x_2 \cdot y} = \frac{.1937 - (.4642)(0)}{\sqrt{1 - (.4642)^2}\ \sqrt{1 - (0)^2}} = .2186$$

**Standard Error of Regression Coefficients**

The standard errors of $\beta_1$, $\beta_2$, . . . , $\beta_n$ measure the expected amount of error that occurs in the estimation of these parameters. These standard errors or $\sigma_{\beta_i}$ are computed for use in determining the significance of the estimated parameters $\beta$'s, and hence the significance of their independent variables.

To calculate the standard error for $\beta_1$ and $\beta_2$ for the previous example, the following two formulas are introduced:

$$\hat{\sigma}_{\beta 1} = \frac{\hat{\sigma}_{yx_1 x_2}}{\sqrt{\Sigma x_1^2(1 - R^2_{x_1 x_2})}}$$

$$\hat{\sigma}_{\beta 2} = \frac{\hat{\sigma}_{yx_1 x_2}}{\sqrt{\Sigma x_2^2(1 - R^2_{x_1 x_2})}}$$

The magnitudes of $\hat{\sigma}_{\beta 1}$ and $\hat{\sigma}_{\beta 2}$ are:

$$\hat{\sigma}_{\beta 1} = \frac{4.443}{\sqrt{398.9(1 - .0375)}} = .2267$$

$$\hat{\sigma}_{\beta 2} = \frac{4.443}{\sqrt{16.9(1 - .0375)}} = .2731$$

The estimated regression equation and the standard error of $\hat{\beta}_1$ and $\hat{\beta}_2$ can be written as follows:

$$\hat{Y} = 7.6869 + .2988X_1 - .3475X_2$$
$$\quad\quad\quad (.2267) \quad\quad (.2731)$$

When compared with the magnitudes of $\hat{\beta}_1$ and $\hat{\sigma}\beta_1$, $\hat{\beta}_1$ is not significantly different from zero. This is also true with respect to $\hat{\beta}_2$. A t-test can be run for each $\hat{\beta}$ as applied in the simple two variable regression, to test for the significance of the $\hat{\beta}_i$. The t-test will show that the $\hat{\beta}_i$ are not significant.

In this example, the magnitude of $R^2$ is very small and the parameter of the model ($\hat{\beta}_1$) is not significantly different from zero, as shown by comparing $\hat{\sigma}_{\beta_1}$ and $\hat{\beta}_1$. These two factors indicate that the estimated equation, or the model, is not reliable for prediction. To improve the reliability of the model, other variables and/or the structure of the model need to be changed. It might be worthwhile to use a nonlinear model rather than the linear model.

## 2. Nonlinear Models:

If the relationship between two or more variables is nonlinear, then any attempt to fit a linear model to the data may fail to produce a reliable, predictive model. An inspection of the scatter diagram is helpful, if not necessary, to decide on the type of model that appropriately fits the data. Even if the decision is made to construct a nonlinear model to represent the nonlinear relationship that exists between the variables, it is important to know the degree of the polynomial from the scatter diagram to properly construct the appropriate model. For example, if the scatter diagram shows that the data follows a polynomial of second degree, then the model becomes:

$$Y = \alpha + \beta_1 X + \beta_2 X^2$$

A higher polynomial curve will determine the elements to be included in the model. In general, any polynomial of second degree or higher can be represented by a nonlinear model. A model for a polynomial of degree n becomes:

$$Y = \alpha + \beta_1 X + \beta_2 X^2 + \beta_3 X^3 + \ldots + \beta_n X^n$$

In addition to polynomials of second degree or higher, growth curves, represented by exponential models, are also nonlinear.

### A. Two-variable Nonlinear Model:

Models that express a nonlinear relationship that exists between an independent and a dependent variable are called two-variable or simple nonlinear model. Exponential models of two variables are here considered to illustrate simple nonlinear models. For example, $Y = \alpha X^\beta$ and $Y = \alpha e^{\beta X}$ are two exponential nonlinear models containing two variables, Y and X.

Exponential two-variable models can be transformed to simple- or two-variable linear models by using logarithms. For the model: $Y = \alpha X^\beta$ the transformed linear model becomes:

$$\log_e Y = \log_e \alpha + \beta \log_e X \tag{1}$$

and $Y = \alpha e^{\beta X}$; a nonlinear exponential model can be transformed to a simple linear model of the form:

$$\log_e Y = \log_e \alpha + \beta X \tag{2}$$

To estimate the parameters of models (1) and (2), the least square method may be used and a system of normal equation can be derived for each model to solve for the unknown parameters. For the first model, equation (1), the system of two-normal equation is:

$$\Sigma \log Y = n \log \hat{\alpha} + \hat{\beta} \Sigma \log X$$
$$\Sigma \log X \log Y = \log \hat{\alpha} \Sigma \log X + \hat{\beta} \Sigma (\log X)^2$$

while the two-normal equation for model (2) is:

$$\Sigma \log Y = n \log \hat{\alpha} + \hat{\beta} \Sigma X$$
$$\Sigma (\log Y) X = \log \hat{\alpha} \Sigma X + \hat{\beta} \Sigma X^2$$

Solve the first and the second sets of equations for log $\hat{\alpha}$ and $\hat{\beta}$; then the antilog of log $\hat{\alpha}$ provides the value of $\hat{\alpha}$.

Having the estimated equations: $\hat{Y} = \hat{\alpha}X^{\beta}$ and $\hat{Y} = \hat{\alpha}e^{\beta X}$ one can follow the same procedure and use the formula presented in the simple linear model to calculate the standard error of estimate $\hat{\sigma}_{yx}$, the standard error of $\beta(\hat{\sigma}_{\beta})$, coefficient of determination $R^2$, and run tests of significance.

B. Multiple Nonlinear Model:

Polynomials of second or higher degree constitute multiple nonlinear models. For example, if the relationship between Y and X can be represented by a polynomial of third degree then the model becomes:

$$Y = \alpha + \beta_1 X + \beta_2 X^2 + \beta_3 X^3$$

The least square method may be applied to estimate the parameters of this model $\alpha$, $\beta_1$, $\beta_2$, and $\beta_3$. The minimization of the sum squared residuals ($\Sigma e^2$) provides us with a system of four-normal equation to solve for the four unknown parameters. The system of normal equation is:

$$\Sigma X \quad = n\hat{\alpha} \quad + \hat{\beta}_1\Sigma X \quad + \hat{\beta}_2\Sigma X^2 + \hat{\beta}_3\Sigma X^3$$

$$\Sigma XY = \hat{\alpha}\Sigma X \quad + \hat{\beta}_1\Sigma X^2 + \hat{\beta}_2\Sigma X^3 + \hat{\beta}_3\Sigma X^4$$

$$\Sigma X^2Y = \hat{\alpha}\Sigma X^2 + \hat{\beta}_1\Sigma X^3 + \hat{\beta}_2\Sigma X^4 + \hat{\beta}_3\Sigma X^5$$

$$\Sigma X^3Y = \hat{\alpha}\Sigma X^3 + \hat{\beta}_1\Sigma X^4 + \hat{\beta}_2\Sigma X^5 + \hat{\beta}_3\Sigma X^6$$

In matrix notation:

$$\begin{bmatrix} \Sigma X \\ \Sigma XY \\ \Sigma X^2Y \\ \Sigma X^3Y \end{bmatrix} = \begin{bmatrix} n & \Sigma X & \Sigma X^2 & \Sigma X^3 \\ \Sigma X & \Sigma X^2 & \Sigma X^3 & \Sigma X^4 \\ \Sigma X^2 & \Sigma X^3 & \Sigma X^4 & \Sigma X^5 \\ \Sigma X^3 & \Sigma X^4 & \Sigma X^5 & \Sigma X^6 \end{bmatrix} \begin{bmatrix} \alpha \\ \beta_1 \\ \beta_2 \\ \beta_3 \end{bmatrix}$$

or $\quad \mathbf{Y = XA}$

To solve for the values of A, find $\mathbf{X}^{-1}$ and multiply:

$$\mathbf{X^{-1}Y = IA}$$

The estimated equation of this model is:

$$\hat{Y} = \hat{\alpha} + \hat{\beta}_1X + \hat{\beta}_2X^2 + \hat{\beta}_3X^3$$

The standard error of estimate, the standard error for the regression coefficients, and the coefficient of determination can be calculated for this model by following the same procedure presented in the multiple linear model.

The multiple nonlinear model can be transformed into the form of a multiple linear model. For example, in the multiple nonlinear model:

$$Y = \alpha + \beta_1 X + \beta_2 X^2 + \beta_3 X^3$$

if we set $X = X_1$

and $X^2 = X_2$

$X^3 = X_3$

**202**

then this multiple nonlinear model becomes a multiple linear model:

$$Y = \alpha + \beta_1 X_1 + \beta_2 X_2 + \beta_3 X_3$$

As the degree of polynomials increases it will be very difficult to use a desk calculator to estimate the parameters and to calculate measures needed to test the predictive ability of the model. Therefore, it is advisable to use computer programs designed for this type of analysis.

## EXERCISES

9.1 Define:
    a. An econometric model
    b. Linear model
    c. Nonlinear model
    d. Multiple linear model
    e. Coefficient of determination

9.2 a. What is the difference between regression and correlation analysis?
    b. Why is the disturbance term included in any econometric model?
    c. What is the difference between the standard error of estimate and the standard error of regression coefficient?

9.3 The following table shows the data for Y and X, where Y represents personal consumption expenditures of durable goods, and X represents total disposable personal income. The data is seasonally adjusted at annual rates, in billions of dollars.

| Year | Y | X |
|------|------|-------|
| 1950 | 30.5 | 206.9 |
| 1951 | 29.6 | 226.6 |
| 1952 | 29.3 | 238.3 |
| 1953 | 33.2 | 252.6 |
| 1954 | 32.8 | 257.4 |
| 1955 | 39.6 | 275.3 |
| 1956 | 38.9 | 293.2 |
| 1957 | 40.8 | 308.5 |
| 1958 | 37.9 | 318.8 |
| 1959 | 44.3 | 337.3 |

Construct an appropriate model for the two variables, and determine the reliability of the model to predict.

9.4 Project: $Y_t = f(X_{t-1})$
$Y_t$ represents personal consumption expenditures on automobiles and parts seasonally adjusted at annual rates for the period 1960 to 1969.
$X_{t-1}$ represents personal savings, seasonally adjusted at annual rates for the period 1959 to 1968. (Notice one year lag.)
Construct a model to represent the relationship between the variables; then test the reliability of the model to predict.

9.5  Project: $Y_t = f(X_{1t}, X_{2t-1})$

$Y_t$ represents total personal consumption on nondurable goods, seasonally adjusted at annual rates for the period 1961 to 1970.

$X_{1t}$ represents personal outlays, seasonally adjusted at annual rates for the period from 1961 to 1970.

$X_{2t-1}$ represents personal savings, seasonally adjusted at annual rates for the period from 1960 to 1969.

Test the reliability of the model to predict.

# APPENDIX

# Matrices

Matrix algebra is being used extensively in quantitative business and economics to solve for the unknowns of a system of linear equations, to calculate the expected pay off values in decision theory, and to estimate parameters of multiple linear or nonlinear econometric models.

This appendix covers selective parts of matrix algebra either applied or mentioned in the chapters of this book. These parts are: determinants, addition, multiplication, and the inverse of a matrix.

Definition of a matrix:

A matrix is a rectangular array of real numbers arranged in m rows and n columns. Let A be a rectangular array, and m by n (m × n) matrix. The elements of matrix A are:

$$A = \begin{bmatrix} a_{11} & a_{12} & a_{13} & \cdots & a_{1n} \\ a_{21} & a_{22} & a_{23} & \cdots & a_{2n} \\ \cdot & \cdot & \cdot & & \cdot \\ \cdot & \cdot & \cdot & & \cdot \\ \cdot & \cdot & \cdot & & \cdot \\ a_{m1} & a_{m2} & a_{m3} & \cdots & a_{mn} \end{bmatrix}_{m,n}$$

or $A = [a_{ij}]_{mn}$; $i = 1,2,3, \ldots ,m$ and $j = 1,2,3, \ldots ,n$.

If the number of rows are equal to the number of columns (m = n), then A becomes a square matrix (n × n).

An array with m rows and one column is called a column vector while a row vector has one row and n columns.

Determinants

A determinant is a single number associated with a square matrix. The determinant of matrix A is denoted by |A|.

$$|A| = \begin{vmatrix} a_{11} & a_{12} & a_{13} & \cdots & a_{1n} \\ a_{21} & a_{22} & a_{23} & \cdots & a_{2n} \\ \cdot & \cdot & \cdot & & \cdot \\ \cdot & \cdot & \cdot & & \cdot \\ \cdot & \cdot & \cdot & & \cdot \\ a_{n1} & a_{n2} & a_{n3} & \cdots & a_{nn} \end{vmatrix}$$

Calculation of determinants

1. Let X be a 2 × 2 square matrix:

$$X = \begin{bmatrix} X_{11} & X_{12} \\ X_{21} & X_{22} \end{bmatrix}$$

then $|X| = X_{11} X_{22} - X_{21} X_{22}$

*Example:*

$$X = \begin{bmatrix} 5 & 9 \\ 4 & 12 \end{bmatrix} \qquad \text{find } |X|$$

**Solution:** $|X| = (5)(12) - (4)(9)$
$$= 24$$

2. Let A be a square matrix of higher order than 2 × 2:

$$A = \begin{bmatrix} a_{11} & a_{12} & a_{13} \\ a_{21} & a_{22} & a_{23} \\ a_{31} & a_{32} & a_{33} \end{bmatrix}$$

$$|A| = \sum_{j=1}^{n} a_{ij} c_{ij}$$

where $c_{ij}$ refers to the co-factors of the minors of A

$$c_{ij} = (-1)^{i+j} |M_{ij}|, \quad |M_{ij}| = \text{determinants of the minor}$$

then $|A| = a_{11}(-1)^{1+1} \begin{vmatrix} a_{22} & a_{23} \\ a_{32} & a_{33} \end{vmatrix} + a_{12}(-1)^{1+2} \begin{vmatrix} a_{21} & a_{23} \\ a_{31} & a_{33} \end{vmatrix}$

$$+ a_{13}(-1)^{1+3} \begin{vmatrix} a_{21} & a_{22} \\ a_{31} & a_{32} \end{vmatrix}$$

or $|A| = a_{11}(-1)^{1+1} \begin{vmatrix} a_{22} & a_{23} \\ a_{32} & a_{33} \end{vmatrix} + a_{21}(-1)^{3+1} \begin{vmatrix} a_{12} & a_{13} \\ a_{32} & a_{33} \end{vmatrix}$

$$+ a_{31}(-1)^{3+1} \begin{vmatrix} a_{12} & a_{13} \\ a_{22} & a_{23} \end{vmatrix}$$

*Example:* Let A be a 3 × 3 matrix:

$$A = \begin{bmatrix} 2 & 4 & 6 \\ 8 & -1 & 3 \\ 5 & 7 & 9 \end{bmatrix} \qquad \text{Find } |A|$$

**Solution:**

$$|A| = (2)(-1)^{1+1} \begin{vmatrix} -1 & 3 \\ 7 & 9 \end{vmatrix} + 4(-1)^{1+2} \begin{vmatrix} 8 & 3 \\ 5 & 9 \end{vmatrix} + 6(-1)^{1+3} \begin{vmatrix} 8 & -1 \\ 5 & 7 \end{vmatrix}$$

$$= (2)(-9 - 21) - 4(72 - 15) + 6(56 - (-5))$$
$$= (2)(-30) - 4(57) + 6(61)$$
$$= -60 - 228 + 366$$
$$= 78$$

or $\quad |A| = (2)(-1)^{1+1} \begin{vmatrix} -1 & 3 \\ 7 & 9 \end{vmatrix} + 8(-1)^{2+1} \begin{vmatrix} 4 & 6 \\ 7 & 9 \end{vmatrix} + 5(-1)^{3+1} \begin{vmatrix} 4 & 6 \\ -1 & 3 \end{vmatrix}$

$$= -60 + 48 + 90$$
$$= 78$$

If the order of a determinant is very large, say 9 or 8, then simplify the determinant order by reducing the rows to zero by some arithmetic operation.

Addition and subtraction of matrices

Matrices of the same order (m,n) are called conformable for addition or subtraction.

*Example:* Add the following two matrices:

$$A = \begin{bmatrix} 5 & 6 & -1 \\ 4 & 2 & 9 \end{bmatrix}_{2,3} \qquad B = \begin{bmatrix} 2 & 5 & 9 \\ -6 & 4 & -2 \end{bmatrix}_{2,3}$$

**Solution:** Let C = A + B

$$C = \begin{bmatrix} 5+2 & 6+5 & -1+9 \\ 4+(-6) & 2+4 & 9+(-2) \end{bmatrix}_{2,3} = \begin{bmatrix} 7 & 11 & 8 \\ -2 & 6 & 7 \end{bmatrix}_{2,3}$$

*Example:* Subtract the matrices B from A in the above example:

**Solution:** Let D = A − B

$$D = \begin{bmatrix} 5-2 & 6-5 & -1-9 \\ 4+6 & 2-4 & 9+2 \end{bmatrix} = \begin{bmatrix} 3 & 1 & -10 \\ 10 & -2 & 11 \end{bmatrix}$$

Multiplication of matrices:

Two matrices can be multiplied if the number of columns in one equals the number of rows in the other, or vice versa. If the number of columns in E is equal to the number of rows in F, then E and F can be multiplied.

If $\quad E = (e_{ij})_{m,n} \qquad$ and $\qquad F = (f_{ij})_{n,p}$

then $\quad EF = X = (x_{ik})_{m,p}$

*Example:* Multiply the following two matrices E and F:

$$E = \begin{bmatrix} 1 & 2 \\ 3 & 4 \end{bmatrix}_{2,2} \qquad F = \begin{bmatrix} 1 & 2 & 3 \\ 3 & 2 & 1 \end{bmatrix}_{2,3}$$

**Solution:** E is 2 × 2 matrix, 2 rows and 2 columns.
F is 2 × 3 matrix, 2 rows and 3 columns.

The number of columns in E = the number of rows in F, then we can multiply E by F to produce X; or

$$X = E\ F$$

$$X = \begin{bmatrix} 1 & 2 \\ 3 & 4 \end{bmatrix}_{2,2} \Bigg| \begin{bmatrix} 1 & 2 & 3 \\ 3 & 2 & 1 \end{bmatrix}_{2,3}$$

$$= \begin{bmatrix} (1)(1) + (2)(3) & (1)(2) + (2)(2) & (1)(3) + (2)(1) \\ (3)(1) + (4)(3) & (3)(2) + (4)(2) & (3)(3) + (4)(1) \end{bmatrix}$$

$$= \begin{bmatrix} 7 & 6 & 5 \\ 15 & 14 & 13 \end{bmatrix}_{2,3}$$

Notice that EF = FE even if the number of columns in F equals the number of rows in E.

Transposition of a Matrix:

The term transpose will be mentioned in finding the inverse of a square matrix. It means rows are changed to columns or vice versa. If A is 2 × 3 matrix, then A transposed ($A^T$ or $A'$) is a matrix of the order 3 × 2.

*Example:* Let A be 3 × 4 matrix, find $A^T$.

$$A = \begin{bmatrix} -2 & 4 & 7 & 8 \\ 3 & 5 & 0 & 9 \\ 2 & 0 & 5 & 7 \end{bmatrix}_{3,4}$$

**Solution:**

$$A^T \text{ or } A' = \begin{bmatrix} -2 & 3 & 2 \\ 4 & 5 & 0 \\ 7 & 0 & 5 \\ 8 & 9 & 7 \end{bmatrix}_{4,3}$$

Inverse of a square matrix

To calculate the inverse of any square matrix, follow these three steps:
1. Find the determinant of X or |X|.
2. Find the adjoint matrix of X or Adj X.
3. $X^{-1} = \dfrac{1}{|X|}$ Adj X.

*Example:* Calculate the inverse of the following matrix:

$$X = \begin{bmatrix} 2 & 3 & 4 \\ 5 & 6 & 7 \\ 1 & 7 & 8 \end{bmatrix}$$

**208**

**Solution:**

1. The determinant of X.

$$|X| = 2(-1)^{1+1} \begin{vmatrix} 6 & 7 \\ 7 & 8 \end{vmatrix} + 3(-1)^{1+2} \begin{vmatrix} 5 & 7 \\ 1 & 8 \end{vmatrix} + 4(-1)^{1+3} \begin{vmatrix} 5 & 6 \\ 1 & 7 \end{vmatrix}$$

$$= -2 - 99 + 116$$

$$= 15$$

2. Calculate the adjoint matrix of X (Adj X). The adjoint matrix of X can be calculated as follows:

a. Calculate the minors. A minor is the determinant obtained from $|X|$ by deleting the ith row and the jth column.

For example:

$$X = \begin{bmatrix} 2 & 3 & 4 \\ 5 & 6 & 7 \\ 1 & 7 & 8 \end{bmatrix} \qquad |X| = \begin{vmatrix} 2 & 3 & 4 \\ 5 & 6 & 7 \\ 1 & 7 & 8 \end{vmatrix}$$

$$\text{The minors of X} = \begin{vmatrix} \begin{vmatrix} 6 & 7 \\ 7 & 8 \end{vmatrix} & \begin{vmatrix} 5 & 7 \\ 1 & 8 \end{vmatrix} & \begin{vmatrix} 5 & 6 \\ 1 & 7 \end{vmatrix} \\ \begin{vmatrix} 3 & 4 \\ 7 & 8 \end{vmatrix} & \begin{vmatrix} 2 & 4 \\ 1 & 8 \end{vmatrix} & \begin{vmatrix} 2 & 3 \\ 1 & 7 \end{vmatrix} \\ \begin{vmatrix} 3 & 4 \\ 6 & 7 \end{vmatrix} & \begin{vmatrix} 2 & 4 \\ 5 & 7 \end{vmatrix} & \begin{vmatrix} 2 & 3 \\ 5 & 6 \end{vmatrix} \end{vmatrix}$$

b. Find the cofactors of X. The cofactor is a signed minor. The sign attached to the minor can be determined by the following rule: $(-1)^{i+j}$ where i denotes the number of the row and j the number of the column.

For example the sign of the minor $\begin{vmatrix} 6 & 7 \\ 7 & 8 \end{vmatrix}$ is:

$(-1)^{1+1}$ because the minor falls in the first row and in the first column, the sign is $(-1)^2$ = +1. The cofactor for this minor is:

$$+ \begin{vmatrix} 6 & 7 \\ 7 & 8 \end{vmatrix} = (6)(8) - (7)(7) = -1$$

Apply the sign rule to the minor $\begin{vmatrix} 5 & 7 \\ 1 & 8 \end{vmatrix}$ and record the cofactor.

The sign is $(-1)^{1+2}$ because the minor falls in the first row and in the second column. The cofactor becomes:

$$(-1)^3 \begin{vmatrix} 5 & 7 \\ 1 & 8 \end{vmatrix} = (-1)\,[(5)(8) - (1)(7)] = -33$$

Use the same steps to calculate the cofactor matrix of X:

$$
\begin{bmatrix}
-1 & -33 & 29 \\
4 & 12 & -11 \\
-3 & 6 & -3
\end{bmatrix}
$$

C. The transpose of the cofactor matrix of X is the Adj X:

$$
\text{Adj } X =
\begin{bmatrix}
-1 & 4 & -3 \\
-33 & 12 & 6 \\
29 & -11 & -3
\end{bmatrix}
$$

3. The inverse of X $(X^{-1})$ is:

$$
X^{-1} = \frac{1}{|X|}
\begin{bmatrix}
-1 & 4 & -3 \\
-33 & 12 & 6 \\
29 & -11 & -3
\end{bmatrix}
$$

$$
= \frac{1}{15}
\begin{bmatrix}
-1 & 4 & -3 \\
-33 & 12 & 6 \\
29 & -11 & -3
\end{bmatrix}
$$

$$
=
\begin{bmatrix}
-1/15 & 4/15 & -3/15 \\
-33/15 & 12/15 & 6/15 \\
29/15 & -11/15 & -3/15
\end{bmatrix}
$$

$$
=
\begin{bmatrix}
-.07 & .27 & -.20 \\
-2.2 & .80 & .40 \\
1.93 & -.73 & -.20
\end{bmatrix}
$$

If we multiply X by its inverse $X^{-1}$, the result will be the unit or identity matrix. The identity matrix contains one in the diagonal of the matrix and zeros everywhere and it serves as one in the scalar system. The multiplication of X by $X^{-1}$ produces:

$$
X \cdot X^{-1} =
\begin{bmatrix}
2 & 3 & 4 \\
5 & 6 & 7 \\
1 & 7 & 8
\end{bmatrix}
\cdot
\begin{bmatrix}
-.07 & .27 & -.20 \\
-2.2 & .80 & .40 \\
1.93 & -.73 & -.20
\end{bmatrix}
$$

$$
=
\begin{bmatrix}
.98 & .02 & 0 \\
-.04 & 1.04 & 0 \\
-.03 & .03 & 1
\end{bmatrix}
\cong
\begin{bmatrix}
1 & 0 & 0 \\
0 & 1 & 0 \\
0 & 0 & 1
\end{bmatrix}
= I
$$

It is also true that $X^{-1} \cdot X = I$, where I refers to the identity matrix.

There are different methods besides this one to find the inverse of a matrix.

# TABLES

## TABLE 1
Binomial Probability Distribution

### n = 1

| r \ p | .01 | .02 | .03 | .04 | .05 | .06 | .07 | .08 | .09 | .10 |
|---|---|---|---|---|---|---|---|---|---|---|
| 0 | .9900 | .9800 | .9700 | .9600 | .9500 | .9400 | .9300 | .9200 | .9100 | .9000 |
| 1 | .0100 | .0200 | .0300 | .0400 | .0500 | .0600 | .0700 | .0800 | .0900 | .1000 |

| r \ p | .11 | .12 | .13 | .14 | .15 | .16 | .17 | .18 | .19 | .20 |
|---|---|---|---|---|---|---|---|---|---|---|
| 0 | .8900 | .8800 | .8700 | .8600 | .8500 | .8400 | .8300 | .8200 | .8100 | .8000 |
| 1 | .1100 | .1200 | .1300 | .1400 | .1500 | .1600 | .1700 | .1800 | .1900 | .2000 |

| r \ p | .21 | .22 | .23 | .24 | .25 | .26 | .27 | .28 | .29 | .30 |
|---|---|---|---|---|---|---|---|---|---|---|
| 0 | .7900 | .7800 | .7700 | .7600 | .7500 | .7400 | .7300 | .7200 | .7100 | .7000 |
| 1 | .2100 | .2200 | .2300 | .2400 | .2500 | .2600 | .2700 | .2800 | .2900 | .3000 |

| r \ p | .31 | .32 | .33 | .34 | .35 | .36 | .37 | .38 | .39 | .40 |
|---|---|---|---|---|---|---|---|---|---|---|
| 0 | .6900 | .6800 | .6700 | .6600 | .6500 | .6400 | .6300 | .6200 | .6100 | .6000 |
| 1 | .3100 | .3200 | .3300 | .3400 | .3500 | .3600 | .3700 | .3800 | .3900 | .4000 |

| r \ p | .41 | .42 | .43 | .44 | .45 | .46 | .47 | .48 | .49 | .50 |
|---|---|---|---|---|---|---|---|---|---|---|
| 0 | .5900 | .5800 | .5700 | .5600 | .5500 | .5400 | .5300 | .5200 | .5100 | .5000 |
| 1 | .4100 | .4200 | .4300 | .4400 | .4500 | .4600 | .4700 | .4800 | .4900 | .5000 |

### n = 2

| r \ p | .01 | .02 | .03 | .04 | .05 | .06 | .07 | .08 | .09 | .10 |
|---|---|---|---|---|---|---|---|---|---|---|
| 0 | .9801 | .9604 | .9409 | .9216 | .9025 | .8836 | .8649 | .8464 | .8281 | .8100 |
| 1 | .0198 | .0392 | .0582 | .0768 | .0950 | .1128 | .1302 | .1472 | .1638 | .1800 |
| 2 | .0001 | .0004 | .0009 | .0016 | .0025 | .0036 | .0049 | .0064 | .0081 | .0100 |

| r \ p | .11 | .12 | .13 | .14 | .15 | .16 | .17 | .18 | .19 | .20 |
|---|---|---|---|---|---|---|---|---|---|---|
| 0 | .7921 | .7744 | .7569 | .7396 | .7225 | .7056 | .6889 | .6724 | .6561 | .6400 |
| 1 | .1958 | .2112 | .2262 | .2408 | .2550 | .2688 | .2822 | .2952 | .3078 | .3200 |
| 2 | .0121 | .0144 | .0169 | .0196 | .0225 | .0256 | .0289 | .0324 | .0361 | .0400 |

| r \ p | .21 | .22 | .23 | .24 | .25 | .26 | .27 | .28 | .29 | .30 |
|---|---|---|---|---|---|---|---|---|---|---|
| 0 | .6241 | .6084 | .5929 | .5776 | .5625 | .5476 | .5329 | .5184 | .5041 | .4900 |
| 1 | .3318 | .3432 | .3542 | 3648 | .3750 | .3848 | .3942 | .4032 | .4118 | .4200 |
| 2 | .0441 | .0484 | .0529 | .0576 | .0625 | .0676 | .0729 | .0784 | .0841 | .0900 |

| r \ p | .31 | .32 | .33 | .34 | .35 | .36 | .37 | .38 | .39 | .40 |
|---|---|---|---|---|---|---|---|---|---|---|
| 0 | .4761 | .4624 | .4489 | .4356 | .4225 | .4096 | .3969 | .3844 | .3721 | .3600 |
| 1 | .4278 | .4352 | .4422 | .4488 | .4550 | .4608 | .4662 | .4712 | .4758 | .4800 |
| 2 | .0961 | .1024 | .1089 | .1156 | .1225 | .1296 | .1369 | .1444 | .1521 | .1600 |

| r \ p | .41 | .42 | .43 | .44 | .45 | .46 | .47 | .48 | .49 | .50 |
|---|---|---|---|---|---|---|---|---|---|---|
| 0 | .3481 | .3364 | .3249 | .3136 | .3025 | .2916 | .2809 | .2704 | .2601 | .2500 |
| 1 | .4838 | .4872 | .4902 | .4928 | .4950 | .4968 | .4982 | .4992 | .4998 | .5000 |
| 2 | .1681 | .1764 | .1849 | .1936 | .2025 | .2116 | .2209 | .2304 | .2401 | .2500 |

Charles Clark and Lawrence Schkade, *Statistical Methods for Business Decisions* (Cincinnati: South-Western Publishing Co., 1969), pp. 77–94, by Special Permission.

TABLE 1—*Continued*

n = 3

| p<br>r | .01 | .02 | .03 | .04 | .05 | .06 | .07 | .08 | .09 | .10 |
|---|---|---|---|---|---|---|---|---|---|---|
| 0 | .9704 | .9412 | .9127 | .8847 | .8574 | .8306 | .8044 | .7787 | .7536 | .7290 |
| 1 | .0294 | .0576 | .0847 | .1106 | .1354 | .1590 | .1816 | .2031 | .2236 | .2430 |
| 2 | .0003 | .0012 | .0026 | .0046 | .0071 | .0102 | .0137 | .0177 | .0221 | .0270 |
| 3 | .0000 | .0000 | .0000 | .0001 | .0001 | .0002 | .0003 | .0005 | .0007 | .0010 |

| | .11 | .12 | .13 | .14 | .15 | .16 | .17 | .18 | .19 | .20 |
|---|---|---|---|---|---|---|---|---|---|---|
| 0 | .7050 | .6815 | .6585 | .6361 | .6141 | .5927 | .5718 | .5514 | .5314 | .5120 |
| 1 | .2614 | .2788 | .2952 | .3106 | .3251 | .3387 | .3513 | .3631 | .3740 | .3840 |
| 2 | .0323 | .0380 | .0441 | .0506 | .0574 | .0645 | .0720 | .0797 | .0877 | .0960 |
| 3 | .0013 | .0017 | .0022 | .0027 | .0034 | .0041 | .0049 | .0058 | .0069 | .0080 |

| | .21 | .22 | .23 | .24 | .25 | .26 | .27 | .28 | .29 | .30 |
|---|---|---|---|---|---|---|---|---|---|---|
| 0 | .4930 | .4746 | .4565 | .4390 | .4219 | .4052 | .3890 | .3732 | .3579 | .3430 |
| 1 | .3932 | .4015 | .4091 | .4159 | .4219 | .4271 | .4316 | .4355 | .4386 | .4410 |
| 2 | .1045 | .1133 | .1222 | .1313 | .1406 | .1501 | .1597 | .1693 | .1791 | .1890 |
| 3 | .0093 | .0106 | .0122 | .0138 | .0156 | .0176 | .0197 | .0220 | .0244 | .0270 |

| | .31 | .32 | .33 | .34 | .35 | .36 | .37 | .38 | .39 | .40 |
|---|---|---|---|---|---|---|---|---|---|---|
| 0 | .3285 | .3144 | .3008 | .2875 | .2746 | .2621 | .2500 | .2383 | .2270 | .2160 |
| 1 | .4428 | .4439 | .4444 | .4443 | .4436 | .4424 | .4406 | .4382 | .4354 | .4320 |
| 2 | .1989 | .2089 | .2189 | .2289 | .2389 | .2488 | .2587 | .2686 | .2783 | .2880 |
| 3 | .0298 | .0328 | .0359 | .0393 | .0429 | .0467 | .0507 | .0549 | .0593 | .0640 |

| | .41 | .42 | .43 | .44 | .45 | .46 | .47 | .48 | .49 | .50 |
|---|---|---|---|---|---|---|---|---|---|---|
| 0 | .2054 | .1951 | .1852 | .1756 | .1664 | .1575 | .1489 | .1406 | .1327 | .1250 |
| 1 | .4282 | .4239 | .4191 | .4140 | .4084 | .4024 | .3961 | .3894 | .3823 | .3750 |
| 2 | .2975 | .3069 | .3162 | .3252 | .3341 | .3428 | .3512 | .3594 | .3674 | .3750 |
| 3 | .0689 | .0741 | .0795 | .0852 | .0911 | .0973 | .1038 | .1106 | .1176 | .1250 |

n = 4

| p<br>r | .01 | .02 | .03 | .04 | .05 | .06 | .07 | .08 | .09 | .10 |
|---|---|---|---|---|---|---|---|---|---|---|
| 0 | .9606 | .9224 | .8853 | .8493 | .8145 | .7807 | .7481 | .7164 | .6857 | .6561 |
| 1 | .0388 | .0753 | .1095 | .1416 | .1715 | .1993 | .2252 | .2492 | .2713 | .2916 |
| 2 | .0006 | .0023 | .0051 | .0088 | .0135 | .0191 | .0254 | .0325 | .0402 | .0486 |
| 3 | .0000 | .0000 | .0001 | .0002 | .0005 | .0008 | .0013 | .0019 | .0027 | .0036 |
| 4 | .0000 | .0000 | .0000 | .0000 | .0000 | .0000 | .0000 | .0000 | .0001 | .0001 |

| | .11 | .12 | .13 | .14 | .15 | .16 | .17 | .18 | .19 | .20 |
|---|---|---|---|---|---|---|---|---|---|---|
| 0 | .6274 | .5997 | .5729 | .5470 | .5220 | .4979 | .4746 | .4521 | .4305 | .4096 |
| 1 | .3102 | .3271 | .3424 | .3562 | .3685 | .3793 | .3888 | .3970 | .4039 | .4096 |
| 2 | .0575 | .0669 | .0767 | .0870 | .0975 | .1084 | .1195 | .1307 | .1421 | .1536 |
| 3 | .0047 | .0061 | .0076 | .0094 | .0115 | .0138 | .0163 | .0191 | .0222 | .0256 |
| 4 | .0001 | .0002 | .0003 | .0004 | .0005 | .0007 | .0008 | .0010 | .0013 | .0016 |

| | .21 | .22 | .23 | .24 | .25 | .26 | .27 | .28 | .29 | .30 |
|---|---|---|---|---|---|---|---|---|---|---|
| 0 | .3895 | .3702 | .3515 | .3336 | .3164 | .2999 | .2840 | .2687 | .2541 | .2401 |
| 1 | .4142 | .4176 | .4200 | .4214 | .4219 | .4214 | .4201 | .4180 | .4152 | .4116 |
| 2 | .1651 | .1767 | .1882 | .1996 | .2109 | .2221 | .2331 | .2439 | .2544 | .2646 |
| 3 | .0293 | .0332 | .0375 | .0420 | .0469 | .0520 | .0575 | .0632 | .0693 | .0756 |
| 4 | .0019 | .0023 | .0028 | .0033 | .0039 | .0046 | .0053 | .0061 | .0071 | .0081 |

| | .31 | .32 | .33 | .34 | .35 | .36 | .37 | .38 | .39 | .40 |
|---|---|---|---|---|---|---|---|---|---|---|
| 0 | .2267 | .2138 | .2015 | .1897 | .1785 | .1678 | .1575 | .1478 | .1385 | .1296 |
| 1 | .4074 | .4025 | .3970 | .3910 | .3845 | .3775 | .3701 | .3623 | .3541 | .3456 |
| 2 | .2745 | .2841 | .2933 | .3021 | .3105 | .3185 | .3260 | .3330 | .3396 | .3456 |
| 3 | .0822 | .0891 | .0963 | .1038 | .1115 | .1194 | .1276 | .1361 | .1447 | .1536 |
| 4 | .0092 | .0105 | .0119 | .0134 | .0150 | .0168 | .0187 | .0209 | .0231 | .0256 |

| | .41 | .42 | .43 | .44 | .45 | .46 | .47 | .48 | .49 | .50 |
|---|---|---|---|---|---|---|---|---|---|---|
| 0 | .1212 | .1132 | .1056 | .0983 | .0915 | .0850 | .0789 | .0731 | .0677 | .0625 |
| 1 | .3368 | .3278 | .3185 | .3091 | .2995 | .2897 | .2799 | .2700 | .2600 | .2500 |
| 2 | .3511 | .3560 | .3604 | .3643 | .3675 | .3702 | .3723 | .3738 | .3747 | .3750 |
| 3 | .1627 | .1719 | .1813 | .1908 | .2005 | .2102 | .2201 | .2300 | .2400 | .2500 |
| 4 | .0283 | .0311 | .0342 | .0375 | .0410 | .0448 | .0488 | .0531 | .0576 | .0625 |

TABLE 1—*Continued*

n = 5

| r \ p | .01 | .02 | .03 | .04 | .05 | .06 | .07 | .08 | .09 | .10 |
|---|---|---|---|---|---|---|---|---|---|---|
| 0 | .9510 | .9039 | .8587 | .8154 | .7738 | .7339 | .6957 | .6591 | .6240 | .5905 |
| 1 | .0480 | .0922 | .1328 | .1699 | .2036 | .2342 | .2618 | .2866 | .3086 | .3280 |
| 2 | .0010 | .0038 | .0082 | .0142 | .0214 | .0299 | .0394 | .0498 | .0610 | .0729 |
| 3 | .0000 | .0001 | .0003 | .0006 | .0011 | .0019 | .0030 | .0043 | .0060 | .0081 |
| 4 | .0000 | .0000 | .0000 | .0000 | .0000 | .0001 | .0001 | .0002 | .0003 | .0004 |

| r \ p | .11 | .12 | .13 | .14 | .15 | .16 | .17 | .18 | .19 | .20 |
|---|---|---|---|---|---|---|---|---|---|---|
| 0 | .5584 | .5277 | .4984 | .4704 | .4437 | .4182 | .3939 | .3707 | .3487 | .3277 |
| 1 | .3451 | .3598 | .3724 | .3829 | .3915 | .3983 | .4034 | .4069 | .4089 | .4096 |
| 2 | .0853 | .0981 | .1113 | .1247 | .1382 | .1517 | .1652 | .1786 | .1919 | .2048 |
| 3 | .0105 | .0134 | .0166 | .0203 | .0244 | .0289 | .0338 | .0392 | .0450 | .0512 |
| 4 | .0007 | .0009 | .0012 | .0017 | .0022 | .0028 | .0035 | .0043 | .0053 | .0064 |
| 5 | .0000 | .0000 | .0000 | .0001 | .0001 | .0001 | .0001 | .0002 | .0002 | .0003 |

| r \ p | .21 | .22 | .23 | .24 | .25 | .26 | .27 | .28 | .29 | .30 |
|---|---|---|---|---|---|---|---|---|---|---|
| 0 | .3077 | .2887 | .2707 | .2536 | .2373 | .2219 | .2073 | .1935 | .1804 | .1681 |
| 1 | .4090 | .4072 | .4043 | .4003 | .3955 | .3898 | .3834 | .3762 | .3685 | .3602 |
| 2 | .2174 | .2297 | .2415 | .2529 | .2637 | .2739 | .2836 | .2926 | .3010 | .3087 |
| 3 | .0578 | .0648 | .0721 | .0798 | .0879 | .0962 | .1049 | .1138 | .1229 | .1323 |
| 4 | .0077 | .0091 | .0108 | .0126 | .0146 | .0169 | .0194 | .0221 | .0251 | .0284 |
| 5 | .0004 | .0005 | .0006 | .0008 | .0010 | .0012 | .0014 | .0017 | .0021 | .0024 |

| r \ p | .31 | .32 | .33 | .34 | .35 | .36 | .37 | .38 | .39 | .40 |
|---|---|---|---|---|---|---|---|---|---|---|
| 0 | .1564 | .1454 | .1350 | .1252 | .1160 | .1074 | .0992 | .0916 | .0845 | .0778 |
| 1 | .3513 | .3421 | .3325 | .3226 | .3124 | .3020 | .2914 | .2808 | .2700 | .2592 |
| 2 | .3157 | .3220 | .3275 | .3323 | .3364 | .3397 | .3423 | .3441 | .3452 | .3456 |
| 3 | .1418 | .1515 | .1613 | .1712 | .1811 | .1911 | .2010 | .2109 | .2207 | .2304 |
| 4 | .0319 | .0357 | .0397 | .0441 | .0488 | .0537 | .0590 | .0646 | .0706 | .0768 |
| 5 | .0029 | .0034 | .0039 | .0045 | .0053 | .0060 | .0069 | .0079 | .0090 | .0102 |

| r \ p | .41 | .42 | .43 | .44 | .45 | .46 | .47 | .48 | .49 | .50 |
|---|---|---|---|---|---|---|---|---|---|---|
| 0 | .0715 | .0656 | .0602 | .0551 | .0503 | .0459 | .0418 | .0380 | .0345 | .0312 |
| 1 | .2484 | .2376 | .2270 | .2164 | .2059 | .1956 | .1854 | .1755 | .1657 | .1562 |
| 2 | .3452 | .3442 | .3424 | .3400 | .3369 | .3332 | .3289 | .3240 | .3185 | .3125 |
| 3 | .2399 | .2492 | .2583 | .2671 | .2757 | .2838 | .2916 | .2990 | .3060 | .3125 |
| 4 | .0834 | .0902 | .0974 | .1049 | .1128 | .1209 | .1293 | .1380 | .1470 | .1562 |
| 5 | .0116 | .0131 | .0147 | .0165 | .0185 | .0206 | .0229 | .0255 | .0282 | .0312 |

n = 6

| r \ p | .01 | .02 | .03 | .04 | .05 | .06 | .07 | .08 | .09 | .10 |
|---|---|---|---|---|---|---|---|---|---|---|
| 0 | .9415 | .8858 | .8330 | .7828 | .7351 | .6899 | .6470 | .6064 | .5679 | .5314 |
| 1 | .0571 | .1085 | .1546 | .1957 | .2321 | .2642 | .2922 | .3164 | .3370 | .3543 |
| 2 | .0014 | .0055 | .0120 | .0204 | .0305 | .0422 | .0550 | .0688 | .0833 | .0984 |
| 3 | .0000 | .0002 | .0005 | .0011 | .0021 | .0036 | .0055 | .0080 | .0110 | .0146 |
| 4 | .0000 | .0000 | .0000 | .0000 | .0001 | .0002 | .0003 | .0005 | .0008 | .0012 |
| 5 | .0000 | .0000 | .0000 | .0000 | .0000 | .0000 | .0000 | .0000 | .0000 | .0001 |

| r \ p | .11 | .12 | .13 | .14 | .15 | .16 | .17 | .18 | .19 | .20 |
|---|---|---|---|---|---|---|---|---|---|---|
| 0 | .4970 | .4644 | .4336 | .4046 | .3771 | .3513 | .3269 | .3040 | .2824 | .2621 |
| 1 | .3685 | .3800 | .3888 | .3952 | .3993 | .4015 | .4018 | .4004 | .3975 | .3932 |
| 2 | .1139 | .1295 | .1452 | .1608 | .1762 | .1912 | .2057 | .2197 | .2331 | .2458 |
| 3 | .0188 | .0236 | .0289 | .0349 | .0415 | .0486 | .0562 | .0643 | .0729 | .0819 |
| 4 | .0017 | .0024 | .0032 | .0043 | .0055 | .0069 | .0086 | .0106 | .0128 | .0154 |
| 5 | .0001 | .0001 | .0002 | .0003 | .0004 | .0005 | .0007 | .0009 | .0012 | .0015 |
| 6 | .0000 | .0000 | .0000 | .0000 | .0000 | .0000 | .0000 | .0000 | .0000 | .0001 |

| r \ p | .21 | .22 | .23 | .24 | .25 | .26 | .27 | .28 | .29 | .30 |
|---|---|---|---|---|---|---|---|---|---|---|
| 0 | .2431 | .2252 | .2084 | .1927 | .1780 | .1642 | .1513 | .1393 | .1281 | .1176 |
| 1 | .3877 | .3811 | .3735 | .3651 | .3560 | .3462 | .3358 | .3251 | .3139 | .3025 |
| 2 | .2577 | .2687 | .2789 | .2882 | .2966 | .3041 | .3105 | .3160 | .3206 | .3241 |
| 3 | .0913 | .1011 | .1111 | .1214 | .1318 | .1424 | .1551 | .1639 | .1746 | .1852 |
| 4 | .0182 | .0214 | .0249 | .0287 | .0330 | .0375 | .0425 | .0478 | .0535 | .0595 |
| 5 | .0019 | .0024 | .0030 | .0036 | .0044 | .0053 | .0063 | .0074 | .0087 | .0102 |
| 6 | .0001 | .0001 | .0001 | .0002 | .0002 | .0003 | .0004 | .0005 | .0006 | .0007 |

TABLE 1—*Continued*

### n = 6 (Continued)

| r＼p | .31 | .32 | .33 | .34 | .35 | .36 | .37 | .38 | .39 | .40 |
|---|---|---|---|---|---|---|---|---|---|---|
| 0 | .1079 | .0989 | .0905 | .0827 | .0754 | .0687 | .0625 | .0568 | .0515 | .0467 |
| 1 | .2909 | .2792 | .2673 | .2555 | .2437 | .2319 | .2203 | .2089 | .1976 | .1866 |
| 2 | .3267 | .3284 | .3292 | .3290 | .3280 | .3261 | .3235 | .3201 | .3159 | .3110 |
| 3 | .1957 | .2061 | .2162 | .2260 | .2355 | .2446 | .2533 | .2616 | .2693 | .2765 |
| 4 | .0660 | .0727 | .0799 | .0873 | .0951 | .1032 | .1116 | .1202 | .1291 | .1382 |
| 5 | .0119 | .0137 | .0157 | .0180 | .0205 | .0232 | .0262 | .0295 | .0330 | .0369 |
| 6 | .0009 | .0011 | .0013 | .0015 | .0018 | .0022 | .0026 | .0030 | .0035 | .0041 |

| r | .41 | .42 | .43 | .44 | .45 | .46 | .47 | .48 | .49 | .50 |
|---|---|---|---|---|---|---|---|---|---|---|
| 0 | .0422 | .0381 | .0343 | .0308 | .0277 | .0248 | .0222 | .0198 | .0176 | .0156 |
| 1 | .1759 | .1654 | .1552 | .1454 | .1359 | .1267 | .1179 | .1095 | .1014 | .0938 |
| 2 | .3055 | .2994 | .2928 | .2856 | .2780 | .2699 | .2615 | .2527 | .2436 | .2344 |
| 3 | .2831 | .2891 | .2945 | .2992 | .3032 | .3065 | .3091 | .3110 | .3121 | .3125 |
| 4 | .1475 | .1570 | .1666 | .1763 | .1861 | .1958 | .2056 | .2153 | .2249 | .2344 |
| 5 | .0410 | .0455 | .0503 | .0554 | .0609 | .0667 | .0729 | .0795 | .0864 | .0938 |
| 6 | .0048 | .0055 | .0063 | .0073 | .0083 | .0095 | .0108 | .0122 | .0138 | .0156 |

### n = 7

| r＼p | .01 | .02 | .03 | .04 | .05 | .06 | .07 | .08 | .09 | .10 |
|---|---|---|---|---|---|---|---|---|---|---|
| 0 | .9321 | .8681 | .8080 | .7514 | .6983 | .6485 | .6017 | .5578 | .5168 | .4783 |
| 1 | .0659 | .1240 | .1749 | .2192 | .2573 | .2897 | .3170 | .3396 | .3578 | .3720 |
| 2 | .0020 | .0076 | .0162 | .0274 | .0406 | .0555 | .0716 | .0886 | .1061 | .1240 |
| 3 | .0000 | .0003 | .0008 | .0019 | .0036 | .0059 | .0090 | .0128 | .0175 | .0230 |
| 4 | .0000 | .0000 | .0000 | .0001 | .0002 | .0004 | .0007 | .0011 | .0017 | .0026 |
| 5 | .0000 | .0000 | .0000 | .0000 | .0000 | .0000 | .0000 | .0001 | .0001 | .0002 |

| r | .11 | .12 | .13 | .14 | .15 | .16 | .17 | .18 | .19 | .20 |
|---|---|---|---|---|---|---|---|---|---|---|
| 0 | .4423 | .4087 | .3773 | .3479 | .3206 | .2951 | .2714 | .2493 | .2288 | .2097 |
| 1 | .3827 | .3901 | .3946 | .3965 | .3960 | .3935 | .3891 | .3830 | .3756 | .3670 |
| 2 | .1419 | .1596 | .1769 | .1936 | .2097 | .2248 | .2391 | .2523 | .2643 | .2753 |
| 3 | .0292 | .0363 | .0441 | .0525 | .0617 | .0714 | .0816 | .0923 | .1033 | .1147 |
| 4 | .0036 | .0049 | .0066 | .0086 | .0109 | .0136 | .0167 | .0203 | .0242 | .0287 |
| 5 | .0003 | .0004 | .0006 | .0008 | .0012 | .0016 | .0021 | .0027 | .0034 | .0043 |
| 6 | .0000 | .0000 | .0000 | .0000 | .0001 | .0001 | .0001 | .0002 | .0003 | .0004 |

| r | .21 | .22 | .23 | .24 | .25 | .26 | .27 | .28 | .29 | .30 |
|---|---|---|---|---|---|---|---|---|---|---|
| 0 | .1920 | .1757 | .1605 | .1465 | .1335 | .1215 | .1105 | .1003 | .0910 | .0824 |
| 1 | .3573 | .3468 | .3356 | .3237 | .3115 | .2989 | .2860 | .2731 | .2600 | .2471 |
| 2 | .2850 | .2935 | .3007 | .3067 | .3115 | .3150 | .3174 | .3186 | .3186 | .3177 |
| 3 | .1263 | .1379 | .1497 | .1614 | .1730 | .1845 | .1956 | .2065 | .2169 | .2269 |
| 4 | .0336 | .0389 | .0447 | .0510 | .0577 | .0648 | .0724 | .0803 | .0886 | .0972 |
| 5 | .0054 | .0066 | .0080 | .0097 | .0115 | .0137 | .0161 | .0187 | .0217 | .0250 |
| 6 | .0005 | .0006 | .0008 | .0010 | .0013 | .0016 | .0020 | .0024 | .0030 | .0036 |
| 7 | .0000 | .0000 | .0000 | .0000 | .0001 | .0001 | .0001 | .0001 | .0002 | .0002 |

| r | .31 | .32 | .33 | .34 | .35 | .36 | .37 | .38 | .39 | .40 |
|---|---|---|---|---|---|---|---|---|---|---|
| 0 | .0745 | .0672 | .0606 | .0546 | .0490 | .0440 | .0394 | .0352 | .0314 | .0280 |
| 1 | .2342 | .2215 | .2090 | .1967 | .1848 | .1732 | .1619 | .1511 | .1407 | .1306 |
| 2 | .3156 | .3127 | .3088 | .3040 | .2985 | .2922 | .2853 | .2778 | .2698 | .2613 |
| 3 | .2363 | .2452 | .2535 | .2610 | .2679 | .2740 | .2793 | .2838 | .2875 | .2903 |
| 4 | .1062 | .1154 | .1248 | .1345 | .1442 | .1541 | .1640 | .1739 | .1838 | .1935 |
| 5 | .0286 | .0326 | .0369 | .0416 | .0466 | .0520 | .0578 | .0640 | .0705 | .0774 |
| 6 | .0043 | .0051 | .0061 | .0071 | .0084 | .0098 | .0113 | .0131 | .0150 | .0172 |
| 7 | .0003 | .0003 | .0004 | .0005 | .0006 | .0008 | .0009 | .0011 | .0014 | .0016 |

| r | .41 | .42 | .43 | .44 | .45 | .46 | .47 | .48 | .49 | .50 |
|---|---|---|---|---|---|---|---|---|---|---|
| 0 | .0249 | .0221 | .0195 | .0173 | .0152 | .0134 | .0117 | .0103 | .0090 | .0078 |
| 1 | .1211 | .1119 | .1032 | .0950 | .0872 | .0798 | .0729 | .0664 | .0604 | .0547 |
| 2 | .2524 | .2431 | .2336 | .2239 | .2140 | .2040 | .1940 | .1840 | .1740 | .1641 |
| 3 | .2923 | .2934 | .2937 | .2932 | .2918 | .2897 | .2867 | .2830 | .2786 | .2734 |
| 4 | .2031 | .2125 | .2216 | .2304 | .2388 | .2468 | .2543 | .2612 | .2676 | .2734 |
| 5 | .0847 | .0923 | .1003 | .1086 | .1172 | .1261 | .1353 | .1447 | .1543 | .1641 |
| 6 | .0196 | .0223 | .0252 | .0284 | .0320 | .0358 | .0400 | .0445 | .0494 | .0547 |
| 7 | .0019 | .0023 | .0027 | .0032 | .0037 | .0044 | .0051 | .0059 | .0068 | .0078 |

TABLE 1—*Continued*

## n = 8

| r \ p | .01 | .02 | .03 | .04 | .05 | .06 | .07 | .08 | .09 | .10 |
|---|---|---|---|---|---|---|---|---|---|---|
| 0 | .9227 | .8508 | .7837 | .7214 | .6634 | .6096 | .5596 | .5132 | .4703 | .4305 |
| 1 | .0746 | .1389 | .1939 | .2405 | .2793 | .3113 | .3370 | .3570 | .3721 | .3826 |
| 2 | .0026 | .0099 | .0210 | .0351 | .0515 | .0695 | .0888 | .1087 | .1288 | .1488 |
| 3 | .0001 | .0004 | .0013 | .0029 | .0054 | .0089 | .0134 | .0189 | .0255 | .0331 |
| 4 | .0000 | .0000 | .0001 | .0002 | .0004 | .0007 | .0013 | .0021 | .0031 | .0046 |
| 5 | .0000 | .0000 | .0000 | .0000 | .0000 | .0000 | .0001 | .0001 | .0002 | .0004 |

| r \ p | .11 | .12 | .13 | .14 | .15 | .16 | .17 | .18 | .19 | .20 |
|---|---|---|---|---|---|---|---|---|---|---|
| 0 | .3937 | .3596 | .3282 | .2992 | .2725 | .2479 | .2252 | .2044 | .1853 | .1678 |
| 1 | .3892 | .3923 | .3923 | .3897 | .3847 | .3777 | .3691 | .3590 | .3477 | .3355 |
| 2 | .1684 | .1872 | .2052 | .2220 | .2376 | .2518 | .2646 | .2758 | .2855 | .2936 |
| 3 | .0416 | .0511 | .0613 | .0723 | .0839 | .0959 | .1084 | .1211 | .1339 | .1468 |
| 4 | .0064 | .0087 | .0115 | .0147 | .0185 | .0228 | .0277 | .0332 | .0393 | .0459 |
| 5 | .0006 | .0009 | .0014 | .0019 | .0026 | .0035 | .0045 | .0058 | .0074 | .0092 |
| 6 | .0000 | .0001 | .0001 | .0002 | .0002 | .0003 | .0005 | .0006 | .0009 | .0011 |
| 7 | .0000 | .0000 | .0000 | .0000 | .0000 | .0000 | .0000 | .0000 | .0001 | .0001 |

| r \ p | .21 | .22 | .23 | .24 | .25 | .26 | .27 | .28 | .29 | .30 |
|---|---|---|---|---|---|---|---|---|---|---|
| 0 | .1517 | .1370 | .1236 | .1113 | .1001 | .0899 | .0806 | .0722 | .0646 | .0576 |
| 1 | .3226 | .3092 | .2953 | .2812 | .2670 | .2527 | .2386 | .2247 | .2110 | .1977 |
| 2 | .3002 | .3052 | .3087 | .3108 | .3115 | .3108 | .3089 | .3058 | .3017 | .2965 |
| 3 | .1596 | .1722 | .1844 | .1963 | .2076 | .2184 | .2285 | .2379 | .2464 | .2541 |
| 4 | .0530 | .0607 | .0689 | .0775 | .0865 | .0959 | .1056 | .1156 | .1258 | .1361 |
| 5 | .0113 | .0137 | .0165 | .0196 | .0231 | .0270 | .0313 | .0360 | .0411 | .0467 |
| 6 | .0015 | .0019 | .0025 | .0031 | .0038 | .0047 | .0058 | .0070 | .0084 | .0100 |
| 7 | .0001 | .0002 | .0002 | .0003 | .0004 | .0005 | .0006 | .0008 | .0010 | .0012 |
| 8 | .0000 | .0000 | .0000 | .0000 | .0000 | .0000 | .0000 | .0000 | .0001 | .0001 |

| r \ p | .31 | .32 | .33 | .34 | .35 | .36 | .37 | .38 | .39 | .40 |
|---|---|---|---|---|---|---|---|---|---|---|
| 0 | .0514 | .0457 | .0406 | .0360 | .0319 | .0281 | .0248 | .0218 | .0192 | .0168 |
| 1 | .1847 | .1721 | .1600 | .1484 | .1373 | .1267 | .1166 | .1071 | .0981 | .0896 |
| 2 | .2904 | .2835 | .2758 | .2675 | .2587 | .2494 | .2397 | .2297 | .2194 | .2090 |
| 3 | .2609 | .2668 | .2717 | .2756 | .2786 | .2805 | .2815 | .2815 | .2806 | .2787 |
| 4 | .1465 | .1569 | .1673 | .1775 | .1875 | .1973 | .2067 | .2157 | .2242 | .2322 |
| 5 | .0527 | .0591 | .0659 | .0732 | .0808 | .0888 | .0971 | .1058 | .1147 | .1239 |
| 6 | .0118 | .0139 | .0162 | .0188 | .0217 | .0250 | .0285 | .0324 | .0367 | .0413 |
| 7 | .0015 | .0019 | .0023 | .0028 | .0033 | .0040 | .0048 | .0057 | .0067 | .0079 |
| 8 | .0001 | .0001 | .0001 | .0002 | .0002 | .0003 | .0004 | .0004 | .0005 | .0007 |

| r \ p | .41 | .42 | .43 | .44 | .45 | .46 | .47 | .48 | .49 | .50 |
|---|---|---|---|---|---|---|---|---|---|---|
| 0 | .0147 | .0128 | .0111 | .0097 | .0084 | .0072 | .0062 | .0053 | .0046 | .0039 |
| 1 | .0816 | .0742 | .0672 | .0608 | .0548 | .0493 | .0442 | .0395 | .0352 | .0312 |
| 2 | .1985 | .1880 | .1776 | .1672 | .1569 | .1489 | .1371 | .1275 | .1183 | .1094 |
| 3 | .2759 | .2723 | .2679 | .2627 | .2568 | .2503 | .2431 | .2355 | .2273 | .2188 |
| 4 | .2397 | .2465 | .2526 | .2580 | .2627 | .2665 | .2695 | .2717 | .2730 | .2734 |
| 5 | .1332 | .1428 | .1525 | .1622 | .1719 | .1816 | .1912 | .2006 | .2098 | .2188 |
| 6 | .0463 | .0517 | .0575 | .0637 | .0703 | .0774 | .0848 | .0926 | .1008 | .1094 |
| 7 | .0092 | .0107 | .0124 | .0143 | .0164 | .0188 | .0215 | .0244 | .0277 | .0312 |
| 8 | .0008 | .0010 | .0012 | .0014 | .0017 | .0020 | .0024 | .0028 | .0033 | .0039 |

## n = 9

| r \ p | .01 | .02 | .03 | .04 | .05 | .06 | .07 | .08 | .09 | .10 |
|---|---|---|---|---|---|---|---|---|---|---|
| 0 | .9135 | .8337 | .7602 | .6925 | .6302 | .5730 | .5204 | .4722 | .4279 | .3874 |
| 1 | .0830 | .1531 | .2116 | .2597 | .2985 | .3292 | .3525 | .3695 | .3809 | .3874 |
| 2 | .0034 | .0125 | .0262 | .0433 | .0629 | .0840 | .1061 | .1285 | .1507 | .1722 |
| 3 | .0001 | .0006 | .0019 | .0042 | .0077 | .0125 | .0186 | .0261 | .0348 | .0446 |
| 4 | .0000 | .0000 | .0001 | .0003 | .0006 | .0012 | .0021 | .0034 | .0052 | .0074 |
| 5 | .0000 | .0000 | .0000 | .0000 | .0000 | .0001 | .0002 | .0003 | .0005 | .0008 |
| 6 | .0000 | .0000 | .0000 | .0000 | .0000 | .0000 | .0000 | .0000 | .0000 | .0001 |

TABLE 1—*Continued*

| n = 9 (Continued) | | | | | | | | | |
|---|---|---|---|---|---|---|---|---|---|

| p / r | .11 | .12 | .13 | .14 | .15 | .16 | .17 | .18 | .19 | .20 |
|---|---|---|---|---|---|---|---|---|---|---|
| 0 | .3504 | .3165 | .2855 | .2573 | .2316 | .2082 | .1869 | .1676 | .1501 | .1342 |
| 1 | .3897 | .3884 | .3840 | .3770 | .3679 | .3569 | .3446 | .3312 | .3169 | .3020 |
| 2 | .1927 | .2119 | .2295 | .2455 | .2597 | .2720 | .2823 | .2908 | .2973 | .3020 |
| 3 | .0556 | .0674 | .0800 | .0933 | .1069 | .1209 | .1349 | .1489 | .1627 | .1762 |
| 4 | .0103 | .0138 | .0179 | .0228 | .0283 | .0345 | .0415 | .0490 | .0573 | .0661 |
| 5 | .0013 | .0019 | .0027 | .0037 | .0050 | .0066 | .0085 | .0108 | .0134 | .0165 |
| 6 | .0001 | .0002 | .0003 | .0004 | .0006 | .0008 | .0012 | .0016 | .0021 | .0028 |
| 7 | .0000 | .0000 | .0000 | .0000 | .0000 | .0001 | .0001 | .0001 | .0002 | .0003 |

| p / r | .21 | .22 | .23 | .24 | .25 | .26 | .27 | .28 | .29 | .30 |
|---|---|---|---|---|---|---|---|---|---|---|
| 0 | .1199 | .1069 | .0952 | .0846 | .0751 | .0665 | .0589 | .0520 | .0458 | .0404 |
| 1 | .2867 | .2713 | .2558 | .2404 | .2253 | .2104 | .1960 | .1820 | .1685 | .1556 |
| 2 | .3049 | .3061 | .3056 | .3037 | .3003 | .2957 | .2899 | .2831 | .2754 | .2668 |
| 3 | .1891 | .2014 | .2130 | .2238 | .2336 | .2424 | .2502 | .2569 | .2624 | .2668 |
| 4 | .0754 | .0852 | .0954 | .1060 | .1168 | .1278 | .1388 | .1499 | .1608 | .1715 |
| 5 | .0200 | .0240 | .0285 | .0335 | .0389 | .0449 | .0513 | .0583 | .0657 | .0735 |
| 6 | .0036 | .0045 | .0057 | .0070 | .0087 | .0105 | .0127 | .0151 | .0179 | .0210 |
| 7 | .0004 | .0005 | .0007 | .0010 | .0012 | .0016 | .0020 | .0025 | .0031 | .0039 |
| 8 | .0000 | .0000 | .0001 | .0001 | .0001 | .0001 | .0002 | .0002 | .0003 | .0004 |

| p / r | .31 | .32 | .33 | .34 | .35 | .36 | .37 | .38 | .39 | .40 |
|---|---|---|---|---|---|---|---|---|---|---|
| 0 | .0355 | .0311 | .0272 | .0238 | .0207 | .0180 | .0156 | .0135 | .0117 | .0101 |
| 1 | .1433 | .1317 | .1206 | .1102 | .1004 | .0912 | .0826 | .0747 | .0673 | .0605 |
| 2 | .2576 | .2478 | .2376 | .2270 | .2162 | .2052 | .1941 | .1831 | .1721 | .1612 |
| 3 | .2701 | .2721 | .2731 | .2729 | .2716 | .2693 | .2660 | .2618 | .2567 | .2508 |
| 4 | .1820 | .1921 | .2017 | .2109 | .2194 | .2272 | .2344 | .2407 | .2462 | .2508 |
| 5 | .0818 | .0904 | .0994 | .1086 | .1181 | .1278 | .1376 | .1475 | .1574 | .1672 |
| 6 | .0245 | .0284 | .0326 | .0373 | .0424 | .0479 | .0539 | .0603 | .0671 | .0743 |
| 7 | .0047 | .0057 | .0069 | .0082 | .0098 | .0116 | .0136 | .0158 | .0184 | .0212 |
| 8 | .0005 | .0007 | .0008 | .0011 | .0013 | .0016 | .0020 | .0024 | .0029 | .0035 |
| 9 | .0000 | .0000 | .0000 | .0001 | .0001 | .0001 | .0001 | .0002 | .0002 | .0003 |

| p / r | .41 | .42 | .43 | .44 | .45 | .46 | .47 | .48 | .49 | .50 |
|---|---|---|---|---|---|---|---|---|---|---|
| 0 | .0087 | .0074 | .0064 | .0054 | .0046 | .0039 | .0033 | .0028 | .0023 | .0020 |
| 1 | .0542 | .0484 | .0431 | .0383 | .0339 | .0299 | .0263 | .0231 | .0202 | .0176 |
| 2 | .1506 | .1402 | .1301 | .1204 | .1110 | .1020 | .0934 | .0853 | .0776 | .0703 |
| 3 | .2442 | .2369 | .2291 | .2207 | .2119 | .2027 | .1933 | .1837 | .1739 | .1641 |
| 4 | .2545 | .2573 | .2592 | .2601 | .2600 | .2590 | .2571 | .2543 | .2506 | .2461 |
| 5 | .1769 | .1863 | .1955 | .2044 | .2128 | .2207 | .2280 | .2347 | .2408 | .2461 |
| 6 | .0819 | .0900 | .0983 | .1070 | .1160 | .1253 | .1348 | .1445 | .1542 | .1641 |
| 7 | .0244 | .0279 | .0318 | .0360 | .0407 | .0458 | .0512 | .0571 | .0635 | .0703 |
| 8 | .0042 | .0051 | .0060 | .0071 | .0083 | .0097 | .0114 | .0132 | .0153 | .0176 |
| 9 | .0003 | .0004 | .0005 | .0006 | .0008 | .0009 | .0011 | .0014 | .0016 | .0020 |

| n = 10 | | | | | | | | | |
|---|---|---|---|---|---|---|---|---|---|

| p / r | .01 | .02 | .03 | .04 | .05 | .06 | .07 | .08 | .09 | .10 |
|---|---|---|---|---|---|---|---|---|---|---|
| 0 | .9044 | .8171 | .7374 | .6648 | .5987 | .5386 | .4840 | .4344 | .3894 | .3487 |
| 1 | .0914 | .1667 | .2281 | .2770 | .3151 | .3438 | .3643 | .3777 | .3851 | .3874 |
| 2 | .0042 | .0153 | .0317 | .0519 | .0746 | .0988 | .1234 | .1478 | .1714 | .1937 |
| 3 | .0001 | .0008 | .0026 | .0058 | .0105 | .0168 | .0248 | .0343 | .0452 | .0574 |
| 4 | .0000 | .0000 | .0001 | .0004 | .0010 | .0019 | .0033 | .0052 | .0078 | .0112 |
| 5 | .0000 | .0000 | .0000 | .0000 | .0001 | .0001 | .0003 | .0005 | .0009 | .0015 |
| 6 | .0000 | .0000 | .0000 | .0000 | .0000 | .0000 | .0000 | .0000 | .0001 | .0001 |

| p / r | .11 | .12 | .13 | .14 | .15 | .16 | .17 | .18 | .19 | .20 |
|---|---|---|---|---|---|---|---|---|---|---|
| 0 | .3118 | .2785 | .2484 | .2213 | .1969 | .1749 | .1552 | .1374 | .1216 | .1074 |
| 1 | .3854 | .3798 | .3712 | .3603 | .3474 | .3331 | .3178 | .3017 | .2852 | .2684 |
| 2 | .2143 | .2330 | .2496 | .2639 | .2759 | .2856 | .2929 | .2980 | .3010 | .3020 |
| 3 | .0706 | .0847 | .0995 | .1146 | .1298 | .1450 | .1600 | .1745 | .1883 | .2013 |
| 4 | .0153 | .0202 | .0260 | .0326 | .0401 | .0483 | .0573 | .0670 | .0773 | .0881 |
| 5 | .0023 | .0033 | .0047 | .0064 | .0085 | .0111 | .0141 | .0177 | .0218 | .0264 |
| 6 | .0002 | .0004 | .0006 | .0009 | .0012 | .0018 | .0024 | .0032 | .0043 | .0055 |
| 7 | .0000 | .0000 | .0000 | .0001 | .0001 | .0002 | .0003 | .0004 | .0006 | .0008 |
| 8 | .0000 | .0000 | .0000 | .0000 | .0000 | .0000 | .0000 | .0000 | .0001 | .0001 |

# TABLE 1—Continued

## n = 10 (Continued)

| p<br>r | .21 | .22 | .23 | .24 | .25 | .26 | .27 | .28 | .29 | .30 |
|---|---|---|---|---|---|---|---|---|---|---|
| 0 | .0947 | .0834 | .0733 | .0643 | .0563 | .0492 | .0430 | .0374 | .0326 | .0282 |
| 1 | .2517 | .2351 | .2188 | .2030 | .1877 | .1730 | .1590 | .1456 | .1330 | .1211 |
| 2 | .3011 | .2984 | .2942 | .2885 | .2816 | .2735 | .2646 | .2548 | .2444 | .2335 |
| 3 | .2134 | .2244 | .2343 | .2429 | .2503 | .2563 | .2609 | .2642 | .2662 | .2668 |
| 4 | .0993 | .1108 | .1225 | .1343 | .1460 | .1576 | .1689 | .1798 | .1903 | .2001 |
| 5 | .0317 | .0375 | .0439 | .0509 | .0584 | .0664 | .0750 | .0839 | .0933 | .1029 |
| 6 | .0070 | .0088 | .0109 | .0134 | .0162 | .0195 | .0231 | .0272 | .0317 | .0368 |
| 7 | .0011 | .0014 | .0019 | .0024 | .0031 | .0039 | .0049 | .0060 | .0074 | .0090 |
| 8 | .0001 | .0002 | .0002 | .0003 | .0004 | .0005 | .0007 | .0009 | .0011 | .0014 |
| 9 | .0000 | .0000 | .0000 | .0000 | .0000 | .0000 | .0001 | .0001 | .0001 | .0001 |

| p<br>r | .31 | .32 | .33 | .34 | .35 | .36 | .37 | .38 | .39 | .40 |
|---|---|---|---|---|---|---|---|---|---|---|
| 0 | .0245 | .0211 | .0182 | .0157 | .0135 | .0115 | .0098 | .0084 | .0071 | .0060 |
| 1 | .1099 | .0995 | .0898 | .0808 | .0725 | .0649 | .0578 | .0514 | .0456 | .0403 |
| 2 | .2222 | .2107 | .1990 | .0873 | .1757 | .1642 | .1529 | .1419 | .1312 | .1209 |
| 3 | .2662 | .2644 | .2614 | .2573 | .2522 | .2462 | .2394 | .2319 | .2237 | .2150 |
| 4 | .2093 | .2177 | .2253 | .2320 | .2377 | .2424 | .2461 | .2487 | .2503 | .2508 |
| 5 | .1128 | .1229 | .1332 | .1434 | .1536 | .1636 | .1734 | .1829 | .1920 | .2007 |
| 6 | .0422 | .0482 | .0547 | .0616 | .0689 | .0767 | .0849 | .0934 | .1023 | .1115 |
| 7 | .0108 | .0130 | .0154 | .0181 | .0212 | .0247 | .0285 | .0327 | .0374 | .0425 |
| 8 | .0018 | .0023 | .0028 | .0035 | .0043 | .0052 | .0063 | .0075 | .0090 | .0106 |
| 9 | .0002 | .0002 | .0003 | .0004 | .0005 | .0006 | .0008 | .0010 | .0013 | .0016 |
| 10 | .0000 | .0000 | .0000 | .0000 | .0000 | .0000 | .0000 | .0001 | .0001 | .0001 |

| p<br>r | .41 | .42 | .43 | .44 | .45 | .46 | .47 | .48 | .49 | .50 |
|---|---|---|---|---|---|---|---|---|---|---|
| 0 | .0051 | .0043 | .0036 | .0030 | .0025 | .0021 | .0017 | .0014 | .0012 | .0010 |
| 1 | .0355 | .0312 | .0273 | .0238 | .0207 | .0180 | .0155 | .0133 | .0114 | .0098 |
| 2 | .1111 | .1017 | .0927 | .0843 | .0763 | .0688 | .0619 | .0554 | .0494 | .0439 |
| 3 | .2058 | .1963 | .1865 | .1765 | .1665 | .1564 | .1464 | .1364 | .1267 | .1172 |
| 4 | .2503 | .2488 | .2462 | .2427 | .2384 | .2331 | .2271 | .2204 | .2130 | .2051 |
| 5 | .2087 | .2162 | .2229 | .2289 | .2340 | .2383 | .2417 | .2441 | .2456 | .2461 |
| 6 | .1209 | .1304 | .1401 | .1499 | .1596 | .1692 | .1786 | .1878 | .1966 | .2051 |
| 7 | .0480 | .0540 | .0604 | .0673 | .0746 | .0824 | .0905 | .0991 | .1080 | .1172 |
| 8 | .0125 | .0147 | .0171 | .0198 | .0229 | .0263 | .0301 | .0343 | .0389 | .0439 |
| 9 | .0019 | .0024 | .0029 | .0035 | .0042 | .0050 | .0059 | .0070 | .0083 | .0098 |
| 10 | .0001 | .0002 | .0002 | .0003 | .0003 | .0004 | .0005 | .0006 | .0008 | .0010 |

## n = 11

| p<br>r | .01 | .02 | .03 | .04 | .05 | .06 | .07 | .08 | .09 | .10 |
|---|---|---|---|---|---|---|---|---|---|---|
| 0 | .8953 | .8007 | .7153 | .6382 | .5688 | .5063 | .4501 | .3996 | .3544 | .3138 |
| 1 | .0995 | .1798 | .2433 | .2925 | .3293 | .3555 | .3727 | .3823 | .3855 | .3835 |
| 2 | .0050 | .0183 | .0376 | .0609 | .0867 | .1135 | .1403 | .1662 | .1906 | .2131 |
| 3 | .0002 | .0011 | .0035 | .0076 | .0137 | .0217 | .0317 | .0434 | .0566 | .0710 |
| 4 | .0000 | .0000 | .0002 | .0006 | .0014 | .0028 | .0048 | .0075 | .0112 | .0158 |
| 5 | .0000 | .0000 | .0000 | .0000 | .0001 | .0002 | .0005 | .0009 | .0015 | .0025 |
| 6 | .0000 | .0000 | .0000 | .0000 | .0000 | .0000 | .0000 | .0001 | .0002 | .0003 |

| p<br>r | .11 | .12 | .13 | .14 | .15 | .16 | .17 | .18 | .19 | .20 |
|---|---|---|---|---|---|---|---|---|---|---|
| 0 | .2775 | .2451 | .2161 | .1903 | .1673 | .1469 | .1288 | .1127 | .0985 | .0859 |
| 1 | .3773 | .3676 | .3552 | .3408 | .3248 | .3078 | .2901 | .2721 | .2541 | .2362 |
| 2 | .2332 | .2507 | .2654 | .2774 | .2866 | .2932 | .2971 | .2987 | .2980 | .2953 |
| 3 | .0865 | .1025 | .1190 | .1355 | .1517 | .1675 | .1826 | .1967 | .2097 | .2215 |
| 4 | .0214 | .0280 | .0356 | .0441 | .0536 | .0638 | .0748 | .0864 | .0984 | .1107 |
| 5 | .0037 | .0053 | .0074 | .0101 | .0132 | .0170 | .0214 | .0265 | .0323 | .0388 |
| 6 | .0005 | .0007 | .0011 | .0016 | .0023 | .0032 | .0044 | .0058 | .0076 | .0097 |
| 7 | .0000 | .0001 | .0001 | .0002 | .0003 | .0004 | .0006 | .0009 | .0013 | .0017 |
| 8 | .0000 | .0000 | .0000 | .0000 | .0000 | .0000 | .0001 | .0001 | .0001 | .0002 |

# TABLE 1—Continued

## n = 11 (Continued)

| p \ r | .21 | .22 | .23 | .24 | .25 | .26 | .27 | .28 | .29 | .30 |
|---|---|---|---|---|---|---|---|---|---|---|
| 0 | .0748 | .0650 | .0564 | .0489 | .0422 | .0364 | .0314 | .0270 | .0231 | .0198 |
| 1 | .2187 | .2017 | .1854 | .1697 | .1549 | .1408 | .1276 | .1153 | .1038 | .0932 |
| 2 | .2907 | .2845 | .2768 | .2680 | .2581 | .2474 | .2360 | .2242 | .2121 | .1998 |
| 3 | .2318 | .2407 | .2481 | .2539 | .2581 | .2608 | .2619 | .2616 | .2599 | .2568 |
| 4 | .1232 | .1358 | .1482 | .1603 | .1721 | .1832 | .1937 | .2035 | .2123 | .2201 |
| 5 | .0459 | .0536 | .0620 | .0709 | .0803 | .0901 | .1003 | .1108 | .1214 | .1321 |
| 6 | .0122 | .0151 | .0185 | .0224 | .0268 | .0317 | .0371 | .0431 | .0496 | .0566 |
| 7 | .0023 | .0030 | .0039 | .0050 | .0064 | .0079 | .0098 | .0120 | .0145 | .0173 |
| 8 | .0003 | .0004 | .0006 | .0008 | .0011 | .0014 | .0018 | .0023 | .0030 | .0037 |
| 9 | .0000 | .0000 | .0001 | .0001 | .0001 | .0002 | .0002 | .0003 | .0004 | .0005 |

| p \ r | .31 | .32 | .33 | .34 | .35 | .36 | .37 | .38 | .39 | .40 |
|---|---|---|---|---|---|---|---|---|---|---|
| 0 | .0169 | .0144 | .0122 | .0104 | .0088 | .0074 | .0062 | .0052 | .0044 | .0036 |
| 1 | .0834 | .0744 | .0662 | .0587 | .0518 | .0457 | .0401 | .0351 | .0306 | .0266 |
| 2 | .1874 | .1751 | .1630 | .1511 | .1395 | .1284 | .1177 | .1075 | .0978 | .0887 |
| 3 | .2526 | .2472 | .2408 | .2335 | .2254 | .2167 | .2074 | .1977 | .1876 | .1774 |
| 4 | .2269 | .2326 | .2372 | .2406 | .2428 | .2438 | .2436 | .2423 | .2399 | .2365 |
| 5 | .1427 | .1533 | .1636 | .1735 | .1830 | .1920 | .2003 | .2079 | .2148 | .2207 |
| 6 | .0641 | .0721 | .0806 | .0894 | .0985 | .1080 | .1176 | .1274 | .1373 | .1471 |
| 7 | .0206 | .0242 | .0283 | .0329 | .0379 | .0434 | .0494 | .0558 | .0627 | .0701 |
| 8 | .0046 | .0057 | .0070 | .0085 | .0102 | .0122 | .0145 | .0171 | .0200 | .0234 |
| 9 | .0007 | .0009 | .0011 | .0015 | .0018 | .0023 | .0028 | .0035 | .0043 | .0052 |
| 10 | .0001 | .0001 | .0001 | .0001 | .0002 | .0003 | .0003 | .0004 | .0005 | .0007 |

| p \ r | .41 | .42 | .43 | .44 | .45 | .46 | .47 | .48 | .49 | .50 |
|---|---|---|---|---|---|---|---|---|---|---|
| 0 | .0030 | .0025 | .0021 | .0017 | .0014 | .0011 | .0009 | .0008 | .0006 | .0005 |
| 1 | .0231 | .0199 | .0171 | .0147 | .0125 | .0107 | .0090 | .0076 | .0064 | .0054 |
| 2 | .0801 | .0721 | .0646 | .0577 | .0513 | .0454 | .0401 | .0352 | .0308 | .0269 |
| 3 | .1670 | .1566 | .1462 | .1359 | .1259 | .1161 | .1067 | .0976 | .0888 | .0806 |
| 4 | .2321 | .2267 | .2206 | .2136 | .2060 | .1978 | .1892 | .1801 | .1707 | .1611 |
| 5 | .2258 | .2299 | .2329 | .2350 | .2360 | .2359 | .2348 | .2327 | .2296 | .2256 |
| 6 | .1569 | .1664 | .1757 | .1846 | .1931 | .2010 | .2083 | .2148 | .2206 | .2256 |
| 7 | .0779 | .0861 | .0947 | .1036 | .1128 | .1223 | .1319 | .1416 | .1514 | .1611 |
| 8 | .0271 | .0312 | .0357 | .0407 | .0462 | .0521 | .0585 | .0654 | .0727 | .0806 |
| 9 | .0063 | .0075 | .0090 | .0107 | .0126 | .0148 | .0173 | .0201 | .0233 | .0269 |
| 10 | .0009 | .0011 | .0014 | .0017 | .0021 | .0025 | .0031 | .0037 | .0045 | .0054 |
| 11 | .0001 | .0001 | .0001 | .0001 | .0002 | .0002 | .0002 | .0003 | .0004 | .0005 |

## n = 12

| p \ r | .01 | .02 | .03 | .04 | .05 | .06 | .07 | .08 | .09 | .10 |
|---|---|---|---|---|---|---|---|---|---|---|
| 0 | .8864 | .7847 | .6938 | .6127 | .5404 | .4759 | .4186 | .3677 | .3225 | .2824 |
| 1 | .1074 | .1922 | .2575 | .3064 | .3413 | .3645 | .3781 | .3837 | .3827 | .3766 |
| 2 | .0060 | .0216 | .0438 | .0702 | .0988 | .1280 | .1565 | .1835 | .2082 | .2301 |
| 3 | .0002 | .0015 | .0045 | .0098 | .0173 | .0272 | .0393 | .0532 | .0686 | .0852 |
| 4 | .0000 | .0001 | .0003 | .0009 | .0021 | .0039 | .0067 | .0104 | .0153 | .0213 |
| 5 | .0000 | .0000 | .0000 | .0001 | .0002 | .0004 | .0008 | .0014 | .0024 | .0038 |
| 6 | .0000 | .0000 | .0000 | .0000 | .0000 | .0000 | .0001 | .0001 | .0003 | .0005 |

| p \ r | .11 | .12 | .13 | .14 | .15 | .16 | .17 | .18 | .19 | .20 |
|---|---|---|---|---|---|---|---|---|---|---|
| 0 | .2470 | .2157 | .1880 | .1637 | .1422 | .1234 | .1069 | .0924 | .0798 | .0687 |
| 1 | .3663 | .3529 | .3372 | .3197 | .3012 | .2821 | .2627 | .2434 | .2245 | .2062 |
| 2 | .2490 | .2647 | .2771 | .2863 | .2924 | .2955 | .2960 | .2939 | .2897 | .2835 |
| 3 | .1026 | .1203 | .1380 | .1553 | .1720 | .1876 | .2021 | .2151 | .2265 | .2362 |
| 4 | .0285 | .0369 | .0464 | .0569 | .0683 | .0804 | .0931 | .1062 | .1195 | .1329 |
| 5 | .0056 | .0081 | .0111 | .0148 | .0193 | .0245 | .0305 | .0373 | .0449 | .0532 |
| 6 | .0008 | .0013 | .0019 | .0028 | .0040 | .0054 | .0073 | .0096 | .0123 | .0155 |
| 7 | .0001 | .0001 | .0002 | .0004 | .0006 | .0009 | .0013 | .0018 | .0025 | .0033 |
| 8 | .0000 | .0000 | .0000 | .0000 | .0001 | .0001 | .0002 | .0002 | .0004 | .0005 |
| 9 | .0000 | .0000 | .0000 | .0000 | .0000 | .00000 | .0000 | .0000 | .0000 | .0001 |

TABLE 1—*Continued*

## n = 12 (Continued)

| r \ p | .21 | .22 | .23 | .24 | .25 | .26 | .27 | .28 | .29 | .30 |
|---|---|---|---|---|---|---|---|---|---|---|
| 0 | .0591 | .0507 | .0434 | .0371 | .0317 | .0270 | .0229 | .0194 | .0164 | .0138 |
| 1 | .1885 | .1717 | .1557 | .1407 | .1267 | .1137 | .1016 | .0906 | .0804 | .0712 |
| 2 | .2756 | .2663 | .2558 | .2444 | .2323 | .2197 | .2068 | .1937 | .1807 | .1678 |
| 3 | .2442 | .2503 | .2547 | .2573 | .2581 | .2573 | .2549 | .2511 | .2460 | .2397 |
| 4 | .1460 | .1589 | .1712 | .1828 | .1936 | .2034 | .2122 | .2197 | .2261 | .2311 |
| 5 | .0621 | .0717 | .0818 | .0924 | .1032 | .1143 | .1255 | .1367 | .1477 | .1585 |
| 6 | .0193 | .0236 | .0285 | .0340 | .0401 | .0469 | .0542 | .0620 | .0704 | .0792 |
| 7 | .0044 | .0057 | .0073 | .0092 | .0115 | .0141 | .0172 | .0207 | .0246 | .0291 |
| 8 | .0007 | .0010 | .0014 | .0018 | .0024 | .0031 | .0040 | .0050 | .0063 | .0078 |
| 9 | .0001 | .0001 | .0002 | .0003 | .0004 | .0005 | .0007 | .0009 | .0011 | .0015 |
| 10 | .0000 | .0000 | .0000 | .0000 | .0000 | .0001 | .0001 | .0001 | .0001 | .0002 |

| r \ p | .31 | .32 | .33 | .34 | .35 | .36 | .37 | .38 | .39 | .40 |
|---|---|---|---|---|---|---|---|---|---|---|
| 0 | .0116 | .0098 | .0082 | .0068 | .0057 | .0047 | .0039 | .0032 | .0027 | .0022 |
| 1 | .0628 | .0552 | .0484 | .0422 | .0368 | .0319 | .0276 | .0237 | .0204 | .0174 |
| 2 | .1552 | .1429 | .1310 | .1197 | .1088 | .0986 | .0890 | .0800 | .0716 | .0639 |
| 3 | .2324 | .2241 | .2151 | .2055 | .1954 | .1849 | .1742 | .1634 | .1526 | .1419 |
| 4 | .2349 | .2373 | .2384 | .2382 | .2367 | .2340 | .2302 | .2254 | .2195 | .2128 |
| 5 | .1688 | .1787 | .1879 | .1963 | .2039 | .2106 | .2163 | .2210 | .2246 | .2270 |
| 6 | .0885 | .0981 | .1079 | .1180 | .1281 | .1382 | .1482 | .1580 | .1675 | .1766 |
| 7 | .0341 | .0396 | .0456 | .0521 | .0591 | .0666 | .0746 | .0830 | .0918 | .1009 |
| 8 | .0096 | .0116 | .0140 | .0168 | .0199 | .0234 | .0274 | .0318 | .0367 | .0420 |
| 9 | .0019 | .0024 | .0031 | .0038 | .0048 | .0059 | .0071 | .0087 | .0104 | .0125 |
| 10 | .0003 | .0003 | .0005 | .0006 | .0008 | .0010 | .0013 | .0016 | .0020 | .0025 |
| 11 | .0000 | .0000 | .0000 | .0001 | .0001 | .0001 | .0001 | .0002 | .0002 | .0003 |

| r \ p | .41 | .42 | .43 | .44 | .45 | .46 | .47 | .48 | .49 | .50 |
|---|---|---|---|---|---|---|---|---|---|---|
| 0 | .0018 | .0014 | .0012 | .0010 | .0008 | .0006 | .0005 | .0004 | .0003 | .0002 |
| 1 | .0148 | .0126 | .0106 | .0090 | .0075 | .0063 | .0052 | .0043 | .0036 | .0029 |
| 2 | .0567 | .0502 | .0442 | .0388 | .0339 | .0294 | .0255 | .0220 | .0189 | .0161 |
| 3 | .1314 | .1211 | .1111 | .1015 | .0923 | .0836 | .0754 | .0676 | .0604 | .0537 |
| 4 | .2054 | .1973 | .1886 | .1794 | .1700 | .1602 | .1504 | .1405 | .1306 | .1208 |
| 5 | .2284 | .2285 | .2276 | .2256 | .2225 | .2184 | .2134 | .2075 | .2008 | .1934 |
| 6 | .1851 | .1931 | .2003 | .2068 | .2124 | .2171 | .2208 | .2234 | .2250 | .2256 |
| 7 | .1103 | .1198 | .1295 | .1393 | .1489 | .1585 | .1678 | .1768 | .1853 | .1934 |
| 8 | .0479 | .0542 | .0611 | .0684 | .0762 | .0844 | .0930 | .1020 | .1113 | .1208 |
| 9 | .0148 | .0175 | .0205 | .0239 | .0277 | .0319 | .0367 | .0418 | .0475 | .0537 |
| 10 | .0031 | .0038 | .0046 | .0056 | .0068 | .0082 | .0098 | .0116 | .0137 | .0161 |
| 11 | .0004 | .0005 | .0006 | .0008 | .0010 | .0013 | .0016 | .0019 | .0024 | .0029 |
| 12 | .0000 | .0000 | .0000 | .0001 | .0001 | .0001 | .0001 | .0001 | .0002 | .0002 |

## n = 13

| r \ p | .01 | .02 | .03 | .04 | .05 | .06 | .07 | .08 | .09 | .10 |
|---|---|---|---|---|---|---|---|---|---|---|
| 0 | .8775 | .7690 | .6730 | .5882 | .5133 | .4474 | .3893 | .3383 | .2935 | .2542 |
| 1 | .1152 | .2040 | .2706 | .3186 | .3512 | .3712 | .3809 | .3824 | .3773 | .3672 |
| 2 | .0070 | .0250 | .0502 | .0797 | .1109 | .1422 | .1720 | .1995 | .2239 | .2448 |
| 3 | .0003 | .0019 | .0057 | .0122 | .0214 | .0333 | .0475 | .0636 | .0812 | .0997 |
| 4 | .0000 | .0001 | .0004 | .0013 | .0028 | .0053 | .0089 | .0138 | .0201 | .0277 |
| 5 | .0000 | .0000 | .0000 | .0001 | .0003 | .0006 | .0012 | .0022 | .0036 | .0055 |
| 6 | .0000 | .0000 | .0000 | .0000 | .0000 | .0001 | .0001 | .0003 | .0005 | .0008 |
| 7 | .0000 | .0000 | .0000 | .0000 | .0000 | .0000 | .0000 | .0000 | .0000 | .0001 |

| r \ p | .11 | .12 | .13 | .14 | .15 | .16 | .17 | .18 | .19 | .20 |
|---|---|---|---|---|---|---|---|---|---|---|
| 0 | .2198 | .1898 | .1636 | .1408 | .1209 | .1037 | .0887 | .0758 | .0646 | .0550 |
| 1 | .3532 | .3364 | .3178 | .2979 | .2774 | .2567 | .2362 | .2163 | .1970 | .1787 |
| 2 | .2619 | .2753 | .2849 | .2910 | .2937 | .2934 | .2903 | .2848 | .2773 | .2680 |
| 3 | .1187 | .1376 | .1561 | .1737 | .1900 | .2049 | .2180 | .2293 | .2385 | .2457 |
| 4 | .0367 | .0469 | .0583 | .0707 | .0838 | .0976 | .1116 | .1258 | .1399 | .1535 |
| 5 | .0082 | .0115 | .0157 | .0207 | .0266 | .0335 | .0412 | .0497 | .0591 | .0691 |
| 6 | .0013 | .0021 | .0031 | .0045 | .0063 | .0085 | .0112 | .0145 | .0185 | .0230 |
| 7 | .0002 | .0003 | .0005 | .0007 | .0011 | .0016 | .0023 | .0032 | .0043 | .0058 |
| 8 | .0000 | .0000 | .0001 | .0001 | .0001 | .0002 | .0004 | .0005 | .0008 | .0011 |
| 9 | .0000 | .0000 | .0000 | .0000 | .0000 | .0000 | .0000 | .0001 | .0001 | .0001 |

# TABLE 1—Continued

## n = 13 (Continued)

| r \ p | .21 | .22 | .23 | .24 | .25 | .26 | .27 | .28 | .29 | .30 |
|---|---|---|---|---|---|---|---|---|---|---|
| 0 | .0467 | .0396 | .0334 | .0282 | .0238 | .0200 | .0167 | .0140 | .0117 | .0097 |
| 1 | .1613 | .1450 | .1299 | .1159 | .1029 | .0911 | .0804 | .0706 | .0619 | .0540 |
| 2 | .2573 | .2455 | .2328 | .2195 | .2059 | .1921 | .1784 | .1648 | .1516 | .1388 |
| 3 | .2508 | .2539 | .2550 | .2542 | .2517 | .2475 | .2419 | .2351 | .2271 | .2181 |
| 4 | .1667 | .1790 | .1904 | .2007 | .2097 | .2174 | .2237 | .2285 | .2319 | .2337 |
| 5 | .0797 | .0909 | .1024 | .1141 | .1258 | .1375 | .1489 | .1600 | .1705 | .1803 |
| 6 | .0283 | .0342 | .0408 | .0480 | .0559 | .0644 | .0734 | .0829 | .0928 | .1030 |
| 7 | .0075 | .0096 | .0122 | .0152 | .0186 | .0226 | .0272 | .0323 | .0379 | .0442 |
| 8 | .0015 | .0020 | .0027 | .0036 | .0047 | .0060 | .0075 | .0094 | .0116 | .0142 |
| 9 | .0002 | .0003 | .0005 | .0006 | .0009 | .0012 | .0015 | .0020 | .0026 | .0034 |
| 10 | .0000 | .0000 | .0001 | .0001 | .0001 | .0002 | .0002 | .0003 | .0004 | .0006 |
| 11 | .0000 | .0000 | .0000 | .0000 | .0000 | .0000 | .0000 | .0000 | .0000 | .0001 |

| r \ p | .31 | .32 | .33 | .34 | .35 | .36 | .37 | .38 | .39 | .40 |
|---|---|---|---|---|---|---|---|---|---|---|
| 0 | .0080 | .0066 | .0055 | .0045 | .0037 | .0030 | .0025 | .0020 | .0016 | .0013 |
| 1 | .0469 | .0407 | .0351 | .0302 | .0259 | .0221 | .0188 | .0159 | .0135 | .0113 |
| 2 | .1265 | .1148 | .1037 | .0933 | .0836 | .0746 | .0663 | .0586 | .0516 | .0453 |
| 3 | .2084 | .1981 | .1874 | .1763 | .1651 | .1538 | .1427 | .1317 | .1210 | .1107 |
| 4 | .2341 | .2331 | .2307 | .2270 | .2222 | .2163 | .2095 | .2018 | .1934 | .1845 |
| 5 | .1893 | .1974 | .2045 | .2105 | .2154 | .2190 | .2215 | .2227 | .2226 | .2214 |
| 6 | .1134 | .1239 | .1343 | .1446 | .1546 | .1643 | .1734 | .1820 | .1898 | .1968 |
| 7 | .0509 | .0583 | .0662 | .0745 | .0833 | .0924 | .1019 | .1115 | .1213 | .1312 |
| 8 | .0172 | .0206 | .0244 | .0288 | .0336 | .0390 | .0449 | .0513 | .0582 | .0656 |
| 9 | .0043 | .0054 | .0067 | .0082 | .0101 | .0122 | .0146 | .0175 | .0207 | .0243 |
| 10 | .0008 | .0010 | .0013 | .0017 | .0022 | .0027 | .0034 | .0043 | .0053 | .0065 |
| 11 | .0001 | .0001 | .0002 | .0002 | .0003 | .0004 | .0006 | .0007 | .0009 | .0012 |
| 12 | .0000 | .0000 | .0000 | .0000 | .0000 | .0000 | .0001 | .0001 | .0001 | .0001 |

| r \ p | .41 | .42 | .43 | .44 | .45 | .46 | .47 | .48 | .49 | .50 |
|---|---|---|---|---|---|---|---|---|---|---|
| 0 | .0010 | .0008 | .0007 | .0005 | .0004 | .0003 | .0003 | .0002 | .0002 | .0001 |
| 1 | .0095 | .0079 | .0066 | .0054 | .0045 | .0037 | .0030 | .0024 | .0020 | .0016 |
| 2 | .0395 | .0344 | .0298 | .0256 | .0220 | .0188 | .0160 | .0135 | .0114 | .0095 |
| 3 | .1007 | .0913 | .0823 | .0739 | .0660 | .0587 | .0519 | .0457 | .0401 | .0349 |
| 4 | .1750 | .1653 | .1553 | .1451 | .1350 | .1250 | .1151 | .1055 | .0962 | .0873 |
| 5 | .2189 | .2154 | .2108 | .2053 | .1989 | .1917 | .1838 | .1753 | .1664 | .1571 |
| 6 | .2029 | .2080 | .2121 | .2151 | .2169 | .2177 | .2173 | .2158 | .2131 | .2095 |
| 7 | .1410 | .1506 | .1600 | .1690 | .1775 | .1854 | .1927 | .1992 | .2048 | .2095 |
| 8 | .0735 | .0818 | .0905 | .0996 | .1089 | .1185 | .1282 | .1379 | .1476 | .1571 |
| 9 | .0284 | .0329 | .0379 | .0435 | .0495 | .0561 | .0631 | .0707 | .0788 | .0873 |
| 10 | .0079 | .0095 | .0114 | .0137 | .0162 | .0191 | .0224 | .0261 | .0303 | .0349 |
| 11 | .0015 | .0019 | .0024 | .0029 | .0036 | .0044 | .0054 | .0066 | .0079 | .0095 |
| 12 | .0002 | .0002 | .0003 | .0004 | .0005 | .0006 | .0008 | .0010 | .0013 | .0016 |
| 13 | .0000 | .0000 | .0000 | .0000 | .0000 | .0000 | .0001 | .0001 | .0001 | .0001 |

## n = 14

| r \ p | .01 | .02 | .03 | .04 | .05 | .06 | .07 | .08 | .09 | .10 |
|---|---|---|---|---|---|---|---|---|---|---|
| 0 | .8687 | .7536 | .6528 | .5647 | .4877 | .4205 | .3620 | .3112 | .2670 | .2288 |
| 1 | .1229 | .2153 | .2827 | .3294 | .3593 | .3758 | .3815 | .3788 | .3698 | .3559 |
| 2 | .0081 | .0286 | .0568 | .0892 | .1229 | .1559 | .1867 | .2141 | .2377 | .2570 |
| 3 | .0003 | .0023 | .0070 | .0149 | .0259 | .0398 | .0562 | .0745 | .0940 | .1142 |
| 4 | .0000 | .0001 | .0006 | .0017 | .0037 | .0070 | .0116 | .0178 | .0256 | .0349 |
| 5 | .0000 | .0000 | .0000 | .0001 | .0004 | .0009 | .0018 | .0031 | .0051 | .0078 |
| 6 | .0000 | .0000 | .0000 | .0000 | .0000 | .0001 | .0002 | .0004 | .0008 | .0013 |
| 7 | .0000 | .0000 | .0000 | .0000 | .0000 | .0000 | .0000 | .0000 | .0001 | .0002 |

| r \ p | .11 | .12 | .13 | .14 | .15 | .16 | .17 | .18 | .19 | .20 |
|---|---|---|---|---|---|---|---|---|---|---|
| 0 | .1956 | .1670 | .1423 | .1211 | .1028 | .0871 | .0736 | .0621 | .0523 | .0440 |
| 1 | .3385 | .3188 | .2977 | .2759 | .2539 | .2322 | .2112 | .1910 | .1719 | .1539 |
| 2 | .2720 | .2826 | .2892 | .2919 | .2912 | .2875 | .2811 | .2725 | .2620 | .2501 |
| 3 | .1345 | .1542 | .1728 | .1901 | .2056 | .2190 | .2303 | .2393 | .2459 | .2501 |
| 4 | .0457 | .0578 | .0710 | .0851 | .0998 | .1147 | .1297 | .1444 | .1586 | .1720 |
| 5 | .0113 | .0158 | .0212 | .0277 | .0352 | .0437 | .0531 | .0634 | .0744 | .0860 |
| 6 | .0021 | .0032 | .0048 | .0068 | .0093 | .0125 | .0163 | .0209 | .0262 | .0322 |
| 7 | .0003 | .0005 | .0008 | .0013 | .0019 | .0027 | .0038 | .0052 | .0070 | .0092 |
| 8 | .0000 | .0001 | .0001 | .0002 | .0003 | .0005 | .0007 | .0010 | .0014 | .0020 |
| 9 | .0000 | .0000 | .0000 | .0000 | .0000 | .0001 | .0001 | .0001 | .0002 | .0003 |

# TABLE 1—*Continued*

## n = 14 (Continued)

| r \ p | .21 | .22 | .23 | .24 | .25 | .26 | .27 | .28 | .29 | .30 |
|---|---|---|---|---|---|---|---|---|---|---|
| 0 | .0369 | .0309 | .0258 | .0214 | .0178 | .0148 | .0122 | .0101 | .0083 | .0068 |
| 1 | .1372 | .1218 | .1077 | .0948 | .0832 | .0726 | .0632 | .0548 | .0473 | .0407 |
| 2 | .2371 | .2234 | .2091 | .1946 | .1802 | .1659 | .1519 | .1385 | .1256 | .1134 |
| 3 | .2521 | .2520 | .2499 | .2459 | .2402 | .2331 | .2248 | .2154 | .2052 | .1943 |
| 4 | .1843 | .1955 | .2052 | .2135 | .2202 | .2252 | .2286 | .2304 | .2305 | .2290 |
| 5 | .0980 | .1103 | .1226 | .1348 | .1468 | .1583 | .1691 | .1792 | .1883 | .1963 |
| 6 | .0391 | .0466 | .0549 | .0639 | .0734 | .0834 | .0938 | .1045 | .1153 | .1262 |
| 7 | .0119 | .0150 | .0188 | .0231 | .0280 | .0335 | .0397 | .0464 | .0538 | .0618 |
| 8 | .0028 | .0037 | .0049 | .0064 | .0082 | .0103 | .0128 | .0158 | .0192 | .0232 |
| 9 | .0005 | .0007 | .0010 | .0013 | .0018 | .0024 | .0032 | .0041 | .0052 | .0066 |
| 10 | .0001 | .0001 | .0001 | .0002 | .0003 | .0004 | .0006 | .0008 | .0011 | .0014 |
| 11 | .0000 | .0000 | .0000 | .0000 | .0000 | .0001 | .0001 | .0001 | .0002 | .0002 |

| r \ p | .31 | .32 | .33 | .34 | .35 | .36 | .37 | .38 | .39 | .40 |
|---|---|---|---|---|---|---|---|---|---|---|
| 0 | .0055 | .0045 | .0037 | .0030 | .0024 | .0019 | .0016 | .0012 | .0010 | .0008 |
| 1 | .0349 | .0298 | .0253 | .0215 | .0181 | .0152 | .0128 | .0106 | .0088 | .0073 |
| 2 | .1018 | .0911 | .0811 | .0719 | .0634 | .0557 | .0487 | .0424 | .0367 | .0317 |
| 3 | .1830 | .1715 | .1598 | .1481 | .1366 | .1253 | .1144 | .1039 | .0940 | .0845 |
| 4 | .2261 | .2219 | .2164 | .2098 | .2022 | .1938 | .1848 | .1752 | .1652 | .1549 |
| 5 | .2032 | .2088 | .2132 | .2161 | .2178 | .2181 | .2170 | .2147 | .2112 | .2066 |
| 6 | .1369 | .1474 | .1575 | .1670 | .1759 | .1840 | .1912 | .1974 | .2026 | .2066 |
| 7 | .0703 | .0793 | .0886 | .0983 | .1082 | .1183 | .1283 | .1383 | .1480 | .1574 |
| 8 | .0276 | .0326 | .0382 | .0443 | .0510 | .0582 | .0659 | .0742 | .0828 | .0918 |
| 9 | .0083 | .0102 | .0125 | .0152 | .0183 | .0218 | .0258 | .0303 | .0353 | .0408 |
| 10 | .0019 | .0024 | .0031 | .0039 | .0049 | .0061 | .0076 | .0093 | .0113 | .0136 |
| 11 | .0003 | .0004 | .0006 | .0007 | .0010 | .0013 | .0016 | .0021 | .0026 | .0033 |
| 12 | .0000 | .0000 | .0001 | .0001 | .0001 | .0002 | .0002 | .0003 | .0004 | .0005 |
| 13 | .0000 | .0000 | .0000 | .0000 | .0000 | .0000 | .0000 | .0000 | .0000 | .0001 |

| r \ p | .41 | .42 | .43 | .44 | .45 | .46 | .47 | .48 | .49 | .50 |
|---|---|---|---|---|---|---|---|---|---|---|
| 0 | .0006 | .0005 | .0004 | .0003 | .0002 | .0002 | .0001 | .0001 | .0001 | .0001 |
| 1 | .0060 | .0049 | .0040 | .0033 | .0027 | .0021 | .0017 | .0014 | .0011 | .0009 |
| 2 | .0272 | .0233 | .0198 | .0168 | .0141 | .0118 | .0099 | .0082 | .0068 | .0056 |
| 3 | .0757 | .0674 | .0597 | .0527 | .0462 | .0403 | .0350 | .0303 | .0260 | .0222 |
| 4 | .1446 | .1342 | .1239 | .1138 | .1040 | .0945 | .0854 | .0768 | .0687 | .0611 |
| 5 | .2009 | .1943 | .1869 | .1788 | .1701 | .1610 | .1515 | .1418 | .1320 | .1222 |
| 6 | .2094 | .2111 | .2115 | .2108 | .2088 | .2057 | .2015 | .1963 | .1902 | .1833 |
| 7 | .1663 | .1747 | .1824 | .1892 | .1952 | .2003 | .2043 | .2071 | .2089 | .2095 |
| 8 | .1011 | .1107 | .1204 | .1301 | .1398 | .1493 | .1585 | .1673 | .1756 | .1833 |
| 9 | .0469 | .0534 | .0605 | .0682 | .0762 | .0848 | .0937 | .1030 | .1125 | .1222 |
| 10 | .0163 | .0193 | .0228 | .0268 | .0312 | .0361 | .0415 | .0475 | .0540 | .0611 |
| 11 | .0041 | .0051 | .0063 | .0076 | .0093 | .0112 | .0134 | .0160 | .0189 | .0222 |
| 12 | .0007 | .0009 | .0012 | .0015 | .0019 | .0024 | .0030 | .0037 | .0045 | .0056 |
| 13 | .0001 | .0001 | .0001 | .0002 | .0002 | .0003 | .0004 | .0005 | .0007 | .0009 |
| 14 | .0000 | .0000 | .0000 | .0000 | .0000 | .0000 | .0000 | .0000 | .0000 | .0001 |

## n = 15

| r \ p | .01 | .02 | .03 | .04 | .05 | .06 | .07 | .08 | .09 | .10 |
|---|---|---|---|---|---|---|---|---|---|---|
| 0 | .8601 | .7386 | .6333 | .5421 | .4633 | .3953 | .3367 | .2863 | .2430 | .2059 |
| 1 | .1303 | .2261 | .2938 | .3388 | .3658 | .3785 | .3801 | .3734 | .3605 | .3432 |
| 2 | .0092 | .0323 | .0636 | .0988 | .1348 | .1691 | .2003 | .2273 | .2496 | .2669 |
| 3 | .0004 | .0029 | .0085 | .0178 | .0307 | .0468 | .0653 | .0857 | .1070 | .1285 |
| 4 | .0000 | .0002 | .0008 | .0022 | .0049 | .0090 | .0148 | .0223 | .0317 | .0428 |
| 5 | .0000 | .0000 | .0001 | .0002 | .0006 | .0013 | .0024 | .0043 | .0069 | .0105 |
| 6 | .0000 | .0000 | .0000 | .0000 | .0000 | .0001 | .0003 | .0006 | .0011 | .0019 |
| 7 | .0000 | .0000 | .0000 | .0000 | .0000 | .0000 | .0000 | .0001 | .0001 | .0003 |

TABLE 1—*Continued*

### n = 15 (Continued)

| r \ p | .11 | .12 | .13 | .14 | .15 | .16 | .17 | .18 | .19 | .20 |
|---|---|---|---|---|---|---|---|---|---|---|
| 0 | .1741 | .1470 | .1238 | .1041 | .0874 | .0731 | .0611 | .0510 | .0424 | .0352 |
| 1 | .3228 | .3006 | .2775 | .2542 | .2312 | .2090 | .1378 | .1678 | .1492 | .1319 |
| 2 | .2793 | .2870 | .2903 | .2897 | .2856 | .2787 | .2692 | .2578 | .2449 | .2309 |
| 3 | .1496 | .1696 | .1880 | .2044 | .2184 | .2300 | .2389 | .2452 | .2489 | .2501 |
| 4 | .0555 | .0694 | .0843 | .0998 | .1156 | .1314 | .1468 | .1615 | .1752 | .1876 |
| 5 | .0151 | .0208 | .0277 | .0357 | .0449 | .0551 | .0662 | .0780 | .0904 | .1032 |
| 6 | .0031 | .0047 | .0069 | .0097 | .0132 | .0175 | .0226 | .0285 | .0353 | .0430 |
| 7 | .0005 | .0008 | .0013 | .0020 | .0030 | .0043 | .0059 | .0081 | .0107 | .0138 |
| 8 | .0001 | .0001 | .0002 | .0003 | .0005 | .0008 | .0012 | .0018 | .0025 | .0035 |
| 9 | .0000 | .0000 | .0000 | .0000 | .0001 | .0001 | .0002 | .0003 | .0005 | .0007 |
| 10 | .0000 | .0000 | .0000 | .0000 | .0000 | .0000 | .0000 | .0000 | .0001 | .0001 |

| r \ p | .21 | .22 | .23 | .24 | .25 | .26 | .27 | .28 | .29 | .30 |
|---|---|---|---|---|---|---|---|---|---|---|
| 0 | .0291 | .0241 | .0198 | .0163 | .0134 | .0109 | .0089 | .0072 | .0059 | .0047 |
| 1 | .1162 | .1018 | .0889 | .0772 | .0668 | .0576 | .0494 | .0423 | .0360 | .0305 |
| 2 | .2162 | .2010 | .1858 | .1707 | .1559 | .1416 | .1280 | .1150 | .1029 | .0916 |
| 3 | .2490 | .2457 | .2405 | .2336 | .2252 | .2156 | .2051 | .1939 | .1821 | .1700 |
| 4 | .1986 | .2079 | .2155 | .2213 | .2252 | .2273 | .2276 | .2262 | .2231 | .2186 |
| 5 | .1161 | .1290 | .1416 | .1537 | .1651 | .1757 | .1852 | .1935 | .2005 | .2061 |
| 6 | .0514 | .0606 | .0705 | .0809 | .0917 | .1029 | .1142 | .1254 | .1365 | .1472 |
| 7 | .0176 | .0220 | .0271 | .0329 | .0393 | .0465 | .0543 | .0627 | .0717 | .0811 |
| 8 | .0047 | .0062 | .0081 | .0104 | .0131 | .0163 | .0201 | .0244 | .0293 | .0348 |
| 9 | .0010 | .0014 | .0019 | .0025 | .0034 | .0045 | .0058 | .0074 | .0093 | .0116 |
| 10 | .0002 | .0002 | .0003 | .0005 | .0007 | .0009 | .0013 | .0017 | .0023 | .0030 |
| 11 | .0000 | .0000 | .0000 | .0001 | .0001 | .0002 | .0002 | .0003 | .0004 | .0006 |
| 12 | .0000 | .0000 | .0000 | .0000 | .0000 | .0000 | .0000 | .0000 | .0001 | .0001 |

| r \ p | .31 | .32 | .33 | .34 | .35 | .36 | .37 | .38 | .39 | .40 |
|---|---|---|---|---|---|---|---|---|---|---|
| 0 | .0038 | .0031 | .0025 | .0020 | .0016 | .0012 | .0010 | .0008 | .0006 | .0005 |
| 1 | .0258 | .0217 | .0182 | .0152 | .0126 | .0104 | .0086 | .0071 | .0058 | .0047 |
| 2 | .0811 | .0715 | .0627 | .0547 | .0476 | .0411 | .0354 | .0303 | .0259 | .0219 |
| 3 | .1579 | .1457 | .1338 | .1222 | .1110 | .1002 | .0901 | .0805 | .0716 | .0634 |
| 4 | .2128 | .2057 | .1977 | .1888 | .1792 | .1692 | .1587 | .1481 | .1374 | .1268 |
| 5 | .210 | .2130 | .2142 | .2140 | .2123 | .2093 | .2051 | .1997 | .1933 | .1859 |
| 6 | .1575 | .1671 | .1759 | .1837 | .1906 | .1963 | .2008 | .2040 | .2059 | .2066 |
| 7 | .0910 | .1011 | .1114 | .1217 | .1319 | .1419 | .1516 | .1608 | .1693 | .1771 |
| 8 | .0409 | .0476 | .0549 | .0627 | .0710 | .0798 | .0890 | .0985 | .1082 | .1181 |
| 9 | .0143 | .0174 | .0210 | .0251 | .0298 | .0349 | .0407 | .0470 | .0538 | .0612 |
| 10 | .0038 | .0049 | .0062 | .0078 | .0096 | .0118 | .0143 | .0173 | .0206 | .0245 |
| 11 | .0008 | .0011 | .0014 | .0018 | .0024 | .0030 | .0038 | .0048 | .0060 | .0074 |
| 12 | .0001 | .0002 | .0002 | .0003 | .0004 | .0006 | .0007 | .0010 | .0013 | .0016 |
| 13 | .0000 | .0000 | .0000 | .0000 | .0001 | .0001 | .0001 | .0001 | .0002 | .0003 |

| r \ p | .41 | .42 | .43 | .44 | .45 | .46 | .47 | .48 | .49 | .50 |
|---|---|---|---|---|---|---|---|---|---|---|
| 0 | .0004 | .0003 | .0002 | .0002 | .0001 | .0001 | .0001 | .0001 | .0000 | .0000 |
| 1 | .0038 | .0031 | .0025 | .0020 | .0016 | .0012 | .0010 | .0008 | .0006 | .0005 |
| 2 | .0185 | .0156 | .0130 | .0108 | .0090 | .0074 | .0060 | .0049 | .0040 | .0032 |
| 3 | .0558 | .0489 | .0426 | .0369 | .0318 | .0272 | .0232 | .0197 | .0166 | .0139 |
| 4 | .1163 | .1061 | .0963 | .0869 | .0780 | .0696 | .0617 | .0545 | .0478 | .0417 |
| 5 | .1778 | .1691 | .1598 | .1502 | .1404 | .1304 | .1204 | .1106 | .1010 | .0916 |
| 6 | .2060 | .2041 | .2010 | .1967 | .1914 | .1851 | .1780 | .1702 | .1617 | .1527 |
| 7 | .1840 | .1900 | .1949 | .1987 | .2013 | .2028 | .2030 | .2020 | .1997 | .1964 |
| 8 | .1279 | .1376 | .1470 | .1561 | .1647 | .1727 | .1800 | .1864 | .1919 | .1964 |
| 9 | .0691 | .0775 | .0863 | .0954 | .1048 | .1144 | .1241 | .1338 | .1434 | .1527 |
| 10 | .0288 | .0337 | .0390 | .0450 | .0515 | .0585 | .0661 | .0741 | .0827 | .0916 |
| 11 | .0091 | .0111 | .0134 | .0161 | .0191 | .0226 | .0266 | .0311 | .0361 | .0417 |
| 12 | .0021 | .0027 | .0034 | .0042 | .0052 | .0064 | .0079 | .0096 | .0116 | .0139 |
| 13 | .0003 | .0004 | .0006 | .0008 | .0010 | .0013 | .0016 | .0020 | .0026 | .0032 |
| 14 | .0000 | .0000 | .0001 | .0001 | .0001 | .0002 | .0002 | .0003 | .0004 | .0005 |

| r \ p | .01 | .02 | .03 | .04 | .05 | .06 | .07 | .08 | .09 | .10 |
|---|---|---|---|---|---|---|---|---|---|---|
| 0 | .8515 | .7238 | .6143 | .5204 | .4401 | .3716 | .3131 | .2634 | .2211 | .1853 |
| 1 | .1376 | .2363 | .3040 | .3469 | .3706 | .3795 | .3771 | .3665 | .3499 | .3294 |
| 2 | .0104 | .0362 | .0705 | .1084 | .1463 | .1817 | .2129 | .2390 | .2596 | .2745 |
| 3 | .0005 | .0034 | .0102 | .0211 | .0359 | .0541 | .0748 | .0970 | .1198 | .1423 |
| 4 | .0000 | .0002 | .0010 | .0029 | .0061 | .0112 | .0183 | .0274 | .0385 | .0514 |
| 5 | .0000 | .0000 | .0001 | .0003 | .0008 | .0017 | .0033 | .0057 | .0091 | .0137 |
| 6 | .0000 | .0000 | .0000 | .0000 | .0001 | .0002 | .0005 | .0009 | .0017 | .0028 |
| 7 | .0000 | .0000 | .0000 | .0000 | .0000 | .0000 | .0000 | .0001 | .0002 | .0004 |
| 8 | .0000 | .0000 | .0000 | .0000 | .0000 | .0000 | .0000 | .0000 | .0000 | .0001 |

| r \ p | .11 | .12 | .13 | .14 | .15 | .16 | .17 | .18 | .19 | .20 |
|---|---|---|---|---|---|---|---|---|---|---|
| 0 | .1550 | .1293 | .1077 | .0895 | .0743 | .0614 | .0507 | .0418 | .0343 | .0281 |
| 1 | .3065 | .2822 | .2575 | .2332 | .2097 | .1873 | .1662 | .1468 | .1289 | .1126 |
| 2 | .2841 | .2886 | .2886 | .2847 | .2775 | .2675 | .2554 | .2416 | .2267 | .2111 |
| 3 | .1638 | .1837 | .2013 | .2163 | .2285 | .2378 | .2441 | .2475 | .2482 | .2463 |
| 4 | .0658 | .0814 | .0977 | .1144 | .1311 | .1472 | .1625 | .1766 | .1892 | .2001 |
| 5 | .0195 | .0266 | .0351 | .0447 | .0555 | .0673 | .0799 | .0930 | .1065 | .1201 |
| 6 | .0044 | .0067 | .0096 | .0133 | .0180 | .0235 | .0300 | .0374 | .0458 | .0550 |
| 7 | .0008 | .0013 | .0020 | .0031 | .0045 | .0064 | .0088 | .0117 | .0153 | .0197 |
| 8 | .0001 | .0002 | .0003 | .0006 | .0009 | .0014 | .0020 | .0029 | .0041 | .0055 |
| 9 | .0000 | .0000 | .0000 | .0001 | .0001 | .0002 | .0004 | .0006 | .0008 | .0012 |
| 10 | .0000 | .0000 | .0000 | .0000 | .0000 | .0000 | .0001 | .0001 | .0001 | .0002 |

| r \ p | .21 | .22 | .23 | .24 | .25 | .26 | .27 | .28 | .29 | .30 |
|---|---|---|---|---|---|---|---|---|---|---|
| 0 | .0230 | .0188 | .0153 | .0124 | .0100 | .0081 | .0065 | .0052 | .0042 | .0033 |
| 1 | .0979 | .0847 | .0730 | .0626 | .0535 | .0455 | .0385 | .0325 | .0273 | .0228 |
| 2 | .1952 | .1792 | .1635 | .1482 | .1336 | .1198 | .1068 | .0947 | .0835 | .0732 |
| 3 | .2421 | .2359 | .2279 | .2185 | .2079 | .1964 | .1843 | .1718 | .1591 | .1465 |
| 4 | .2092 | .2162 | .2212 | .2242 | .2252 | .2243 | .2215 | .2171 | .2112 | .2040 |
| 5 | .1334 | .1464 | .1586 | .1699 | .1802 | .1891 | .1966 | .2026 | .2071 | .2099 |
| 6 | .0650 | .0757 | .0869 | .0984 | .1101 | .1218 | .1333 | .1445 | .1551 | .1649 |
| 7 | .0247 | .0305 | .0371 | .0444 | .0524 | .0611 | .0704 | .0803 | .0905 | .1010 |
| 8 | .0074 | .0097 | .0125 | .0158 | .0197 | .0242 | .0293 | .0351 | .0416 | .0487 |
| 9 | .0017 | .0024 | .0033 | .0044 | .0058 | .0075 | .0096 | .0121 | .0151 | .0185 |
| 10 | .0003 | .0005 | .0007 | .0010 | .0014 | .0019 | .0025 | .0033 | .0043 | .0056 |
| 11 | .0000 | .0001 | .0001 | .0002 | .0002 | .0004 | .0005 | .0007 | .0010 | .0013 |
| 12 | .0000 | .0000 | .0000 | .0000 | .0000 | .0001 | .0001 | .0001 | .0002 | .0002 |

| r \ p | .31 | .32 | .33 | .34 | .35 | .36 | .37 | .38 | .39 | .40 |
|---|---|---|---|---|---|---|---|---|---|---|
| 0 | .0026 | .0021 | .0016 | .0013 | .0010 | .0008 | .0006 | .0005 | .0004 | .0003 |
| 1 | .0190 | .0157 | .0130 | .0107 | .0087 | .0071 | .0058 | .0047 | .0038 | .0030 |
| 2 | .0639 | .0555 | .0480 | .0413 | .0353 | .0301 | .0255 | .0215 | .0180 | .0150 |
| 3 | .1341 | .1220 | .1103 | .0992 | .0888 | .0790 | .0699 | .0615 | .0538 | .0468 |
| 4 | .1958 | .1865 | .1766 | .1662 | .1553 | .1444 | .1333 | .1224 | .1118 | .1014 |
| 5 | .2111 | .2107 | .2088 | .2054 | .2008 | .1949 | .1879 | .1801 | .1715 | .1623 |
| 6 | .1739 | .1818 | .1885 | .1940 | .1982 | .2010 | .2024 | .2024 | .2010 | .1983 |
| 7 | .1116 | .1222 | .1326 | .1428 | .1524 | .1615 | .1698 | .1772 | .1836 | .1889 |
| 8 | .0564 | .0647 | .0735 | .0827 | .0923 | .1022 | .1122 | .1222 | .1320 | .1417 |
| 9 | .0225 | .0271 | .0322 | .0379 | .0442 | .1511 | .0586 | .0666 | .0750 | .0840 |
| 10 | .0071 | .0089 | .0111 | .0137 | .0167 | .0201 | .0241 | .0286 | .0336 | .0392 |
| 11 | .0017 | .0023 | .0030 | .0038 | .0049 | .0062 | .0077 | .0095 | .0117 | .0142 |
| 12 | .0003 | .0004 | .0006 | .0008 | .0011 | .0014 | .0019 | .0024 | .0031 | .0040 |
| 13 | .0000 | .0001 | .0001 | .0001 | .0002 | .0003 | .0003 | .0005 | .0006 | .0008 |
| 14 | .0000 | .0000 | .0000 | .0000 | .0000 | .0000 | .0000 | .0001 | .0001 | .0001 |

| r \ p | .41 | .42 | .43 | .44 | .45 | .46 | .47 | .48 | .49 | .50 |
|---|---|---|---|---|---|---|---|---|---|---|
| 0 | .0002 | .0002 | .0001 | .0001 | .0001 | .0001 | .0000 | .0000 | .0000 | .0000 |
| 1 | .0024 | .0019 | .0015 | .0012 | .0009 | .0007 | .0005 | .0004 | .0003 | .0002 |
| 2 | .0125 | .0103 | .0085 | .0069 | .0056 | .0046 | .0037 | .0029 | .0023 | .0018 |
| 3 | .0405 | .0349 | .0299 | .0254 | .0215 | .0181 | .0151 | .0126 | .0104 | .0085 |
| 4 | .0915 | .0821 | .0732 | .0649 | .0572 | .0501 | .0436 | .0378 | .0325 | .0278 |
| 5 | .1526 | .1426 | .1325 | .1224 | .1123 | .1024 | .0929 | .0837 | .0749 | .0667 |
| 6 | .1944 | .1894 | .1833 | .1762 | .1684 | .1600 | .1510 | .1416 | .1319 | .1222 |
| 7 | .1930 | .1959 | .1975 | .1978 | .1969 | .1947 | .1912 | .1867 | .1811 | .1746 |
| 8 | .1509 | .1596 | .1676 | .1749 | .1812 | .1865 | .1908 | .1939 | .1958 | .1964 |
| 9 | .0932 | .1027 | .1124 | .1221 | .1318 | .1413 | .1504 | .1591 | .1672 | .1746 |
| 10 | .0453 | .0521 | .0594 | .0672 | .0755 | .0842 | .0934 | .1028 | .1124 | .1222 |
| 11 | .0172 | .0206 | .0244 | .0288 | .0337 | .0391 | .0452 | .0518 | .0589 | .0667 |
| 12 | .0050 | .0062 | .0077 | .0094 | .0115 | .0139 | .0167 | .0199 | .0236 | .0278 |
| 13 | .0011 | .0014 | .0018 | .0023 | .0029 | .0036 | .0046 | .0057 | .0070 | .0085 |
| 14 | .0002 | .0002 | .0003 | .0004 | .0005 | .0007 | .0009 | .0011 | .0014 | .0018 |
| 15 | .0000 | .0000 | .0000 | .0000 | .0001 | .0001 | .0001 | .0001 | .0002 | .0002 |

TABLE 1—*Continued*

n = 17

| r \ p | .01 | .02 | .03 | .04 | .05 | .06 | .07 | .08 | .09 | .10 |
|---|---|---|---|---|---|---|---|---|---|---|
| 0 | .8429 | .7093 | .5958 | .4996 | .4181 | .3493 | .2912 | .2423 | .2012 | .1668 |
| 1 | .1447 | .2461 | .3133 | .3539 | .3741 | .3790 | .3726 | .3582 | .3383 | .3150 |
| 2 | .0117 | .0402 | .0775 | .1180 | .1575 | .1935 | .2244 | .2492 | .2677 | .2800 |
| 3 | .0006 | .0041 | .0120 | .0246 | .0415 | .0618 | .0844 | .1083 | .1324 | .1556 |
| 4 | .0000 | .0003 | .0013 | .0036 | .0076 | .0138 | .0222 | .0330 | .0458 | .0605 |
| 5 | .0000 | .0000 | .0001 | .0004 | .0010 | .0023 | .0044 | .0075 | .0118 | .0175 |
| 6 | .0000 | .0000 | .0000 | .0000 | .0001 | .0003 | .0007 | .0013 | .0023 | .0039 |
| 7 | .0000 | .0000 | .0000 | .0000 | .0000 | .0000 | .0001 | .0002 | .0004 | .0007 |
| 8 | .0000 | .0000 | .0000 | .0000 | .0000 | .0000 | .0000 | .0000 | .0000 | .0001 |

| r \ p | .11 | .12 | .13 | .14 | .15 | .16 | .17 | .18 | .19 | .20 |
|---|---|---|---|---|---|---|---|---|---|---|
| 0 | .1379 | .1138 | .0937 | .0770 | .0631 | .0516 | .0421 | .0343 | .0278 | .0225 |
| 1 | .2898 | .2638 | .2381 | .2131 | .1893 | .1671 | .1466 | .1279 | .1109 | .0957 |
| 2 | .2865 | .2878 | .2846 | .2775 | .2673 | .2547 | .2402 | .2245 | .2081 | .1914 |
| 3 | .1771 | .1963 | .2126 | .2259 | .2359 | .2425 | .2460 | .2464 | .2441 | .2393 |
| 4 | .0766 | .0937 | .1112 | .1287 | .1457 | .1617 | .1764 | .1893 | .2004 | .2093 |
| 5 | .0246 | .0332 | .0432 | .0545 | .0668 | .0801 | .0939 | .1081 | .1222 | .1361 |
| 6 | .0061 | .0091 | .0129 | .0177 | .0236 | .0305 | .0385 | .0474 | .0573 | .0680 |
| 7 | .0012 | .0019 | .0030 | .0045 | .0065 | .0091 | .0124 | .0164 | .0211 | .0267 |
| 8 | .0002 | .0003 | .0006 | .0009 | .0014 | .0022 | .0032 | .0045 | .0062 | .0084 |
| 9 | .0000 | .0000 | .0001 | .0002 | .0003 | .0004 | .0006 | .0010 | .0015 | .0021 |
| 10 | .0000 | .0000 | .0000 | .0000 | .0000 | .0001 | .0001 | .0002 | .0003 | .0004 |
| 11 | .0000 | .0000 | .0000 | .0000 | .0000 | .0000 | .0000 | .0000 | .0000 | .0001 |

| r \ p | .21 | .22 | .23 | .24 | .25 | .26 | .27 | .28 | .29 | .30 |
|---|---|---|---|---|---|---|---|---|---|---|
| 0 | .0182 | .0146 | .0118 | .0094 | .0075 | .0060 | .0047 | .0038 | .0030 | .0023 |
| 1 | .0822 | .0702 | .0597 | .0505 | .0426 | .0357 | .0299 | .0248 | .0206 | .0169 |
| 2 | .1747 | .1584 | .1427 | .1277 | .1136 | .1005 | .0883 | .0772 | .0672 | .0581 |
| 3 | .2322 | .2234 | .2131 | .2016 | .1893 | .1765 | .1634 | .1502 | .1372 | .1245 |
| 4 | .2161 | .2205 | .2228 | .2228 | .2209 | .2170 | .2115 | .2044 | .1961 | .1868 |
| 5 | .1493 | .1617 | .1730 | .1830 | .1914 | .1982 | .2033 | .2067 | .2083 | .2081 |
| 6 | .0794 | .0912 | .1034 | .1156 | .1276 | .1393 | .1504 | .1608 | .1701 | .1784 |
| 7 | .0332 | .0404 | .0485 | .0573 | .0668 | .0769 | .0874 | .0982 | .1092 | .1201 |
| 8 | .0110 | .0143 | .0181 | .0226 | .0279 | .0338 | .0404 | .0478 | .0558 | .0644 |
| 9 | .0029 | .0040 | .0054 | .0071 | .0093 | .0119 | .0150 | .0186 | .0228 | .0276 |
| 10 | .0006 | .0009 | .0013 | .0018 | .0025 | .0033 | .0044 | .0058 | .0074 | .0095 |
| 11 | .0001 | .0002 | .0002 | .0004 | .0005 | .0007 | .0010 | .0014 | .0019 | .0026 |
| 12 | .0000 | .0000 | .0000 | .0001 | .0001 | .0001 | .0002 | .0003 | .0004 | .0006 |
| 13 | .0000 | .0000 | .0000 | .0000 | .0000 | .0000 | .0000 | .0000 | .0001 | .0001 |

| r \ p | .31 | .32 | .33 | .34 | .35 | .36 | .37 | .38 | .39 | .40 |
|---|---|---|---|---|---|---|---|---|---|---|
| 0 | .0018 | .0014 | .0011 | .0009 | .0007 | .0005 | .0004 | .0003 | .0002 | .0002 |
| 1 | .0139 | .0114 | .0093 | .0075 | .0060 | .0048 | .0039 | .0031 | .0024 | .0019 |
| 2 | .0500 | .0428 | .0364 | .0309 | .0260 | .0218 | .0182 | .0151 | .0125 | .0102 |
| 3 | .1123 | .1007 | .0898 | .0795 | .0701 | .0614 | .0534 | .0463 | .0398 | .0341 |
| 4 | .1766 | .1659 | .1547 | .1434 | .1320 | .1208 | .1099 | .0993 | .0892 | .0796 |
| 5 | .2063 | .2030 | .1982 | .1921 | .1849 | .1767 | .1677 | .1582 | .1482 | .1379 |
| 6 | .1854 | .1910 | .1952 | .1979 | .1991 | .1988 | .1970 | .1939 | .1895 | .1839 |
| 7 | .1309 | .1413 | .1511 | .1602 | .1685 | .1757 | .1818 | .1868 | .1904 | .1927 |
| 8 | .0735 | .0831 | .0930 | .1032 | .1134 | .1235 | .1335 | .1431 | .1521 | .1606 |
| 9 | .0330 | .0391 | .0458 | .0531 | .0611 | .0695 | .0784 | .0877 | .0973 | .1070 |
| 10 | .0119 | .0147 | .0181 | .0219 | .0263 | .0313 | .0368 | .0430 | .0498 | .0571 |
| 11 | .0034 | .0044 | .0057 | .0072 | .0090 | .0112 | .0138 | .0168 | .0202 | .0242 |
| 12 | .0008 | .0010 | .0014 | .0018 | .0024 | .0031 | .0040 | .0051 | .0065 | .0081 |
| 13 | .0001 | .0002 | .0003 | .0004 | .0005 | .0007 | .0009 | .0012 | .0016 | .0021 |
| 14 | .0000 | .0000 | .0000 | .0001 | .0001 | .0001 | .0002 | .0002 | .0003 | .0004 |
| 15 | .0000 | .0000 | .0000 | .0000 | .0000 | .0000 | .0000 | .0000 | .0000 | .0001 |

TABLE 1—*Continued*

| n = 17 (Continued) | | | | | | | | | |

| r \ p | .41 | .42 | .43 | .44 | .45 | .46 | .47 | .48 | .49 | .50 |
|---|---|---|---|---|---|---|---|---|---|---|
| 0 | .0001 | .0001 | .0001 | .0001 | .0000 | .0000 | .0000 | .0000 | .0000 | .0000 |
| 1 | .0015 | .0012 | .0009 | .0007 | .0005 | .0004 | .0003 | .0002 | .0002 | .0001 |
| 2 | .0084 | .0068 | .0055 | .0044 | .0035 | .0028 | .0022 | .0017 | .0013 | .0010 |
| 3 | .0290 | .0246 | .0207 | .0173 | .0144 | .0119 | .0097 | .0079 | .0064 | .0052 |
| 4 | .0706 | .0622 | .0546 | .0475 | .0411 | .0354 | .0302 | .0257 | .0217 | .0182 |
| 5 | .1276 | .1172 | .1070 | .0971 | .0875 | .0784 | .0697 | .0616 | .0541 | .0472 |
| 6 | .1773 | .1697 | .1614 | .1525 | .1432 | .1335 | .1237 | .1138 | .1040 | .0944 |
| 7 | .1936 | .1932 | .1914 | .1883 | .1841 | .1787 | .1723 | .1650 | .1570 | .1484 |
| 8 | .1682 | .1748 | .1805 | .1850 | .1883 | .1903 | .1910 | .1904 | .1886 | .1855 |
| 9 | .1169 | .1266 | .1361 | .1453 | .1540 | .1621 | .1694 | .1758 | .1812 | .1855 |
| 10 | .0650 | .0733 | .0822 | .0914 | .1008 | .1105 | .1202 | .1298 | .1393 | .1484 |
| 11 | .0287 | .0338 | .0394 | .0457 | .0525 | .0599 | .0678 | .0763 | .0851 | .0944 |
| 12 | .0100 | .0122 | .0149 | .0179 | .0215 | .0255 | .0301 | .0352 | .0409 | .0472 |
| 13 | .0027 | .0034 | .0043 | .0054 | .0068 | .0084 | .0103 | .0125 | .0151 | .0182 |
| 14 | .0005 | .0007 | .0009 | .0012 | .0016 | .0020 | .0026 | .0033 | .0041 | .0052 |
| 15 | .0001 | .0001 | .0001 | .0002 | .0003 | .0003 | .0005 | .0006 | .0008 | .0010 |
| 16 | .0000 | .0000 | .0000 | .0000 | .0000 | .0000 | .0001 | .0001 | .0001 | .0001 |

| n = 18 | | | | | | | | | |

| r \ p | .01 | .02 | .03 | .04 | .05 | .06 | .07 | .08 | .09 | .10 |
|---|---|---|---|---|---|---|---|---|---|---|
| 0 | .8345 | .6951 | .5780 | .4796 | .3972 | .3283 | .2708 | .2229 | .1831 | .1501 |
| 1 | .1517 | .2554 | .3217 | .3597 | .3763 | .3772 | .3669 | .3489 | .3260 | .3002 |
| 2 | .0130 | .0443 | .0846 | .1274 | .1683 | .2047 | .2348 | .2579 | .2741 | .2835 |
| 3 | .0007 | .0048 | .0140 | .0283 | .0473 | .0697 | .0942 | .1196 | .1446 | .1680 |
| 4 | .0000 | .0004 | .0016 | .0044 | .0093 | .0167 | .0266 | .0390 | .0536 | .0700 |
| 5 | .0000 | .0000 | .0001 | .0005 | .0014 | .0030 | .0056 | .0095 | .0148 | .0218 |
| 6 | .0000 | .0000 | .0000 | .0000 | .0002 | .0004 | .0009 | .0018 | .0032 | .0052 |
| 7 | .0000 | .0000 | .0000 | .0000 | .0000 | .0000 | .0001 | .0003 | .0005 | .0010 |
| 8 | .0000 | .0000 | .0000 | .0000 | .0000 | .0000 | .0000 | .0000 | .0001 | .0002 |

| r \ p | .11 | .12 | .13 | .14 | .15 | .16 | .17 | .18 | .19 | .20 |
|---|---|---|---|---|---|---|---|---|---|---|
| 0 | .1227 | .1002 | .0815 | .0662 | .0536 | .0434 | .0349 | .0281 | .0225 | .0180 |
| 1 | .2731 | .2458 | .2193 | .1940 | .1704 | .1486 | .1288 | .1110 | .0951 | .0811 |
| 2 | .2869 | .2850 | .2785 | .2685 | .2556 | .2407 | .2243 | .2071 | .1897 | .1723 |
| 3 | .1891 | .2072 | .2220 | .2331 | .2406 | .2445 | .2450 | .2425 | .2373 | .2297 |
| 4 | .0877 | .1060 | .1244 | .1423 | .1592 | .1746 | .1882 | .1996 | .2087 | .2153 |
| 5 | .0303 | .0405 | .0520 | .0649 | .0787 | .0931 | .1079 | .1227 | .1371 | .1507 |
| 6 | .0081 | .0120 | .0168 | .0229 | .0301 | .0384 | .0479 | .0584 | .0697 | .0816 |
| 7 | .0017 | .0028 | .0043 | .0064 | .0091 | .0126 | .0168 | .0220 | .0280 | .0350 |
| 8 | .0003 | .0005 | .0009 | .0014 | .0022 | .0033 | .0047 | .0066 | .0090 | .0120 |
| 9 | .0000 | .0001 | .0001 | .0003 | .0004 | .0007 | .0011 | .0016 | .0024 | .0033 |
| 10 | .0000 | .0000 | .0000 | .0000 | .0001 | .0001 | .0002 | .0003 | .0005 | .0008 |
| 11 | .0000 | .0000 | .0000 | .0000 | .0000 | .0000 | .0000 | .0001 | .0001 | .0001 |

| r \ p | .21 | .22 | .23 | .24 | .25 | .26 | .27 | .28 | .29 | .30 |
|---|---|---|---|---|---|---|---|---|---|---|
| 0 | .0144 | .0114 | .0091 | .0072 | .0056 | .0044 | .0035 | .0027 | .0021 | .0016 |
| 1 | .0687 | .0580 | .0487 | .0407 | .0338 | .0280 | .0231 | .0189 | .0155 | .0126 |
| 2 | .1553 | .1390 | .1236 | .1092 | .0958 | .0836 | .0725 | .0626 | .0537 | .0458 |
| 3 | .2202 | .2091 | .1969 | .1839 | .1704 | .1567 | .1431 | .1298 | .1169 | .1046 |
| 4 | .2195 | .2212 | .2205 | .2177 | .2130 | .2065 | .1985 | .1892 | .1790 | .1681 |
| 5 | .1634 | .1747 | .1845 | .1925 | .1988 | .2031 | .2055 | .2061 | .2048 | .2017 |
| 6 | .0941 | .1067 | .1194 | .1317 | .1436 | .1546 | .1647 | .1736 | .1812 | .1873 |
| 7 | .0429 | .0516 | .0611 | .0713 | .0820 | .0931 | .1044 | .1157 | .1269 | .1376 |
| 8 | .0157 | .0200 | .0251 | .0310 | .0376 | .0450 | .0531 | .0619 | .0713 | .0811 |
| 9 | .0046 | .0063 | .0083 | .0109 | .0139 | .0176 | .0218 | .0267 | .0323 | .0386 |
| 10 | .0011 | .0016 | .0022 | .0031 | .0042 | .0056 | .0073 | .0094 | .0119 | .0149 |
| 11 | .0002 | .0003 | .0005 | .0007 | .0010 | .0014 | .0020 | .0026 | .0035 | .0046 |
| 12 | .0000 | .0001 | .0001 | .0001 | .0002 | .0003 | .0004 | .0006 | .0008 | .0012 |
| 13 | .0000 | .0000 | .0000 | .0000 | .0000 | .0000 | .0001 | .0001 | .0002 | .0002 |

TABLE 1—*Continued*

## n = 18 (Continued)

| r \ p | .31 | .32 | .33 | .34 | .35 | .36 | .37 | .38 | .39 | .40 |
|---|---|---|---|---|---|---|---|---|---|---|
| 0 | .0013 | .0010 | .0007 | .0006 | .0004 | .0003 | .0002 | .0002 | .0001 | .0001 |
| 1 | .0102 | .0082 | .0066 | .0052 | .0042 | .0033 | .0026 | .0020 | .0016 | .0012 |
| 2 | .0388 | .0327 | .0275 | .0229 | .0190 | .0157 | .0129 | .0105 | .0086 | .0069 |
| 3 | .0930 | .0822 | .0722 | .0630 | .0547 | .0471 | .0404 | .0344 | .0292 | .0246 |
| 4 | .1567 | .1450 | .1333 | .1217 | .1104 | .0994 | .0890 | .0791 | .0699 | .0614 |
| 5 | .1971 | .1911 | .1838 | .1755 | .1664 | .1566 | .1463 | .1358 | .1252 | .1146 |
| 6 | .1919 | .1948 | .1962 | .1959 | .1941 | .1908 | .1862 | .1803 | .1734 | .1655 |
| 7 | .1478 | .1572 | .1656 | .1730 | .1792 | .1840 | .1875 | .1895 | .1900 | .1892 |
| 8 | .0913 | .1017 | .1122 | .1226 | .1327 | .1423 | .1514 | .1597 | .1671 | .1734 |
| 9 | .0456 | .0532 | .0614 | .0701 | .0794 | .0890 | .0988 | .1087 | .1187 | .1284 |
| 10 | .0184 | .0225 | .0272 | .0325 | .0385 | .0450 | .0522 | .0600 | .0683 | .0771 |
| 11 | .0060 | .0077 | .0097 | .0122 | .0151 | .0184 | .0223 | .0267 | .0318 | .0374 |
| 12 | .0016 | .0021 | .0028 | .0037 | .0047 | .0060 | .0076 | .0096 | .0118 | .0145 |
| 13 | .0003 | .0005 | .0006 | .0009 | .0012 | .0016 | .0021 | .0027 | .0035 | .0045 |
| 14 | .0001 | .0001 | .0001 | .0002 | .0002 | .0003 | .0004 | .0006 | .0008 | .0011 |
| 15 | .0000 | .0000 | .0000 | .0000 | .0000 | .0000 | .0001 | .0001 | .0001 | .0002 |

| r \ p | .41 | .42 | .43 | .44 | .45 | .46 | .47 | .48 | .49 | .50 |
|---|---|---|---|---|---|---|---|---|---|---|
| 0 | .0001 | .0001 | .0000 | .0000 | .0000 | .0000 | .0000 | .0000 | .0000 | .0000 |
| 1 | .0009 | .0007 | .0005 | .0004 | .0003 | .0002 | .0002 | .0001 | .0001 | .0001 |
| 2 | .0055 | .0044 | .0035 | .0028 | .0022 | .0017 | .0013 | .0010 | .0008 | .0006 |
| 3 | .0206 | .0171 | .0141 | .0116 | .0095 | .0077 | .0062 | .0050 | .0039 | .0031 |
| 4 | .0536 | .0464 | .0400 | .0342 | .0291 | .0246 | .0206 | .0172 | .0142 | .0117 |
| 5 | .1042 | .0941 | .0844 | .0753 | .0666 | .0586 | .0512 | .0444 | .0382 | .0327 |
| 6 | .1569 | .1477 | .1380 | .1281 | .1181 | .1081 | .0983 | .0887 | .0796 | .0708 |
| 7 | .1869 | .1833 | .1785 | .1726 | .1657 | .1579 | .1494 | .1404 | .1310 | .1214 |
| 8 | .1786 | .1825 | .1852 | .1864 | .1864 | .1850 | .1822 | .1782 | .1731 | .1669 |
| 9 | .1379 | .1469 | .1552 | .1628 | .1694 | .1751 | .1795 | .1828 | .1848 | .1855 |
| 10 | .0862 | .0957 | .1054 | .1151 | .1248 | .1342 | .1433 | .1519 | .1598 | .1669 |
| 11 | .0436 | .0504 | .0578 | .0658 | .0742 | .0831 | .0924 | .1020 | .1117 | .1214 |
| 12 | .0177 | .0213 | .0254 | .0301 | .0354 | .0413 | .0478 | .1549 | .0626 | .0708 |
| 13 | .0057 | .0071 | .0089 | .0109 | .0134 | .0162 | .0196 | .0234 | .0278 | .0327 |
| 14 | .0014 | .0018 | .0024 | .0031 | .0039 | .0049 | .0062 | .0077 | .0095 | .0117 |
| 15 | .0003 | .0004 | .0005 | .0006 | .0009 | .0011 | .0015 | .0019 | .0024 | .0031 |
| 16 | .0000 | .0000 | .0001 | .0001 | .0001 | .0002 | .0002 | .0003 | .0004 | .0006 |
| 17 | .0000 | .0000 | .0000 | .0000 | .0000 | .0000 | .0000 | .0000 | .0000 | .0001 |

## n = 19

| r \ p | .01 | .02 | .03 | .04 | .05 | .06 | .07 | .08 | .09 | .10 |
|---|---|---|---|---|---|---|---|---|---|---|
| 0 | .8262 | .6812 | .5606 | .4604 | .3774 | .3086 | .2519 | .2051 | .1666 | .1351 |
| 1 | .1586 | .2642 | .3294 | .3645 | .3774 | .3743 | .3602 | .3389 | .3131 | .2852 |
| 2 | .0144 | .0485 | .0917 | .1367 | .1787 | .2150 | .2440 | .2652 | .2787 | .2852 |
| 3 | .0008 | .0056 | .0161 | .0323 | .0533 | .0778 | .1041 | .1307 | .1562 | .1796 |
| 4 | .0000 | .0005 | .0020 | .0054 | .0112 | .0199 | .0313 | .0455 | .0618 | .0798 |
| 5 | .0000 | .0000 | .0002 | .0007 | .0018 | .0038 | .0071 | .0119 | .0183 | .0266 |
| 6 | .0000 | .0000 | .0000 | .0001 | .0002 | .0006 | .0012 | .0024 | .0042 | .0069 |
| 7 | .0000 | .0000 | .0000 | .0000 | .0000 | .0001 | .0002 | .0004 | .0008 | .0014 |
| 8 | .0000 | .0000 | .0000 | .0000 | .0000 | .0000 | .0000 | .0001 | .0001 | .0002 |

| r \ p | .11 | .12 | .13 | .14 | .15 | .16 | .17 | .18 | .19 | .20 |
|---|---|---|---|---|---|---|---|---|---|---|
| 0 | .1092 | .0881 | .0709 | .0569 | .0456 | .0364 | .0290 | .0230 | .0182 | .0144 |
| 1 | .2565 | .2284 | .2014 | .1761 | .1529 | .1318 | .1129 | .0961 | .0813 | .0685 |
| 2 | .2854 | .2803 | .2708 | .2581 | .2428 | .2259 | .2081 | .1898 | .1717 | .1540 |
| 3 | .1999 | .2166 | .2293 | .2381 | .2428 | .2439 | .2415 | .2361 | .2282 | .2182 |
| 4 | .0988 | .1181 | .1371 | .1550 | .1714 | .1858 | .1979 | .2073 | .2141 | .2182 |
| 5 | .0366 | .0483 | .0614 | .0757 | .0907 | .1062 | .1216 | .1365 | .1507 | .1636 |
| 6 | .0106 | .0154 | .0214 | .0288 | .0374 | .0472 | .0581 | .0699 | .0825 | .0955 |
| 7 | .0024 | .0039 | .0059 | .0087 | .0122 | .0167 | .0221 | .0285 | .0359 | .0443 |
| 8 | .0004 | .0008 | .0013 | .0021 | .0032 | .0048 | .0068 | .0094 | .0126 | .0166 |
| 9 | .0001 | .0001 | .0002 | .0004 | .0007 | .0011 | .0017 | .0025 | .0036 | .0051 |
| 10 | .0000 | .0000 | .0000 | .0001 | .0001 | .0002 | .0003 | .0006 | .0009 | .0013 |
| 11 | .0000 | .0000 | .0000 | .0000 | .0000 | .0000 | .0001 | .0001 | .0002 | .0003 |

**227**

TABLE 1—Continued

### n = 19 (Continued)

| r \ p | .21 | .22 | .23 | .24 | .25 | .26 | .27 | .28 | .29 | .30 |
|---|---|---|---|---|---|---|---|---|---|---|
| 0 | .0113 | .0089 | .0070 | .0054 | .0042 | .0033 | .0025 | .0019 | .0015 | .0011 |
| 1 | .0573 | .0477 | .0396 | .0326 | .0268 | .0219 | .0178 | .0144 | .0116 | .0093 |
| 2 | .1371 | .1212 | .1064 | .0927 | .0803 | .0692 | .0592 | .0503 | .0426 | .0358 |
| 3 | .2065 | .1937 | .1800 | .1659 | .1517 | .1377 | .1240 | .1109 | .0985 | .0869 |
| 4 | .2196 | .2185 | .2151 | .2096 | .2023 | .1935 | .1835 | .1726 | .1610 | .1491 |
| 5 | .1751 | .1849 | .1928 | .1986 | .2023 | .2040 | .2036 | .2013 | .1973 | .1916 |
| 6 | .1086 | .1217 | .1343 | .1463 | .1574 | .1672 | .1757 | .1827 | .1880 | .1916 |
| 7 | .0536 | .0637 | .0745 | .0858 | .0974 | .1091 | .1207 | .1320 | .1426 | .1525 |
| 8 | .0214 | .0270 | .0334 | .0406 | .0487 | .0575 | .0670 | .0770 | .0874 | .0981 |
| 9 | .0069 | .0093 | .0122 | .0157 | .0198 | .0247 | .0303 | .0366 | .0436 | .0514 |
| 10 | .0018 | .0026 | .0036 | .0050 | .0066 | .0087 | .0112 | .0142 | .0178 | .0220 |
| 11 | .0004 | .0006 | .0009 | .0013 | .0018 | .0025 | .0034 | .0045 | .0060 | .0077 |
| 12 | .0001 | .0001 | .0002 | .0003 | .0004 | .0006 | .0008 | .0012 | .0016 | .0022 |
| 13 | .0000 | .0000 | .0000 | .0000 | .0001 | .0001 | .0002 | .0002 | .0004 | .0005 |
| 14 | .0000 | .0000 | .0000 | .0000 | .0000 | .0000 | .0000 | .0000 | .0001 | .0001 |

| r \ p | .31 | .32 | .33 | .34 | .35 | .36 | .37 | .38 | .39 | .40 |
|---|---|---|---|---|---|---|---|---|---|---|
| 0 | .0009 | .0007 | .0005 | .0004 | .0003 | .0002 | .0002 | .0001 | .0001 | .0001 |
| 1 | .0074 | .0059 | .0046 | .0036 | .0029 | .0022 | .0017 | .0013 | .0010 | .0008 |
| 2 | .0299 | .0249 | .0206 | .0169 | .0138 | .0112 | .0091 | .0073 | .0058 | .0046 |
| 3 | .0762 | .0664 | .0574 | .0494 | .0422 | .0358 | .0302 | .0253 | .0211 | .0175 |
| 4 | .1370 | .1249 | .1131 | .1017 | .0909 | .0806 | .0710 | .0621 | .0540 | .0467 |
| 5 | .1846 | .1764 | .1672 | .1572 | .1468 | .1360 | .1251 | .1143 | .1036 | .0933 |
| 6 | .1935 | .1936 | .1921 | .1890 | .1844 | .1785 | .1714 | .1634 | .1546 | .1451 |
| 7 | .1615 | .1692 | .1757 | .1808 | .1844 | .1865 | .1870 | .1860 | .1835 | .1797 |
| 8 | .1088 | .1195 | .1298 | .1397 | .1489 | .1573 | .1647 | .1710 | .1760 | .1797 |
| 9 | .0597 | .0687 | .0782 | .0880 | .0980 | .1082 | .1182 | .1281 | .1375 | .1464 |
| 10 | .0268 | .0323 | .0385 | .0453 | .0528 | .0608 | .0694 | .0785 | .0879 | .0976 |
| 11 | .0099 | .0124 | .0155 | .0191 | .0233 | .0280 | .0334 | .0394 | .0460 | .0532 |
| 12 | .0030 | .0039 | .0051 | .0066 | .0083 | .0105 | .0131 | .0161 | .0196 | .0237 |
| 13 | .0007 | .0010 | .0014 | .0018 | .0024 | .0032 | .0041 | .0053 | .0067 | .0085 |
| 14 | .0001 | .0002 | .0003 | .0004 | .0006 | .0008 | .0010 | .0014 | .0018 | .0024 |
| 15 | .0000 | .0000 | .0000 | .0001 | .0001 | .0001 | .0002 | .0003 | .0004 | .0005 |
| 16 | .0000 | .0000 | .0000 | .0000 | .0000 | .0000 | .0000 | .0000 | .0001 | .0001 |

| r \ p | .41 | .42 | .43 | .44 | .45 | .46 | .47 | .48 | .49 | .50 |
|---|---|---|---|---|---|---|---|---|---|---|
| 0 | .0000 | .0000 | .0000 | .0000 | .0000 | .0000 | .0000 | .0000 | .0000 | .0000 |
| 1 | .0006 | .0004 | .0003 | .0002 | .0002 | .0001 | .0001 | .0001 | .0001 | .0000 |
| 2 | .0037 | .0029 | .0022 | .0017 | .0013 | .0010 | .0008 | .0006 | .0004 | .0003 |
| 3 | .0144 | .0118 | .0096 | .0077 | .0062 | .0049 | .0039 | .0031 | .0024 | .0018 |
| 4 | .0400 | .0341 | .0289 | .0243 | .0203 | .0168 | .0138 | .0113 | .0092 | .0074 |
| 5 | .0834 | .0741 | .0653 | .0572 | .0497 | .0429 | .0368 | .0313 | .0265 | .0222 |
| 6 | .1353 | .1252 | .1150 | .1049 | .0949 | .0853 | .0751 | .0674 | .0593 | .0518 |
| 7 | .1746 | .1683 | .1611 | .1530 | .1443 | .1350 | .1254 | .1156 | .1058 | .0961 |
| 8 | .1820 | .1829 | .1823 | .1803 | .1771 | .1725 | .1668 | .1601 | .1525 | .1442 |
| 9 | .1546 | .1618 | .1681 | .1732 | .1771 | .1796 | .1808 | .1806 | .1791 | .1762 |
| 10 | .1074 | .1172 | .1268 | .1361 | .1449 | .1530 | .1603 | .1667 | .1721 | .1762 |
| 11 | .0611 | .0694 | .0783 | .0875 | .0970 | .1066 | .1163 | .1259 | .1352 | .1442 |
| 12 | .0283 | .0335 | .0394 | .0458 | .0529 | .0606 | .0688 | .0775 | .0866 | .0961 |
| 13 | .0106 | .0131 | .0160 | .0194 | .0233 | .0278 | .0328 | .0385 | .0448 | .0518 |
| 14 | .0032 | .0041 | .0052 | .0065 | .0082 | .0101 | .0125 | .0152 | .0185 | .0222 |
| 15 | .0007 | .0010 | .0013 | .0017 | .0022 | .0029 | .0037 | .0047 | .0059 | .0074 |
| 16 | .0001 | .0002 | .0002 | .0003 | .0005 | .0006 | .0008 | .0011 | .0014 | .0018 |
| 17 | .0000 | .0000 | .0000 | .0000 | .0001 | .0001 | .0001 | .0002 | .0002 | .0003 |

### n = 20

| r \ p | .01 | .02 | .03 | .04 | .05 | .06 | .07 | .08 | .09 | .10 |
|---|---|---|---|---|---|---|---|---|---|---|
| 0 | .8179 | .6676 | .5438 | .4420 | .3585 | .2901 | .2342 | .1887 | .1516 | .1216 |
| 1 | .1652 | .2725 | .3364 | .3683 | .3774 | .3703 | .3526 | .3282 | .3000 | .2702 |
| 2 | .0159 | .0528 | .0988 | .1458 | .1887 | .2246 | .2521 | .2711 | .2818 | .2852 |
| 3 | .0010 | .0065 | .0183 | .0364 | .0596 | .0860 | .1139 | .1414 | .1672 | .1901 |
| 4 | .0000 | .0006 | .0024 | .0065 | .0133 | .0233 | .0364 | .0523 | .0703 | .0898 |
| 5 | .0000 | .0000 | .0002 | .0009 | .0022 | .0048 | .0088 | .0145 | .0222 | .0319 |
| 6 | .0000 | .0000 | .0000 | .0001 | .0003 | .0008 | .0017 | .0032 | .0055 | .0089 |
| 7 | .0000 | .0000 | .0000 | .0000 | .0000 | .0001 | .0002 | .0005 | .0011 | .0020 |
| 8 | .0000 | .0000 | .0000 | .0000 | .0000 | .0000 | .0000 | .0001 | .0002 | .0004 |
| 9 | .0000 | .0000 | .0000 | .0000 | .0000 | .0000 | .0000 | .0000 | .0000 | .0001 |

# TABLE 1—*Continued*

## n = 20 (Continued)

| r \ p | .11 | .12 | .13 | .14 | .15 | .16 | .17 | .18 | .19 | .20 |
|---|---|---|---|---|---|---|---|---|---|---|
| 0 | .0972 | .0776 | .0617 | .0490 | .0388 | .0306 | .0241 | .0189 | .0148 | .0115 |
| 1 | .2403 | .2115 | .1844 | .1595 | .1368 | .1165 | .0986 | .0829 | .0693 | .0576 |
| 2 | .2822 | .2740 | .2618 | .2466 | .2293 | .2109 | .1919 | .1730 | .1545 | .1369 |
| 3 | .2093 | .2242 | .2347 | .2409 | .2428 | .2410 | .2358 | .2278 | .2175 | .2054 |
| 4 | .1099 | .1299 | .1491 | .1666 | .1821 | .1951 | .2053 | .2125 | .2168 | .2182 |
| 5 | .0435 | .0567 | .0713 | .0868 | .1028 | .1189 | .1345 | .1493 | .1627 | .1746 |
| 6 | .0134 | .0193 | .0266 | .0353 | .0454 | .0566 | .0689 | .0819 | .0954 | .1091 |
| 7 | .0033 | .0053 | .0080 | .0115 | .0160 | .0216 | .0282 | .0360 | .0448 | .0545 |
| 8 | .0007 | .0012 | .0019 | .0030 | .0046 | .0067 | .0094 | .0128 | .0171 | .0222 |
| 9 | .0001 | .0002 | .0004 | .0007 | .0011 | .0017 | .0026 | .0038 | .0053 | .0074 |
| 10 | .0000 | .0000 | .0001 | .0001 | .0002 | .0004 | .0006 | .0009 | .0014 | .0020 |
| 11 | .0000 | .0000 | .0000 | .0000 | .0000 | .0001 | .0001 | .0002 | .0003 | .0005 |
| 12 | .0000 | .0000 | .0000 | .0000 | .0000 | .0000 | .0000 | .0000 | .0001 | .0001 |

| r \ p | .21 | .22 | .23 | .24 | .25 | .26 | .27 | .28 | .29 | .30 |
|---|---|---|---|---|---|---|---|---|---|---|
| 0 | .0090 | .0069 | .0054 | .0041 | .0032 | .0024 | .0018 | .0014 | .0011 | .0008 |
| 1 | .0477 | .0392 | .0321 | .0261 | .0211 | .0170 | .0137 | .0109 | .0087 | .0068 |
| 2 | .1204 | .1050 | .0910 | .0783 | .0669 | .0569 | .0480 | .0403 | .0336 | .0278 |
| 3 | .1920 | .1777 | .1631 | .1484 | .1339 | .1199 | .1065 | .0940 | .0823 | .0716 |
| 4 | .2169 | .2131 | .2070 | .1991 | .1897 | .1790 | .1675 | .1553 | .1429 | .1304 |
| 5 | .1845 | .1923 | .1979 | .2012 | .2023 | .2013 | .1982 | .1933 | .1868 | .1789 |
| 6 | .1226 | .1356 | .1478 | .1589 | .1686 | .1768 | .1833 | .1879 | .1907 | .1916 |
| 7 | .0652 | .0765 | .0883 | .1003 | .1124 | .1242 | .1356 | .1462 | .1558 | .1643 |
| 8 | .0282 | .0351 | .0429 | .0515 | .0609 | .0709 | .0815 | .0924 | .1034 | .1144 |
| 9 | .0100 | .0132 | .0171 | .0217 | .0271 | .0332 | .0402 | .0479 | .0563 | .0654 |
| 10 | .0029 | .0041 | .0056 | .0075 | .0099 | .0128 | .0163 | .0205 | .0253 | .0308 |
| 11 | .0007 | .0010 | .0015 | .0022 | .0030 | .0041 | .0055 | .0072 | .0094 | .0120 |
| 12 | .0001 | .0002 | .0003 | .0005 | .0008 | .0011 | .0015 | .0021 | .0029 | .0039 |
| 13 | .0000 | .0000 | .0001 | .0001 | .0002 | .0002 | .0003 | .0005 | .0007 | .0010 |
| 14 | .0000 | .0000 | .0000 | .0000 | .0000 | .0000 | .0001 | .0001 | .0001 | .0002 |

| r \ p | .31 | .32 | .33 | .34 | .35 | .36 | .37 | .38 | .39 | .40 |
|---|---|---|---|---|---|---|---|---|---|---|
| 0 | .0006 | .0004 | .0003 | .0002 | .0002 | .0001 | .0001 | .0001 | .0001 | .0000 |
| 1 | .0054 | .0042 | .0033 | .0025 | .0020 | .0015 | .0011 | .0009 | .0007 | .0005 |
| 2 | .0229 | .0188 | .0153 | .0124 | .0100 | .0080 | .0064 | .0050 | .0040 | .0031 |
| 3 | .0619 | .0531 | .0453 | .0383 | .0323 | .0270 | .0224 | .0185 | .0152 | .0123 |
| 4 | .1181 | .1062 | .0947 | .0839 | .0738 | .0645 | .0559 | .0482 | .0412 | .0350 |
| 5 | .1698 | .1599 | .1493 | .1384 | .1272 | .1161 | .1051 | .0945 | .0843 | .0746 |
| 6 | .1907 | .1881 | .1839 | .1782 | .1712 | .1632 | .1543 | .1447 | .1347 | .1244 |
| 7 | .1714 | .1770 | .1811 | .1836 | .1844 | .1836 | .1812 | .1774 | .1722 | .1659 |
| 8 | .1251 | .1354 | .1450 | .1537 | .1614 | .1678 | .1730 | .1767 | .1790 | .1797 |
| 9 | .0750 | .0849 | .0952 | .1056 | .1158 | .1259 | .1354 | .1444 | .1526 | .1597 |
| 10 | .0370 | .0440 | .0516 | .0598 | .0686 | .0779 | .0875 | .0974 | .1073 | .1171 |
| 11 | .0151 | .0188 | .0231 | .0280 | .0336 | .0398 | .0467 | .0542 | .0624 | .0710 |
| 12 | .0051 | .0066 | .0085 | .0108 | .0136 | .0168 | .0206 | .0249 | .0299 | .0355 |
| 13 | .0014 | .0019 | .0026 | .0034 | .0045 | .0058 | .0074 | .0094 | .0118 | .0146 |
| 14 | .0003 | .0005 | .0006 | .0009 | .0012 | .0016 | .0022 | .0029 | .0038 | .0049 |
| 15 | .0001 | .0001 | .0001 | .0002 | .0003 | .0004 | .0005 | .0007 | .0010 | .0013 |
| 16 | .0000 | .0000 | .0000 | .0000 | .0000 | .0001 | .0001 | .0001 | .0002 | .0003 |

| r \ p | .41 | .42 | .43 | .44 | .45 | .46 | .47 | .48 | .49 | .50 |
|---|---|---|---|---|---|---|---|---|---|---|
| 0 | .0000 | .0000 | .0000 | .0000 | .0000 | .0000 | .0000 | .0000 | .0000 | .0000 |
| 1 | .0004 | .0003 | .0002 | .0001 | .0001 | .0001 | .0001 | .0000 | .0000 | .0000 |
| 2 | .0024 | .0018 | .0014 | .0011 | .0008 | .0006 | .0005 | .0003 | .0002 | .0002 |
| 3 | .0100 | .0080 | .0064 | .0051 | .0040 | .0031 | .0024 | .0019 | .0014 | .0011 |
| 4 | .0295 | .0247 | .0206 | .0170 | .0139 | .0113 | .0092 | .0074 | .0059 | .0046 |
| 5 | .0656 | .0573 | .0496 | .0427 | .0365 | .0309 | .0260 | .0217 | .0180 | .0148 |
| 6 | .1140 | .1037 | .0936 | .0839 | .0746 | .0658 | .0577 | .0501 | .0432 | .0370 |
| 7 | .1585 | .1502 | .1413 | .1318 | .1221 | .1122 | .1023 | .0925 | .0830 | .0739 |
| 8 | .1790 | .1768 | .1732 | .1683 | .1623 | .1553 | .1474 | .1388 | .1296 | .1201 |
| 9 | .1658 | .1707 | .1742 | .1763 | .1771 | .1763 | .1742 | .1708 | .1661 | .1602 |
| 10 | .1268 | .1359 | .1446 | .1524 | .1593 | .1652 | .1700 | .1734 | .1755 | .1762 |
| 11 | .0801 | .0895 | .0991 | .1089 | .1185 | .1280 | .1370 | .1455 | .1533 | .1602 |
| 12 | .0417 | .0486 | .0561 | .0642 | .0727 | .0818 | .0911 | .1007 | .1105 | .1201 |
| 13 | .0178 | .0217 | .0260 | .0310 | .0366 | .0429 | .0497 | .0572 | .0653 | .0739 |
| 14 | .0062 | .0078 | .0098 | .0122 | .0150 | .0183 | .0221 | .0264 | .0314 | .0370 |
| 15 | .0017 | .0023 | .0030 | .0038 | .0049 | .0062 | .0078 | .0098 | .0121 | .0148 |
| 16 | .0004 | .0005 | .0007 | .0009 | .0013 | .0017 | .0022 | .0028 | .0036 | .0046 |
| 17 | .0001 | .0001 | .0001 | .0002 | .0002 | .0003 | .0005 | .0006 | .0008 | .0011 |
| 18 | .0000 | .0000 | .0000 | .0000 | .0000 | .0000 | .0001 | .0001 | .0001 | .0002 |

## TABLE 2
### Poisson Probability Distribution

| r | 0.10 | 0.20 | 0.30 | 0.40 | λ<br>0.50 | 0.60 | 0.70 | 0.80 | 0.90 | 1.00 |
|---|---|---|---|---|---|---|---|---|---|---|
| 0 | .9048 | .8187 | .7408 | .6703 | .6066 | .5488 | .4966 | .4493 | .4066 | .3679 |
| 1 | .0905 | .1637 | .2222 | .2681 | .3033 | .3293 | .3476 | .3595 | .3659 | .3679 |
| 2 | .0045 | .0164 | .0333 | .0536 | .0758 | .0988 | .1217 | .1438 | .1647 | .1839 |
| 3 | .0002 | .0011 | .0033 | .0072 | .0126 | .0198 | .0284 | .0383 | .0494 | .0613 |
| 4 | .0000 | .0001 | .0003 | .0007 | .0016 | .0030 | .0050 | .0077 | .0111 | .0153 |
| 5 | .0000 | .0000 | .0000 | .0001 | .0002 | .0004 | .0007 | .0012 | .0020 | .0031 |
| 6 | .0000 | .0000 | .0000 | .0000 | .0000 | .0000 | .0001 | .0002 | .0003 | .0005 |
| 7 | .0000 | .0000 | .0000 | .0000 | .0000 | .0000 | .0000 | .0000 | .0000 | .0001 |

| r | 1.10 | 1.20 | 1.30 | 1.40 | λ<br>1.50 | 1.60 | 1.70 | 1.80 | 1.90 | 2.00 |
|---|---|---|---|---|---|---|---|---|---|---|
| 0 | .3329 | .3012 | .2725 | .2466 | .2231 | .2019 | .1827 | .1653 | .1496 | .1353 |
| 1 | .3662 | .3614 | .3543 | .3452 | .3347 | .3230 | .3106 | .2975 | .2842 | .2707 |
| 2 | .2014 | .2169 | .2303 | .2417 | .2510 | .2584 | .2640 | .2678 | .2700 | .2707 |
| 3 | .0738 | .0867 | .0998 | .1128 | .1255 | .1378 | .1496 | .1607 | .1710 | .1804 |
| 4 | .0203 | .0260 | .0324 | .0395 | .0471 | .0551 | .0636 | .0723 | .0812 | .0902 |
| 5 | .0045 | .0062 | .0084 | .0111 | .0141 | .0176 | .0216 | .0260 | .0309 | .0361 |
| 6 | .0008 | .0012 | .0018 | .0026 | .0035 | .0047 | .0061 | .0078 | .0098 | .0120 |
| 7 | .0001 | .0002 | .0003 | .0005 | .0008 | .0011 | .0015 | .0020 | .0027 | .0034 |
| 8 | .0000 | .0000 | .0001 | .0001 | .0001 | .0002 | .0003 | .0005 | .0006 | .0009 |
| 9 | .0000 | .0000 | .0000 | .0000 | .0000 | .0000 | .0000 | .0001 | .0001 | .0002 |

| r | 2.10 | 2.20 | 2.30 | 2.40 | λ<br>2.50 | 2.60 | 2.70 | 2.80 | 2.90 | 3.00 |
|---|---|---|---|---|---|---|---|---|---|---|
| 0 | .1225 | .1108 | .1003 | .0907 | .0821 | .0743 | .0672 | .0608 | .0550 | .0498 |
| 1 | .2572 | .2438 | .2306 | .2177 | .2052 | .1931 | .1815 | .1703 | .1596 | .1494 |
| 2 | .2700 | .2681 | .2652 | .2613 | .2565 | .2510 | .2450 | .2384 | .2314 | .2240 |
| 3 | .1890 | .1966 | .2033 | .2090 | .2138 | .2176 | .2205 | .2225 | .2237 | .2240 |
| 4 | .0992 | .1082 | .1169 | .1254 | .1336 | .1414 | .1488 | .1557 | .1622 | .1680 |
| 5 | .0417 | .0476 | .0538 | .0602 | .0668 | .0735 | .0804 | .0872 | .0940 | .1008 |
| 6 | .0146 | .0174 | .0206 | .0241 | .0278 | .0319 | .0362 | .0407 | .0455 | .0504 |
| 7 | .0044 | .0055 | .0068 | .0083 | .0099 | .0118 | .0139 | .0163 | .0188 | .0216 |
| 8 | .0011 | .0015 | .0019 | .0025 | .0031 | .0038 | .0047 | .0057 | .0068 | .0081 |
| 9 | .0003 | .0004 | .0005 | .0007 | .0009 | .0011 | .0014 | .0018 | .0022 | .0027 |
| 10 | .0001 | .0001 | .0001 | .0002 | .0002 | .0003 | .0004 | .0005 | .0006 | .0008 |
| 11 | .0000 | .0000 | .0000 | .0000 | .0000 | .0001 | .0001 | .0001 | .0002 | .0002 |
| 12 | .0000 | .0000 | .0000 | .0000 | .0000 | .0000 | .0000 | .0000 | .0000 | .0001 |

| r | 3.10 | 3.20 | 3.30 | 3.40 | λ<br>3.50 | 3.60 | 3.70 | 3.80 | 3.90 | 4.00 |
|---|---|---|---|---|---|---|---|---|---|---|
| 0 | .0450 | .0408 | .0369 | .0334 | .0302 | .0273 | .0247 | .0224 | .0202 | .0183 |
| 1 | .1397 | .1304 | .1217 | .1135 | .1057 | .0984 | .0915 | .0850 | .0789 | .0733 |
| 2 | .2165 | .2087 | .2008 | .1929 | .1850 | .1771 | .1692 | .1615 | .1539 | .1465 |
| 3 | .2237 | .2226 | .2209 | .2186 | .2158 | .2125 | .2087 | .2046 | .2001 | .1954 |
| 4 | .1733 | .1781 | .1823 | .1858 | .1888 | .1912 | .1931 | .1944 | .1951 | .1954 |

Charles Clark and Lawrence Schkade, *Statistical Methods for Business Decisions* (Cincinnati: South-Western Publishing Co., 1969), pp. 141–144, by special permission.

# TABLE 2—Continued

| r | 3.10 | 3.20 | 3.30 | 3.40 | λ 3.50 | 3.60 | 3.70 | 3.80 | 3.90 | 4.00 |
|---|------|------|------|------|------|------|------|------|------|------|
| 5 | .1075 | .1140 | .1203 | .1264 | .1322 | .1377 | .1429 | .1477 | .1522 | .1563 |
| 6 | .0555 | .0608 | .0662 | .0716 | .0771 | .0826 | .0881 | .0936 | .0989 | .1042 |
| 7 | .0246 | .0278 | .0312 | .0348 | .0385 | .0425 | .0466 | .0508 | .0551 | .0595 |
| 8 | .0095 | .0111 | .0129 | .0148 | .0169 | .0191 | .0215 | .0241 | .0269 | .0298 |
| 9 | .0033 | .0040 | .0047 | .0056 | .0066 | .0076 | .0089 | .0102 | .0116 | .0132 |
| 10 | .0010 | .0013 | .0016 | .0019 | .0023 | .0028 | .0033 | .0039 | .0045 | .0053 |
| 11 | .0003 | .0004 | .0005 | .0006 | .0007 | .0009 | .0011 | .0013 | .0016 | .0019 |
| 12 | .0001 | .0001 | .0001 | .0002 | .0002 | .0003 | .0003 | .0004 | .0005 | .0006 |
| 13 | .0000 | .0000 | .0000 | .0000 | .0001 | .0001 | .0001 | .0001 | .0002 | .0002 |
| 14 | .0000 | .0000 | .0000 | .0000 | .0000 | .0000 | .0000 | .0000 | .0000 | .0001 |

| r | 4.10 | 4.20 | 4.30 | 4.40 | λ 4.50 | 4.60 | 4.70 | 4.80 | 4.90 | 5.00 |
|---|------|------|------|------|------|------|------|------|------|------|
| 0 | .0166 | .0150 | .0136 | .0123 | .0111 | .0101 | .0091 | .0082 | .0074 | .0067 |
| 1 | .0679 | .0630 | .0583 | .0540 | .0500 | .0462 | .0427 | .0395 | .0365 | .0337 |
| 2 | .1393 | .1323 | .1254 | .1188 | .1125 | .1063 | .1005 | .0948 | .0894 | .0842 |
| 3 | .1904 | .1852 | .1798 | .1743 | .1687 | .1631 | .1574 | .1517 | .1460 | .1404 |
| 4 | .1951 | .1944 | .1933 | .1917 | .1898 | .1875 | .1849 | .1820 | .1789 | .1755 |
| 5 | .1600 | .1633 | .1662 | .1687 | .1708 | .1725 | .1738 | .1747 | .1753 | .1755 |
| 6 | .1093 | .1143 | .1191 | .1237 | .1281 | .1323 | .1362 | .1398 | .1432 | .1462 |
| 7 | .0640 | .0686 | .0732 | .0778 | .0824 | .0869 | .0914 | .0959 | .1002 | .1044 |
| 8 | .0328 | .0360 | .0393 | .0428 | .0463 | .0500 | .0537 | .0575 | .0614 | .0653 |
| 9 | .0150 | .0168 | .0188 | .0209 | .0232 | .0255 | .0281 | .0307 | .0334 | .0363 |
| 10 | .0061 | .0071 | .0081 | .0092 | .0104 | .0118 | .0132 | .0147 | .0164 | .0181 |
| 11 | .0023 | .0027 | .0032 | .0037 | .0043 | .0049 | .0056 | .0064 | .0073 | .0082 |
| 12 | .0008 | .0009 | .0011 | .0013 | .0016 | .0019 | .0022 | .0026 | .0030 | .0034 |
| 13 | .0002 | .0003 | .0004 | .0005 | .0006 | .0007 | .0008 | .0009 | .0011 | .0013 |
| 14 | .0001 | .0001 | .0001 | .0001 | .0002 | .0002 | .0003 | .0003 | .0004 | .0005 |
| 15 | .0000 | .0000 | .0000 | .0000 | .0001 | .0001 | .0001 | .0001 | .0001 | .0002 |

| r | 5.10 | 5.20 | 5.30 | 5.40 | λ 5.50 | 5.60 | 5.70 | 5.80 | 5.90 | 6.00 |
|---|------|------|------|------|------|------|------|------|------|------|
| 0 | .0061 | .0055 | .0050 | .0045 | .0041 | .0037 | .0033 | .0030 | .0027 | .0025 |
| 1 | .0311 | .0287 | .0265 | .0244 | .0225 | .0207 | .0191 | .0176 | .0162 | .0149 |
| 2 | .0793 | .0746 | .0701 | .0659 | .0618 | .0580 | .0544 | .0509 | .0477 | .0446 |
| 3 | .1348 | .1293 | .1239 | .1185 | .1133 | .1082 | .1033 | .0985 | .0938 | .0892 |
| 4 | .1719 | .1681 | .1641 | .1600 | .1558 | .1515 | .1472 | .1428 | .1383 | .1339 |
| 5 | .1753 | .1748 | .1740 | .1728 | .1714 | .1697 | .1678 | .1656 | .1632 | .1606 |
| 6 | .1490 | .1515 | .1537 | .1555 | .1571 | .1584 | .1594 | .1601 | .1605 | .1606 |
| 7 | .1086 | .1125 | .1163 | .1200 | .1234 | .1267 | .1298 | .1326 | .1353 | .1377 |
| 8 | .0692 | .0731 | .0771 | .0810 | .0849 | .0887 | .0925 | .0962 | .0998 | .1033 |
| 9 | .0392 | .0423 | .0454 | .0486 | .0519 | .0552 | .0586 | .0620 | .0654 | .0688 |
| 10 | .0200 | .0220 | .0241 | .0262 | .0285 | .0309 | .0334 | .0359 | .0386 | .0413 |
| 11 | .0093 | .0104 | .0116 | .0129 | .0143 | .0157 | .0173 | .0190 | .0207 | .0225 |
| 12 | .0039 | .0045 | .0051 | .0058 | .0065 | .0073 | .0082 | .0092 | .0102 | .0113 |
| 13 | .0015 | .0018 | .0021 | .0024 | .0028 | .0032 | .0036 | .0041 | .0046 | .0052 |
| 14 | .0006 | .0007 | .0008 | .0009 | .0011 | .0013 | .0015 | .0017 | .0019 | .0022 |
| 15 | .0002 | .0002 | .0003 | .0003 | .0004 | .0005 | .0006 | .0007 | .0008 | .0009 |
| 16 | .0001 | .0001 | .0001 | .0001 | .0001 | .0002 | .0002 | .0002 | .0003 | .0003 |
| 17 | .0000 | .0000 | .0000 | .0000 | .0000 | .0001 | .0001 | .0001 | .0001 | .0001 |

| r | 6.10 | 6.20 | 6.30 | 6.40 | λ 6.50 | 6.60 | 6.70 | 6.80 | 6.90 | 7.00 |
|---|------|------|------|------|------|------|------|------|------|------|
| 0 | .0022 | .0020 | .0018 | .0017 | .0015 | .0014 | .0012 | .0011 | .0010 | .0009 |
| 1 | .0137 | .0126 | .0116 | .0106 | .0098 | .0090 | .0082 | .0076 | .0070 | .0064 |
| 2 | .0417 | .0390 | .0364 | .0340 | .0318 | .0296 | .0276 | .0258 | .0240 | .0223 |
| 3 | .0848 | .0806 | .0765 | .0726 | .0688 | .0652 | .0617 | .0584 | .0552 | .0521 |
| 4 | .1294 | .1249 | .1205 | .1161 | .1118 | .1076 | .1034 | .0992 | .0952 | .0912 |
| 5 | .1579 | .1549 | .1519 | .1487 | .1454 | .1420 | .1385 | .1349 | .1314 | .1277 |
| 6 | .1605 | .1601 | .1595 | .1586 | .1575 | .1562 | .1546 | .1529 | .1511 | .1490 |
| 7 | .1399 | .1418 | .1435 | .1450 | .1462 | .1472 | .1480 | .1486 | .1489 | .1490 |
| 8 | .1066 | .1099 | .1130 | .1160 | .1188 | .1215 | .1240 | .1263 | .1284 | .1304 |
| 9 | .0723 | .0757 | .0791 | .0825 | .0858 | .0891 | .0923 | .0954 | .0985 | .1014 |
| 10 | .0441 | .0469 | .0498 | .0528 | .0558 | .0588 | .0618 | .0649 | .0679 | .0710 |
| 11 | .0244 | .0265 | .0285 | .0307 | .0330 | .0353 | .0377 | .0401 | .0426 | .0452 |
| 12 | .0124 | .0137 | .0150 | .0164 | .0179 | .0194 | .0210 | .0227 | .0245 | .0263 |
| 13 | .0058 | .0065 | .0073 | .0081 | .0089 | .0099 | .0108 | .0119 | .0130 | .0142 |
| 14 | .0025 | .0029 | .0033 | .0037 | .0041 | .0046 | .0052 | .0058 | .0064 | .0071 |

# TABLE 2—*Continued*

| r | 6.10 | 6.20 | 6.30 | 6.40 | λ 6.50 | 6.60 | 6.70 | 6.80 | 6.90 | 7.00 |
|---|------|------|------|------|------|------|------|------|------|------|
| 15 | .0010 | .0012 | .0014 | .0016 | .0018 | .0020 | .0023 | .0026 | .0029 | .0033 |
| 16 | .0004 | .0005 | .0005 | .0006 | .0007 | .0008 | .0010 | .0011 | .0013 | .0014 |
| 17 | .0001 | .0002 | .0002 | .0002 | .0003 | .0003 | .0004 | .0004 | .0005 | .0006 |
| 18 | .0000 | .0001 | .0001 | .0001 | .0001 | .0001 | .0001 | .0002 | .0002 | .0002 |
| 19 | .0000 | .0000 | .0000 | .0000 | .0000 | .0000 | .0001 | .0001 | .0001 | .0001 |

| r | 7.10 | 7.20 | 7.30 | 7.40 | λ 7.50 | 7.60 | 7.70 | 7.80 | 7.90 | 8.00 |
|---|------|------|------|------|------|------|------|------|------|------|
| 0 | .0008 | .0007 | .0007 | .0006 | .0006 | .0005 | .0005 | .0004 | .0004 | .0003 |
| 1 | .0059 | .0054 | .0049 | .0045 | .0041 | .0038 | .0035 | .0032 | .0029 | .0027 |
| 2 | .0208 | .0194 | .0180 | .0167 | .0156 | .0145 | .0134 | .0125 | .0116 | .0107 |
| 3 | .0492 | .0464 | .0438 | .0413 | .0389 | .0366 | .0345 | .0324 | .0305 | .0286 |
| 4 | .0874 | .0836 | .0799 | .0764 | .0729 | .0696 | .0663 | .0632 | .0602 | .0573 |
| 5 | .1241 | .1204 | .1167 | .1130 | .1094 | .1057 | .1021 | .0986 | .0951 | .0916 |
| 6 | .1468 | .1445 | .1420 | .1394 | .1367 | .1339 | .1311 | .1282 | .1252 | .1221 |
| 7 | .1489 | .1486 | .1481 | .1474 | .1465 | .1454 | .1442 | .1428 | .1413 | .1396 |
| 8 | .1321 | .1337 | .1351 | .1363 | .1373 | .1381 | .1388 | .1392 | .1395 | .1396 |
| 9 | .1042 | .1070 | .1096 | .1121 | .1144 | .1167 | .1187 | .1207 | .1224 | .1241 |
| 10 | .0740 | .0770 | .0800 | .0829 | .0858 | .0887 | .0914 | .0941 | .0967 | .0993 |
| 11 | .0478 | .0504 | .0531 | .0558 | .0585 | .0613 | .0640 | .0667 | .0695 | .0722 |
| 12 | .0283 | .0303 | .0323 | .0344 | .0366 | .0388 | .0411 | .0434 | .0457 | .0481 |
| 13 | .0154 | .0168 | .0181 | .0196 | .0211 | .0227 | .0243 | .0260 | .0278 | .0296 |
| 14 | .0078 | .0086 | .0095 | .0104 | .0113 | .0123 | .0134 | .0145 | .0157 | .0169 |
| 15 | .0037 | .0041 | .0046 | .0051 | .0057 | .0062 | .0069 | .0075 | .0083 | .0090 |
| 16 | .0016 | .0019 | .0021 | .0024 | .0026 | .0030 | .0033 | .0037 | .0041 | .0045 |
| 17 | .0007 | .0008 | .0009 | .0010 | .0012 | .0013 | .0015 | .0017 | .0019 | .0021 |
| 18 | .0003 | .0003 | .0004 | .0004 | .0005 | .0006 | .0006 | .0007 | .0008 | .0009 |
| 19 | .0001 | .0001 | .0001 | .0002 | .0002 | .0002 | .0003 | .0003 | .0003 | .0004 |
| 20 | .0000 | .0000 | .0001 | .0001 | .0001 | .0001 | .0001 | .0001 | .0001 | .0002 |
| 21 | .0000 | .0000 | .0000 | .0000 | .0000 | .0000 | .0000 | .0000 | .0001 | .0001 |

| r | 8.10 | 8.20 | 8.30 | 8.40 | λ 8.50 | 8.60 | 8.70 | 8.80 | 8.90 | 9.00 |
|---|------|------|------|------|------|------|------|------|------|------|
| 0 | .0003 | .0003 | .0002 | .0002 | .0002 | .0002 | .0002 | .0002 | .0001 | .0001 |
| 1 | .0025 | .0023 | .0021 | .0019 | .0017 | .0016 | .0014 | .0013 | .0012 | .0011 |
| 2 | .0100 | .0092 | .0086 | .0079 | .0074 | .0068 | .0063 | .0058 | .0054 | .0050 |
| 3 | .0269 | .0252 | .0237 | .0222 | .0208 | .0195 | .0183 | .0171 | .0160 | .0150 |
| 4 | .0544 | .0517 | .0491 | .0466 | .0443 | .0420 | .0398 | .0377 | .0357 | .0337 |
| 5 | .0882 | .0849 | .0816 | .0784 | .0752 | .0722 | .0692 | .0663 | .0635 | .0607 |
| 6 | .1191 | .1160 | .1128 | .1097 | .1066 | .1034 | .1003 | .0972 | .0941 | .0911 |
| 7 | .1378 | .1358 | .1338 | .1317 | .1294 | .1271 | .1247 | .1222 | .1197 | .1171 |
| 8 | .1395 | .1392 | .1388 | .1382 | .1375 | .1366 | .1356 | .1344 | .1332 | .1318 |
| 9 | .1256 | .1269 | .1280 | .1290 | .1299 | .1306 | .1311 | .1315 | .1317 | .1318 |
| 10 | .1017 | .1040 | .1063 | .1084 | .1104 | .1123 | .1140 | .1157 | .1172 | .1186 |
| 11 | .0749 | .0776 | .0802 | .0828 | .0853 | .0878 | .0902 | .0925 | .0948 | .0970 |
| 12 | .0505 | .0530 | .0555 | .0579 | .0604 | .0629 | .0654 | .0679 | .0703 | .0728 |
| 13 | .0315 | .0334 | .0354 | .0374 | .0395 | .0416 | .0438 | .0459 | .0481 | .0504 |
| 14 | .0182 | .0196 | .0210 | .0225 | .0240 | .0256 | .0272 | .0289 | .0306 | .0324 |
| 15 | .0098 | .0107 | .0116 | .0126 | .0136 | .0147 | .0158 | .0169 | .0182 | .0194 |
| 16 | .0050 | .0055 | .0060 | .0066 | .0072 | .0079 | .0086 | .0093 | .0101 | .0109 |
| 17 | .0024 | .0026 | .0029 | .0033 | .0036 | .0040 | .0044 | .0048 | .0053 | .0058 |
| 18 | .0011 | .0012 | .0014 | .0015 | .0017 | .0019 | .0021 | .0024 | .0026 | .0029 |
| 19 | .0005 | .0005 | .0006 | .0007 | .0008 | .0009 | .0010 | .0011 | .0012 | .0014 |
| 20 | .0002 | .0002 | .0002 | .0003 | .0003 | .0004 | .0004 | .0005 | .0005 | .0006 |
| 21 | .0001 | .0001 | .0001 | .0001 | .0001 | .0002 | .0002 | .0002 | .0002 | .0003 |
| 22 | .0000 | .0000 | .0000 | .0000 | .0001 | .0001 | .0001 | .0001 | .0001 | .0001 |

| r | 9.10 | 9.20 | 9.30 | 9.40 | λ 9.50 | 9.60 | 9.70 | 9.80 | 9.90 | 10.00 |
|---|------|------|------|------|------|------|------|------|------|------|
| 0 | .0001 | .0001 | .0001 | .0001 | .0001 | .0001 | .0001 | .0001 | .0001 | .0000 |
| 1 | .0010 | .0009 | .0009 | .0008 | .0007 | .0007 | .0006 | .0005 | .0005 | .0005 |
| 2 | .0046 | .0043 | .0040 | .0037 | .0034 | .0031 | .0029 | .0027 | .0025 | .0023 |
| 3 | .0140 | .0131 | .0123 | .0115 | .0107 | .0100 | .0093 | .0087 | .0081 | .0076 |
| 4 | .0319 | .0302 | .0285 | .0269 | .0254 | .0240 | .0226 | .0213 | .0201 | .0189 |
| 5 | .0581 | .0555 | .0530 | .0506 | .0483 | .0460 | .0439 | .0418 | .0398 | .0378 |
| 6 | .0881 | .0851 | .0822 | .0793 | .0764 | .0736 | .0709 | .0682 | .0656 | .0631 |
| 7 | .1145 | .1118 | .1091 | .1064 | .1037 | .1010 | .0982 | .0955 | .0928 | .0901 |
| 8 | .1302 | .1286 | .1269 | .1251 | .1232 | .1212 | .1191 | .1170 | .1148 | .1126 |
| 9 | .1317 | .1315 | .1311 | .1306 | .1300 | .1293 | .1284 | .1274 | .1263 | .1251 |

TABLE 2—*Continued*

| r | 9.10 | 9.20 | 9.30 | 9.40 | λ 9.50 | 9.60 | 9.70 | 9.80 | 9.90 | 10.00 |
|---|------|------|------|------|--------|------|------|------|------|-------|
| 10 | .1198 | .1210 | .1219 | .1228 | .1235 | .1241 | .1245 | .1249 | .1250 | .1251 |
| 11 | .0991 | .1012 | .1031 | .1049 | .1067 | .1083 | .1098 | .1112 | .1125 | .1137 |
| 12 | .0752 | .0776 | .0799 | .0822 | .0844 | .0866 | .0888 | .0908 | .0928 | .0948 |
| 13 | .0526 | .0549 | .0572 | .0594 | .0617 | .0640 | .0662 | .0685 | .0707 | .0729 |
| 14 | .0342 | .0361 | .0380 | .0399 | .0419 | .0439 | .0459 | .0479 | .0500 | .0521 |
| 15 | .0208 | .0221 | .0235 | .0250 | .0265 | .0281 | .0297 | .0313 | .0330 | .0347 |
| 16 | .0118 | .0127 | .0137 | .0147 | .0157 | .0168 | .0180 | .0192 | .0204 | .0217 |
| 17 | .0063 | .0069 | .0075 | .0081 | .0088 | .0095 | .0103 | .0111 | .0119 | .0128 |
| 18 | .0032 | .0035 | .0039 | .0042 | .0046 | .0051 | .0055 | .0060 | .0065 | .0071 |
| 19 | .0015 | .0017 | .0019 | .0021 | .0023 | .0026 | .0028 | .0031 | .0034 | .0037 |
| 20 | .0007 | .0008 | .0009 | .0010 | .0011 | .0012 | .0014 | .0015 | .0017 | .0019 |
| 21 | .0003 | .0003 | .0004 | .0004 | .0005 | .0006 | .0006 | .0007 | .0008 | .0009 |
| 22 | .0001 | .0001 | .0002 | .0002 | .0002 | .0002 | .0003 | .0003 | .0004 | .0004 |
| 23 | .0000 | .0001 | .0001 | .0001 | .0001 | .0001 | .0001 | .0001 | .0002 | .0002 |
| 24 | .0000 | .0000 | .0000 | .0000 | .0000 | .0000 | .0000 | .0001 | .0001 | .0001 |

| r | 11. | 12. | 13. | 14. | λ 15. | 16. | 17. | 18. | 19. | 20. |
|---|-----|-----|-----|-----|-------|-----|-----|-----|-----|-----|
| 0 | .0000 | .0000 | .0000 | .0000 | .0000 | .0000 | .0000 | .0000 | .0000 | .0000 |
| 1 | .0002 | .0001 | .0000 | .0000 | .0000 | .0000 | .0000 | .0000 | .0000 | .0000 |
| 2 | .0010 | .0004 | .0002 | .0001 | .0000 | .0000 | .0000 | .0000 | .0000 | .0000 |
| 3 | .0037 | .0018 | .0008 | .0004 | .0002 | .0001 | .0000 | .0000 | .0000 | .0000 |
| 4 | .0102 | .0053 | .0027 | .0013 | .0006 | .0003 | .0001 | .0001 | .0000 | .0000 |
| 5 | .0224 | .0127 | .0070 | .0037 | .0019 | .0010 | .0005 | .0002 | .0001 | .0001 |
| 6 | .0411 | .0255 | .0152 | .0087 | .0048 | .0026 | .0014 | .0007 | .0004 | .0002 |
| 7 | .0646 | .0437 | .0281 | .0174 | .0104 | .0060 | .0034 | .0019 | .0010 | .0005 |
| 8 | .0888 | .0655 | .0457 | .0304 | .0194 | .0120 | .0072 | .0042 | .0024 | .0013 |
| 9 | .1085 | .0874 | .0661 | .0473 | .0324 | .0213 | .0135 | .0083 | .0050 | .0029 |
| 10 | .1194 | .1048 | .0859 | .0663 | .0486 | .0341 | .0230 | .0150 | .0095 | .0058 |
| 11 | .1194 | .1144 | .1015 | .0844 | .0663 | .0496 | .0355 | .0245 | .0164 | .0106 |
| 12 | .1094 | .1144 | .1099 | .0984 | .0829 | .0661 | .0504 | .0259 | .0176 | .0176 |
| 13 | .0926 | .1056 | .1099 | .1060 | .0956 | .0814 | .0658 | .0509 | .0378 | .0271 |
| 14 | .0728 | .0905 | .1021 | .1060 | .1024 | .0930 | .0800 | .0655 | .0514 | .0387 |
| 15 | .0534 | .0724 | .0885 | .0989 | .1024 | .0992 | .0906 | .0786 | .0650 | .0516 |
| 16 | .0367 | .0543 | .0719 | .0866 | .0960 | .0992 | .0963 | .0884 | .0772 | .0646 |
| 17 | .0237 | .0383 | .0550 | .0713 | .0847 | .0934 | .0963 | .0936 | .0863 | .0760 |
| 18 | .0145 | .0256 | .0397 | .0554 | .0706 | .0830 | .0909 | .0936 | .0911 | .0844 |
| 19 | .0084 | .0161 | .0272 | .0409 | .0557 | .0699 | .0814 | .0887 | .0911 | .0888 |
| 20 | .0046 | .0097 | .0177 | .0286 | .0418 | .0559 | .0692 | .0798 | .0866 | .0888 |
| 21 | .0024 | .0055 | .0109 | .0191 | .0299 | .0426 | .0560 | .0684 | .0783 | .0846 |
| 22 | .0012 | .0030 | .0065 | .0121 | .0204 | .0310 | .0433 | .0560 | .0676 | .0769 |
| 23 | .0006 | .0016 | .0037 | .0074 | .0133 | .0216 | .0320 | .0438 | .0559 | .0669 |
| 24 | .0003 | .0008 | .0020 | .0043 | .0083 | .0144 | .0226 | .0329 | .0442 | .0557 |
| 25 | .0001 | .0004 | .0010 | .0024 | .0050 | .0092 | .0154 | .0237 | .0336 | .0446 |
| 26 | .0000 | .0002 | .0005 | .0013 | .0029 | .0057 | .0101 | .0164 | .0246 | .0343 |
| 27 | .0000 | .0001 | .0002 | .0007 | .0016 | .0034 | .0063 | .0109 | .0173 | .0254 |
| 28 | .0000 | .0000 | .0001 | .0003 | .0009 | .0019 | .0038 | .0070 | .0117 | .0181 |
| 29 | .0000 | .0000 | .0001 | .0002 | .0004 | .0011 | .0023 | .0044 | .0077 | .0125 |
| 30 | .0000 | .0000 | .0000 | .0001 | .0002 | .0006 | .0013 | .0026 | .0049 | .0083 |
| 31 | .0000 | .0000 | .0000 | .0000 | .0001 | .0003 | .0007 | .0015 | .0030 | .0054 |
| 32 | .0000 | .0000 | .0000 | .0000 | .0001 | .0001 | .0004 | .0009 | .0018 | .0034 |
| 33 | .0000 | .0000 | .0000 | .0000 | .0000 | .0001 | .0002 | .0005 | .0010 | .0020 |
| 34 | .0000 | .0000 | .0000 | .0000 | .0000 | .0000 | .0001 | .0002 | .0006 | .0012 |
| 35 | .0000 | .0000 | .0000 | .0000 | .0000 | .0000 | .0000 | .0001 | .0003 | .0007 |
| 36 | .0000 | .0000 | .0000 | .0000 | .0000 | .0000 | .0000 | .0001 | .0002 | .0004 |
| 37 | .0000 | .0000 | .0000 | .0000 | .0000 | .0000 | .0000 | .0000 | .0001 | .0002 |
| 38 | .0000 | .0000 | .0000 | .0000 | .0000 | .0000 | .0000 | .0000 | .0000 | .0001 |
| 39 | .0000 | .0000 | .0000 | .0000 | .0000 | .0000 | .0000 | .0000 | .0000 | .0001 |

**TABLE 3**
Normal Probability Distribution

| z | .00 | .01 | .02 | .03 | .04 | .05 | .06 | .07 | .08 | .09 |
|---|---|---|---|---|---|---|---|---|---|---|
| 0.0 | .0000 | .0040 | .0080 | .0120 | .0160 | .0199 | .0239 | .0279 | .0319 | .0359 |
| 0.1 | .0398 | .0438 | .0478 | .0517 | .0557 | .0596 | .0636 | .0675 | .0714 | .0753 |
| 0.2 | .0793 | .0832 | .0871 | .0910 | .0948 | .0987 | .1026 | .1064 | .1103 | .1141 |
| 0.3 | .1179 | .1217 | .1255 | .1293 | .1331 | .1368 | .1406 | .1443 | .1480 | .1517 |
| 0.4 | .1554 | .1591 | .1628 | .1664 | .1700 | .1736 | .1772 | .1808 | .1844 | .1879 |
| 0.5 | .1915 | .1950 | .1985 | .2019 | .2054 | .2088 | .2123 | .2157 | .2190 | .2224 |
| 0.6 | .2257 | .2291 | .2324 | .2357 | .2389 | .2422 | .2454 | .2486 | .2517 | .2549 |
| 0.7 | .2580 | .2611 | .2642 | .2673 | .2703 | .2734 | .2764 | .2794 | .2823 | .2852 |
| 0.8 | .2881 | .2910 | .2939 | .2967 | .2995 | .3023 | .3051 | .3078 | .3106 | .3133 |
| 0.9 | .3159 | .3186 | .3212 | .3238 | .3264 | .3289 | .3315 | .3340 | .3365 | .3389 |
| 1.0 | .3413 | .3438 | .3461 | .3485 | .3508 | .3531 | .3554 | .3577 | .3599 | .3621 |
| 1.1 | .3643 | .3665 | .3686 | .3708 | .3729 | .3749 | .3770 | .3790 | .3810 | .3830 |
| 1.2 | .3849 | .3869 | .3888 | .3907 | .3925 | .3944 | .3962 | .3980 | .3997 | .4015 |
| 1.3 | .4032 | .4049 | .4066 | .4082 | .4099 | .4115 | .4131 | .4147 | .4162 | .4177 |
| 1.4 | .4192 | .4207 | .4222 | .4236 | .4251 | .4265 | .4279 | .4292 | .4306 | .4319 |
| 1.5 | .4332 | .4345 | .4357 | .4370 | .4382 | .4394 | .4406 | .4418 | .4429 | .4441 |
| 1.6 | .4452 | .4463 | .4474 | .4484 | .4495 | .4505 | .4515 | .4525 | .4535 | .4545 |
| 1.7 | .4554 | .4564 | .4573 | .4582 | .4591 | .4599 | .4608 | .4616 | .4625 | .4633 |
| 1.8 | .4641 | .4649 | .4656 | .4664 | .4671 | .4678 | .4686 | .4693 | .4699 | .4706 |
| 1.9 | .4713 | .4719 | .4726 | .4732 | .4738 | .4744 | .4750 | .4756 | .4761 | .4767 |
| 2.0 | .4772 | .4778 | .4783 | .4788 | .4793 | .4798 | .4803 | .4808 | .4812 | .4817 |
| 2.1 | .4821 | .4826 | .4830 | .4834 | .4838 | .4842 | .4846 | .4850 | .4854 | .4857 |
| 2.2 | .4861 | .4864 | .4868 | .4871 | .4875 | .4878 | .4881 | .4884 | .4887 | .4890 |
| 2.3 | .4893 | .4896 | .4898 | .4901 | .4904 | .4906 | .4909 | .4911 | .4913 | .4916 |
| 2.4 | .4918 | .4920 | .4922 | .4925 | .4927 | .4929 | .4931 | .4932 | .4934 | .4936 |
| 2.5 | .4938 | .4940 | .4941 | .4943 | .4945 | .4946 | .4948 | .4949 | .4951 | .4952 |
| 2.6 | .4953 | .4955 | .4956 | .4957 | .4959 | .4960 | .4961 | .4962 | .4963 | .4964 |
| 2.7 | .4965 | .4966 | .4967 | .4968 | .4969 | .4970 | .4971 | .4972 | .4973 | .4974 |
| 2.8 | .4974 | .4975 | .4976 | .4977 | .4977 | .4978 | .4979 | .4979 | .4980 | .4981 |
| 2.9 | .4981 | .4982 | .4982 | .4983 | .4984 | .4984 | .4985 | .4985 | .4986 | .4986 |
| 3.0 | .4987 | .4987 | .4987 | .4988 | .4988 | .4989 | .4989. | .4989 | .4990 | .4990 |

Paul G. Hoel and Raymond J. Jessen, *Basic Statistics for Business and Economics* (New York: John Wiley and Sons, Inc., 1971), p. 412 by permission of the publishers.

**TABLE 4**
Student Distribution (t-distribution)

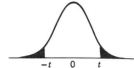

| $\nu$ \ P $\alpha$ | 0.50 | 0.25 | 0.10 | 0.05 | 0.025 | 0.01 | 0.005 |
|---|---|---|---|---|---|---|---|
| 1 | 1.00000 | 2.4142 | 6.3138 | 12.706 | 25.452 | 63.657 | 127.32 |
| 2 | 0.81650 | 1.6036 | 2.9200 | 4.3027 | 6.2053 | 9.9248 | 14.089 |
| 3 | 0.76489 | 1.4226 | 2.3534 | 3.1825 | 4.1765 | 5.8409 | 7.4533 |
| 4 | 0.74070 | 1.3444 | 2.1318 | 2.7764 | 3.4954 | 4.6041 | 5.5976 |
| 5 | 0.72669 | 1.3009 | 2.0150 | 2.5706 | 3.1634 | 4.0321 | 4.7733 |
| 6 | 0.71756 | 1.2733 | 1.9432 | 2.4469 | 2.9687 | 3.7074 | 4.3168 |
| 7 | 0.71114 | 1.2543 | 1.8946 | 2.3646 | 2.8412 | 3.4995 | 4.0293 |
| 8 | 0.70639 | 1.2403 | 1.8595 | 2.3060 | 2.7515 | 3.3554 | 3.8325 |
| 9 | 0.70272 | 1.2297 | 1.8331 | 2.2622 | 2.6850 | 3.2498 | 3.6897 |
| 10 | 0.69981 | 1.2213 | 1.8125 | 2.2281 | 2.6338 | 3.1693 | 3.5814 |
| 11 | 0.69745 | 1.2145 | 1.7959 | 2.2010 | 2.5931 | 3.1058 | 3.4966 |
| 12 | 0.69548 | 1.2089 | 1.7823 | 2.1788 | 2.5600 | 3.0545 | 3.4284 |
| 13 | 0.69384 | 1.2041 | 1.7709 | 2.1604 | 2.5326 | 3.0123 | 3.3725 |
| 14 | 0.69242 | 1.2001 | 1.7613 | 2.1448 | 2.5096 | 2.9768 | 3.3257 |
| 15 | 0.69120 | 1.1967 | 1.7530 | 2.1315 | 2.4899 | 2.9467 | 3.2860 |
| 16 | 0.69013 | 1.1937 | 1.7459 | 2.1199 | 2.4729 | 2.9208 | 3.2520 |
| 17 | 0.68919 | 1.1910 | 1.7396 | 2.1098 | 2.4581 | 2.8982 | 3.2225 |
| 18 | 0.68837 | 1.1887 | 1.7341 | 2.1009 | 2.4450 | 2.8784 | 3.1966 |
| 19 | 0.68763 | 1.1866 | 1.7291 | 2.0930 | 2.4334 | 2.8609 | 3.1737 |
| 20 | 0.68696 | 1.1848 | 1.7247 | 2.0860 | 2.4231 | 2.8453 | 3.1534 |
| 21 | 0.68635 | 1.1831 | 1.7207 | 2.0796 | 2.4138 | 2.8314 | 3.1352 |
| 22 | 0.68580 | 1.1816 | 1.7171 | 2.0739 | 2.4055 | 2.8188 | 3.1188 |
| 23 | 0.68531 | 1.1802 | 1.7139 | 2.0687 | 2.3979 | 2.8073 | 3.1040 |
| 24 | 0.68485 | 1.1789 | 1.7109 | 2.0639 | 2.3910 | 2.7969 | 3.0905 |
| 25 | 0.68443 | 1.1777 | 1.7081 | 2.0595 | 2.3846 | 2.7874 | 3.0782 |
| 26 | 0.68405 | 1.1766 | 1.7056 | 2.0555 | 2.3788 | 2.7787 | 3.0669 |
| 27 | 0.68370 | 1.1757 | 1.7033 | 2.0518 | 2.3734 | 2.7707 | 3.0565 |
| 28 | 0.68335 | 1.1748 | 1.7011 | 2.0484 | 2.3685 | 2.7633 | 3.0469 |
| 29 | 0.68304 | 1.1739 | 1.6991 | 2.0452 | 2.3638 | 2.7564 | 3.0380 |
| 30 | 0.68276 | 1.1731 | 1.6973 | 2.0423 | 2.3596 | 2.7500 | 3.0298 |
| 40 | 0.68066 | 1.1673 | 1.6839 | 2.0211 | 2.3289 | 2.7045 | 2.9712 |
| 60 | 0.67862 | 1.1616 | 1.6707 | 2.0003 | 2.2991 | 2.6603 | 2.9146 |
| 120 | 0.67656 | 1.1559 | 1.6577 | 1.9799 | 2.2699 | 2.6174 | 2.8599 |
| $\infty$ | 0.67449 | 1.1503 | 1.6449 | 1.9600 | 2.2414 | 2.5758 | 2.8070 |

Paul G. Hoel and Raymond J. Jessen, *Basic Statistics for Business and Economics* (New York: John Wiley and Sons, Inc., 1971), p. 413 by permission of the publishers.

# TABLE 5
## Chi-square Distribution

| $\nu$ \ P | 0.995 | 0.975 | 0.050 | 0.025 | 0.010 | 0.005 |
|---|---|---|---|---|---|---|
| 1 | 0.0$^4$3927 | 0.0$^3$9821 | 3.84146 | 5.02389 | 6.63490 | 7.87944 |
| 2 | 0.010025 | 0.050636 | 5.99147 | 7.37776 | 9.21034 | 10.5966 |
| 3 | 0.071721 | 0.215795 | 7.81473 | 9.34840 | 11.3449 | 12.8381 |
| 4 | 0.206990 | 0.484419 | 9.48773 | 11.1433 | 13.2767 | 14.8602 |
| 5 | 0.411740 | 0.831211 | 11.0705 | 12.8325 | 15.0863 | 16.7496 |
| 6 | 0.675727 | 1.237347 | 12.5916 | 14.4494 | 16.8119 | 18.5476 |
| 7 | 0.989265 | 1.68987 | 14.0671 | 16.0128 | 18.4753 | 20.2777 |
| 8 | 1.344419 | 2.17973 | 15.5073 | 17.5346 | 20.0902 | 21.9550 |
| 9 | 1.734926 | 2.70039 | 16.9190 | 19.0228 | 21.6660 | 23.5893 |
| 10 | 2.15585 | 3.24697 | 18.3070 | 20.4831 | 23.2093 | 25.1882 |
| 11 | 2.60321 | 3.81575 | 19.6751 | 21.9200 | 24.7250 | 26.7569 |
| 12 | 3.07382 | 4.40379 | 21.0261 | 23.3367 | 26.2170 | 28.2995 |
| 13 | 3.56503 | 5.00874 | 22.3621 | 24.7356 | 27.6883 | 29.8194 |
| 14 | 4.07468 | 5.62872 | 23.6848 | 26.1190 | 29.1413 | 31.3193 |
| 15 | 4.60094 | 6.26214 | 24.9958 | 27.4884 | 30.5779 | 32.8013 |
| 16 | 5.14224 | 6.90766 | 26.2962 | 28.8454 | 31.9999 | 34.2672 |
| 17 | 5.69724 | 7.56418 | 27.5871 | 30.1910 | 33.4087 | 35.7185 |
| 18 | 6.26481 | 8.23075 | 28.8693 | 31.5264 | 34.8053 | 37.1564 |
| 19 | 6.84398 | 8.90655 | 30.1435 | 32.8523 | 36.1908 | 38.5822 |
| 20 | 7.43386 | 9.59083 | 31.4104 | 34.1696 | 37.5662 | 39.9968 |
| 21 | 8.03366 | 10.28293 | 32.6705 | 35.4789 | 38.9321 | 41.4010 |
| 22 | 8.64272 | 10.9823 | 33.9244 | 36.7807 | 40.2894 | 42.7956 |
| 23 | 9.26042 | 11.6885 | 35.1725 | 30.0757 | 41.6384 | 44.1813 |
| 24 | 9.88623 | 12.4001 | 36.4151 | 39.3641 | 42.9798 | 45.5585 |
| 25 | 10.5197 | 13.1197 | 37.6525 | 40.6465 | 44.3141 | 46.9278 |
| 26 | 11.1603 | 13.8439 | 38.8852 | 41.9232 | 45.6417 | 48.2899 |
| 27 | 11.8076 | 14.5733 | 40.1133 | 43.1944 | 46.9630 | 49.6449 |
| 28 | 12.4613 | 15.3079 | 41.3372 | 44.4607 | 48.2782 | 50.9933 |
| 29 | 13.1211 | 16.0471 | 42.5569 | 45.7222 | 49.5879 | 52.3356 |
| 30 | 13.7867 | 16.7908 | 43.7729 | 46.9792 | 50.8922 | 53.6720 |
| 40 | 20.7065 | 24.4331 | 55.7585 | 59.3417 | 63.6907 | 66.7659 |
| 50 | 27.9907 | 32.3574 | 67.5048 | 71.4202 | 76.1539 | 79.4900 |
| 60 | 35.5346 | 40.4817 | 79.0819 | 83.2976 | 88.3794 | 91.9517 |
| 70 | 43.2752 | 48.7576 | 90.5312 | 95.0231 | 100.425 | 104.215 |
| 80 | 51.1720 | 57.1532 | 101.879 | 106.629 | 112.329 | 116.321 |
| 90 | 59.1963 | 65.6466 | 113.145 | 118.136 | 124.116 | 128.299 |
| 100 | 67.3276 | 74.2219 | 124.342 | 129.561 | 135.807 | 140.169 |

Paul G. Hoel and Raymond J. Jessen, *Basic Statistics for Business and Economics* (New York: John Wiley and Sons, Inc., 1971), p. 415 by permission of the publishers.

# TABLE 6
## F Distribution

## 5% (Roman Type) and 1% (Boldface Type) Points for the Distribution of F

Degrees of freedom for numerator ($v_1$)

| Degrees of freedom for denominator ($v_2$) | 1 | 2 | 3 | 4 | 5 | 6 | 7 | 8 | 9 | 10 | 11 | 12 | 14 | 16 | 20 | 24 | 30 | 40 | 50 | 75 | 100 | 200 | 500 | ∞ |
|---|---|---|---|---|---|---|---|---|---|---|---|---|---|---|---|---|---|---|---|---|---|---|---|---|
| 1 | 161 / **4052** | 200 / **4999** | 216 / **5403** | 225 / **5625** | 230 / **5764** | 234 / **5859** | 237 / **5928** | 239 / **5981** | 241 / **6022** | 242 / **6056** | 243 / **6082** | 244 / **6106** | 245 / **6142** | 246 / **6169** | 248 / **6208** | 249 / **6234** | 250 / **6258** | 251 / **6286** | 252 / **6302** | 253 / **6323** | 253 / **6334** | 254 / **6352** | 254 / **6361** | 254 / **6366** |
| 2 | 18.51 / **98.49** | 19.00 / **99.01** | 19.16 / **99.17** | 19.25 / **99.25** | 19.30 / **99.30** | 19.33 / **99.33** | 19.36 / **99.34** | 19.37 / **99.36** | 19.38 / **99.38** | 19.39 / **99.40** | 19.40 / **99.41** | 19.41 / **99.42** | 19.42 / **99.43** | 19.43 / **99.44** | 19.44 / **99.45** | 19.45 / **99.46** | 19.46 / **99.47** | 19.47 / **99.48** | 19.47 / **99.48** | 19.48 / **99.49** | 19.49 / **99.49** | 19.49 / **99.49** | 19.50 / **99.50** | 19.50 / **99.50** |
| 3 | 10.13 / **34.12** | 9.55 / **30.81** | 9.28 / **29.46** | 9.12 / **28.71** | 9.01 / **28.24** | 8.94 / **27.91** | 8.88 / **27.67** | 8.84 / **27.49** | 8.81 / **27.34** | 8.78 / **27.23** | 8.76 / **27.13** | 8.74 / **27.05** | 8.71 / **26.92** | 8.69 / **26.83** | 8.66 / **26.69** | 8.64 / **26.60** | 8.62 / **26.50** | 8.60 / **26.41** | 8.58 / **26.30** | 8.57 / **26.27** | 8.56 / **26.23** | 8.54 / **26.18** | 8.54 / **26.14** | 8.53 / **26.12** |
| 4 | 7.71 / **21.20** | 6.94 / **18.00** | 6.59 / **16.69** | 6.39 / **15.98** | 6.26 / **15.52** | 6.16 / **15.21** | 6.09 / **14.98** | 6.04 / **14.80** | 6.00 / **14.66** | 5.96 / **14.54** | 5.93 / **14.45** | 5.91 / **14.37** | 5.87 / **14.24** | 5.84 / **14.15** | 5.80 / **14.02** | 5.77 / **13.93** | 5.74 / **13.83** | 5.71 / **13.74** | 5.70 / **13.69** | 5.68 / **13.61** | 5.66 / **13.57** | 5.65 / **13.52** | 5.64 / **13.48** | 5.63 / **13.46** |
| 5 | 6.61 / **16.26** | 5.79 / **13.27** | 5.41 / **12.06** | 5.19 / **11.39** | 5.05 / **10.97** | 4.95 / **10.67** | 4.88 / **10.45** | 4.82 / **10.27** | 4.78 / **10.15** | 4.74 / **10.05** | 4.70 / **9.96** | 4.68 / **9.89** | 4.64 / **9.77** | 4.60 / **9.68** | 4.56 / **9.55** | 4.53 / **9.47** | 4.50 / **9.38** | 4.46 / **9.29** | 4.44 / **9.24** | 4.42 / **9.17** | 4.40 / **9.13** | 4.38 / **9.07** | 4.37 / **9.04** | 4.36 / **9.02** |
| 6 | 5.99 / **13.74** | 5.14 / **10.92** | 4.76 / **9.78** | 4.53 / **9.15** | 4.39 / **8.75** | 4.28 / **8.47** | 4.21 / **8.26** | 4.15 / **8.10** | 4.10 / **7.98** | 4.06 / **7.87** | 4.03 / **7.79** | 4.00 / **7.72** | 3.96 / **7.60** | 3.92 / **7.52** | 3.87 / **7.39** | 3.84 / **7.31** | 3.81 / **7.23** | 3.77 / **7.14** | 3.75 / **7.09** | 3.72 / **7.02** | 3.71 / **6.99** | 3.69 / **6.94** | 3.68 / **6.90** | 3.67 / **6.88** |
| 7 | 5.59 / **12.25** | 4.74 / **9.55** | 4.35 / **8.45** | 4.12 / **7.85** | 3.97 / **7.46** | 3.87 / **7.19** | 3.79 / **7.00** | 3.73 / **6.84** | 3.68 / **6.71** | 3.63 / **6.62** | 3.60 / **6.54** | 3.57 / **6.47** | 3.52 / **6.35** | 3.49 / **6.27** | 3.44 / **6.15** | 3.41 / **6.07** | 3.38 / **5.98** | 3.34 / **5.90** | 3.32 / **5.85** | 3.29 / **5.78** | 3.28 / **5.75** | 3.25 / **5.70** | 3.24 / **5.67** | 3.23 / **5.65** |
| 8 | 5.32 / **11.26** | 4.46 / **8.65** | 4.07 / **7.59** | 3.84 / **7.01** | 3.69 / **6.63** | 3.58 / **6.37** | 3.50 / **6.19** | 3.44 / **6.03** | 3.39 / **5.91** | 3.34 / **5.82** | 3.31 / **5.74** | 3.28 / **5.67** | 3.23 / **5.56** | 3.20 / **5.48** | 3.15 / **5.36** | 3.12 / **5.28** | 3.08 / **5.20** | 3.05 / **5.11** | 3.03 / **5.06** | 3.00 / **5.00** | 2.98 / **4.96** | 2.96 / **4.91** | 2.94 / **4.88** | 2.93 / **4.86** |
| 9 | 5.12 / **10.56** | 4.26 / **8.02** | 3.86 / **6.99** | 3.63 / **6.42** | 3.48 / **6.06** | 3.37 / **5.80** | 3.29 / **5.62** | 3.23 / **5.47** | 3.18 / **5.35** | 3.13 / **5.26** | 3.10 / **5.18** | 3.07 / **5.11** | 3.02 / **5.00** | 2.98 / **4.92** | 2.93 / **4.80** | 2.90 / **4.73** | 2.86 / **4.64** | 2.82 / **4.56** | 2.80 / **4.51** | 2.77 / **4.45** | 2.76 / **4.41** | 2.73 / **4.36** | 2.72 / **4.33** | 2.71 / **4.31** |

Paul G. Hoel and Raymond J. Jessen, *Basic Statistics for Business and Economics* (New York: John Wiley and Sons, Inc., 1971), pp. 416–419 by permission of the publishers.

TABLE 6—Continued

| | | | | | | | | | | | | | | | | | | | | | | | | |
|---|---|---|---|---|---|---|---|---|---|---|---|---|---|---|---|---|---|---|---|---|---|---|---|---|
| 10 | 4.96 / 10.04 | 4.10 / 7.56 | 3.71 / 6.55 | 3.48 / 5.99 | 3.33 / 5.64 | 3.22 / 5.39 | 3.14 / 5.21 | 3.07 / 5.06 | 3.02 / 4.95 | 2.97 / 4.85 | 2.94 / 4.78 | 2.91 / 4.71 | 2.86 / 4.60 | 2.82 / 4.52 | 2.77 / 4.41 | 2.74 / 4.33 | 2.70 / 4.25 | 2.67 / 4.17 | 2.64 / 4.12 | 2.61 / 4.05 | 2.59 / 4.01 | 2.56 / 3.96 | 2.55 / 3.93 | 2.54 / 3.91 |
| 11 | 4.84 / 9.65 | 3.98 / 7.20 | 3.59 / 6.22 | 3.36 / 5.67 | 3.20 / 5.32 | 3.09 / 5.07 | 3.01 / 4.88 | 2.95 / 4.74 | 2.90 / 4.63 | 2.86 / 4.54 | 2.82 / 4.46 | 2.79 / 4.40 | 2.74 / 4.29 | 2.70 / 4.21 | 2.65 / 4.10 | 2.61 / 4.02 | 2.57 / 3.94 | 2.53 / 3.86 | 2.50 / 3.80 | 2.47 / 3.74 | 2.45 / 3.70 | 2.42 / 3.66 | 2.41 / 3.62 | 2.40 / 3.60 |
| 12 | 4.75 / 9.33 | 3.88 / 6.93 | 3.49 / 5.95 | 3.26 / 5.41 | 3.11 / 5.06 | 3.00 / 4.82 | 2.92 / 4.65 | 2.85 / 4.50 | 2.80 / 4.39 | 2.76 / 4.30 | 2.72 / 4.22 | 2.69 / 4.16 | 2.64 / 4.05 | 2.60 / 3.98 | 2.54 / 3.86 | 2.50 / 3.78 | 2.46 / 3.70 | 2.42 / 3.61 | 2.40 / 3.56 | 2.36 / 3.49 | 2.35 / 3.46 | 2.32 / 3.41 | 2.31 / 3.38 | 2.30 / 3.36 |
| 13 | 4.67 / 9.07 | 3.80 / 6.70 | 3.41 / 5.74 | 3.18 / 5.20 | 3.02 / 4.86 | 2.92 / 4.62 | 2.84 / 4.44 | 2.77 / 4.30 | 2.72 / 4.19 | 2.67 / 4.10 | 2.63 / 4.02 | 2.60 / 3.96 | 2.55 / 3.85 | 2.51 / 3.78 | 2.46 / 3.67 | 2.42 / 3.59 | 2.38 / 3.51 | 2.34 / 3.42 | 2.32 / 3.37 | 2.28 / 3.30 | 2.26 / 3.27 | 2.24 / 3.21 | 2.22 / 3.18 | 2.21 / 3.16 |
| 14 | 4.60 / 8.86 | 3.74 / 6.51 | 3.34 / 5.56 | 3.11 / 5.03 | 2.96 / 4.69 | 2.85 / 4.46 | 2.77 / 4.28 | 2.70 / 4.14 | 2.65 / 4.03 | 2.60 / 3.94 | 2.56 / 3.86 | 2.53 / 3.80 | 2.48 / 3.70 | 2.44 / 3.62 | 2.39 / 3.51 | 2.35 / 3.43 | 2.31 / 3.34 | 2.27 / 3.26 | 2.24 / 3.21 | 2.21 / 3.14 | 2.19 / 3.11 | 2.16 / 3.06 | 2.14 / 3.02 | 2.13 / 3.00 |
| 15 | 4.54 / 8.68 | 3.68 / 6.36 | 3.29 / 5.42 | 3.06 / 4.89 | 2.90 / 4.56 | 2.79 / 4.32 | 2.70 / 4.14 | 2.64 / 4.00 | 2.59 / 3.89 | 2.55 / 3.80 | 2.51 / 3.73 | 2.48 / 3.67 | 2.43 / 3.56 | 2.39 / 3.48 | 2.33 / 3.36 | 2.29 / 3.29 | 2.25 / 3.20 | 2.21 / 3.12 | 2.18 / 3.07 | 2.15 / 3.00 | 2.12 / 2.97 | 2.10 / 2.92 | 2.08 / 2.89 | 2.07 / 2.87 |
| 16 | 4.49 / 8.53 | 3.63 / 6.23 | 3.24 / 5.29 | 3.01 / 4.77 | 2.85 / 4.44 | 2.74 / 4.20 | 2.66 / 4.03 | 2.59 / 3.89 | 2.54 / 3.78 | 2.49 / 3.69 | 2.45 / 3.61 | 2.42 / 3.55 | 2.37 / 3.45 | 2.33 / 3.37 | 2.28 / 3.25 | 2.24 / 3.18 | 2.20 / 3.10 | 2.16 / 3.01 | 2.13 / 2.96 | 2.09 / 2.89 | 2.07 / 2.86 | 2.04 / 2.80 | 2.02 / 2.77 | 2.01 / 2.75 |
| 17 | 4.45 / 8.40 | 3.59 / 6.11 | 3.20 / 5.18 | 2.96 / 4.67 | 2.81 / 4.34 | 2.70 / 4.10 | 2.62 / 3.93 | 2.55 / 3.79 | 2.50 / 3.68 | 2.45 / 3.59 | 2.41 / 3.52 | 2.38 / 3.45 | 2.33 / 3.35 | 2.29 / 3.27 | 2.23 / 3.16 | 2.19 / 3.08 | 2.15 / 3.00 | 2.11 / 2.92 | 2.08 / 2.86 | 2.04 / 2.79 | 2.02 / 2.76 | 1.99 / 2.70 | 1.97 / 2.67 | 1.96 / 2.65 |
| 18 | 4.41 / 8.28 | 3.55 / 6.01 | 3.16 / 5.09 | 2.93 / 4.58 | 2.77 / 4.25 | 2.66 / 4.01 | 2.58 / 3.85 | 2.51 / 3.71 | 2.46 / 3.60 | 2.41 / 3.51 | 2.37 / 3.44 | 2.34 / 3.37 | 2.29 / 3.27 | 2.25 / 3.19 | 2.19 / 3.07 | 2.15 / 3.00 | 2.11 / 2.91 | 2.07 / 2.83 | 2.04 / 2.78 | 2.00 / 2.71 | 1.98 / 2.68 | 1.95 / 2.62 | 1.93 / 2.59 | 1.92 / 2.57 |
| 19 | 4.38 / 8.18 | 3.52 / 5.93 | 3.13 / 5.01 | 2.90 / 4.50 | 2.74 / 4.17 | 2.63 / 3.94 | 2.55 / 3.77 | 2.48 / 3.63 | 2.43 / 3.52 | 2.38 / 3.43 | 2.34 / 3.36 | 2.31 / 3.30 | 2.26 / 3.19 | 2.21 / 3.12 | 2.15 / 3.00 | 2.11 / 2.92 | 2.07 / 2.84 | 2.02 / 2.76 | 2.00 / 2.70 | 1.96 / 2.63 | 1.94 / 2.60 | 1.91 / 2.54 | 1.90 / 2.51 | 1.88 / 2.49 |
| 20 | 4.35 / 8.10 | 3.49 / 5.85 | 3.10 / 4.94 | 2.87 / 4.43 | 2.71 / 4.10 | 2.60 / 3.87 | 2.52 / 3.71 | 2.45 / 3.56 | 2.40 / 3.45 | 2.35 / 3.37 | 2.31 / 3.30 | 2.28 / 3.23 | 2.23 / 3.13 | 2.18 / 3.05 | 2.12 / 2.94 | 2.08 / 2.86 | 2.04 / 2.77 | 1.99 / 2.69 | 1.96 / 2.63 | 1.92 / 2.56 | 1.90 / 2.53 | 1.87 / 2.47 | 1.85 / 2.44 | 1.84 / 2.42 |
| 21 | 4.32 / 8.02 | 3.47 / 5.78 | 3.07 / 4.87 | 2.84 / 4.37 | 2.68 / 4.04 | 2.57 / 3.81 | 2.49 / 3.65 | 2.42 / 3.51 | 2.37 / 3.40 | 2.32 / 3.31 | 2.28 / 3.24 | 2.25 / 3.17 | 2.20 / 3.07 | 2.15 / 2.99 | 2.09 / 2.88 | 2.05 / 2.80 | 2.00 / 2.72 | 1.96 / 2.63 | 1.93 / 2.58 | 1.89 / 2.51 | 1.87 / 2.47 | 1.84 / 2.42 | 1.82 / 2.38 | 1.81 / 2.36 |
| 22 | 4.30 / 7.94 | 3.44 / 5.72 | 3.05 / 4.82 | 2.82 / 4.31 | 2.66 / 3.99 | 2.55 / 3.76 | 2.47 / 3.59 | 2.40 / 3.45 | 2.35 / 3.35 | 2.30 / 3.26 | 2.26 / 3.18 | 2.23 / 3.12 | 2.18 / 3.02 | 2.13 / 2.94 | 2.07 / 2.83 | 2.03 / 2.75 | 1.98 / 2.67 | 1.93 / 2.58 | 1.91 / 2.53 | 1.87 / 2.46 | 1.84 / 2.42 | 1.81 / 2.37 | 1.80 / 2.33 | 1.78 / 2.31 |
| 23 | 4.28 / 7.88 | 3.42 / 5.66 | 3.03 / 4.76 | 2.80 / 4.26 | 2.64 / 3.94 | 2.53 / 3.71 | 2.45 / 3.54 | 2.38 / 3.41 | 2.32 / 3.30 | 2.28 / 3.21 | 2.24 / 3.14 | 2.20 / 3.07 | 2.14 / 2.97 | 2.10 / 2.89 | 2.04 / 2.78 | 2.00 / 2.70 | 1.96 / 2.62 | 1.91 / 2.53 | 1.88 / 2.48 | 1.84 / 2.41 | 1.82 / 2.37 | 1.79 / 2.32 | 1.77 / 2.28 | 1.76 / 2.26 |
| 24 | 4.26 / 7.82 | 3.40 / 5.61 | 3.01 / 4.72 | 2.78 / 4.22 | 2.62 / 3.90 | 2.51 / 3.67 | 2.43 / 3.50 | 2.36 / 3.36 | 2.30 / 3.25 | 2.26 / 3.17 | 2.22 / 3.09 | 2.18 / 3.03 | 2.13 / 2.93 | 2.09 / 2.85 | 2.02 / 2.74 | 1.98 / 2.66 | 1.94 / 2.58 | 1.89 / 2.49 | 1.86 / 2.44 | 1.82 / 2.36 | 1.80 / 2.33 | 1.76 / 2.27 | 1.74 / 2.23 | 1.73 / 2.21 |
| 25 | 4.24 / 7.77 | 3.38 / 5.57 | 2.99 / 4.68 | 2.76 / 4.18 | 2.60 / 3.86 | 2.49 / 3.63 | 2.41 / 3.46 | 2.34 / 3.32 | 2.28 / 3.21 | 2.24 / 3.13 | 2.20 / 3.05 | 2.16 / 2.99 | 2.11 / 2.89 | 2.06 / 2.81 | 2.00 / 2.70 | 1.96 / 2.62 | 1.92 / 2.54 | 1.87 / 2.45 | 1.84 / 2.40 | 1.80 / 2.32 | 1.77 / 2.29 | 1.74 / 2.23 | 1.72 / 2.19 | 1.71 / 2.17 |

# TABLE 6—*Continued*
## 5% (Roman Type) and 1% (Boldface Type) Points for the Distribution of F

Degrees of freedom for numerator ($v_1$)

Each cell shows the 5% point (Roman) / 1% point (Boldface).

| $v_2$ | 1 | 2 | 3 | 4 | 5 | 6 | 7 | 8 | 9 | 10 | 11 | 12 | 14 | 16 | 20 | 24 | 30 | 40 | 50 | 75 | 100 | 200 | 500 | ∞ |
|---|---|---|---|---|---|---|---|---|---|---|---|---|---|---|---|---|---|---|---|---|---|---|---|---|
| 26 | 4.22/**7.72** | 3.37/**5.53** | 2.89/**4.64** | 2.74/**4.14** | 2.59/**3.82** | 2.47/**3.59** | 2.39/**3.42** | 2.32/**3.29** | 2.27/**3.17** | 2.22/**3.09** | 2.18/**3.02** | 2.15/**2.96** | 2.10/**2.86** | 2.05/**2.77** | 1.99/**2.66** | 1.95/**2.58** | 1.90/**2.50** | 1.85/**2.41** | 1.82/**2.36** | 1.78/**2.28** | 1.76/**2.25** | 1.72/**2.19** | 1.70/**2.15** | 1.69/**2.13** |
| 27 | 4.21/**7.68** | 3.35/**5.49** | 2.96/**4.60** | 2.73/**4.11** | 2.57/**3.79** | 2.46/**3.56** | 2.37/**3.39** | 2.30/**3.26** | 2.25/**3.14** | 2.20/**3.06** | 2.16/**2.98** | 2.13/**2.93** | 2.08/**2.83** | 2.03/**2.74** | 1.97/**2.63** | 1.93/**2.55** | 1.88/**2.47** | 1.84/**2.38** | 1.80/**2.33** | 1.76/**2.25** | 1.74/**2.21** | 1.71/**2.16** | 1.68/**2.12** | 1.67/**2.10** |
| 28 | 4.20/**7.64** | 3.34/**5.45** | 2.95/**4.57** | 2.71/**4.07** | 2.56/**3.76** | 2.44/**3.53** | 2.36/**3.36** | 2.29/**3.23** | 2.24/**3.11** | 2.19/**3.03** | 2.15/**2.95** | 2.12/**2.90** | 2.06/**2.80** | 2.02/**2.71** | 1.96/**2.60** | 1.91/**2.52** | 1.87/**2.44** | 1.81/**2.35** | 1.78/**2.30** | 1.75/**2.22** | 1.72/**2.18** | 1.69/**2.13** | 1.67/**2.09** | 1.65/**2.06** |
| 29 | 4.18/**7.60** | 3.33/**5.52** | 2.93/**4.54** | 2.70/**4.04** | 2.54/**3.73** | 2.43/**3.50** | 2.35/**3.33** | 2.28/**3.20** | 2.22/**3.08** | 2.18/**3.00** | 2.14/**2.92** | 2.10/**2.87** | 2.05/**2.77** | 2.00/**2.68** | 1.94/**2.57** | 1.90/**2.49** | 1.85/**2.41** | 1.80/**2.32** | 1.77/**2.27** | 1.73/**2.19** | 1.71/**2.15** | 1.68/**2.10** | 1.65/**2.06** | 1.64/**2.03** |
| 30 | 4.17/**7.56** | 3.32/**5.39** | 2.92/**4.51** | 2.69/**4.02** | 2.53/**3.70** | 2.42/**3.47** | 2.34/**3.30** | 2.27/**3.17** | 2.21/**3.06** | 2.16/**2.98** | 2.12/**2.90** | 2.09/**2.84** | 2.04/**2.74** | 1.99/**2.66** | 1.93/**2.55** | 1.89/**2.47** | 1.84/**2.38** | 1.79/**2.29** | 1.76/**2.24** | 1.72/**2.16** | 1.69/**2.13** | 1.66/**2.07** | 1.64/**2.03** | 1.62/**2.01** |
| 32 | 4.15/**7.50** | 3.30/**5.34** | 2.90/**4.46** | 2.67/**3.97** | 2.51/**3.66** | 2.40/**3.42** | 2.32/**3.25** | 2.25/**3.12** | 2.19/**3.01** | 2.14/**2.94** | 2.10/**2.86** | 2.07/**2.80** | 2.02/**2.70** | 1.97/**2.62** | 1.91/**2.51** | 1.86/**2.42** | 1.82/**2.34** | 1.76/**2.25** | 1.74/**2.20** | 1.69/**2.12** | 1.67/**2.08** | 1.64/**2.02** | 1.61/**1.98** | 1.59/**1.96** |
| 34 | 4.13/**7.44** | 3.28/**5.29** | 2.88/**4.42** | 2.65/**3.93** | 2.49/**3.61** | 2.38/**3.38** | 2.30/**3.21** | 2.23/**3.08** | 2.17/**2.97** | 2.12/**2.89** | 2.08/**2.82** | 2.05/**2.76** | 2.00/**2.66** | 1.95/**2.58** | 1.89/**2.47** | 1.84/**2.38** | 1.80/**2.30** | 1.74/**2.21** | 1.71/**2.15** | 1.67/**2.08** | 1.64/**2.04** | 1.61/**1.98** | 1.59/**1.94** | 1.57/**1.91** |
| 36 | 4.11/**7.39** | 3.26/**5.25** | 2.86/**4.38** | 2.63/**3.89** | 2.48/**3.58** | 2.36/**3.35** | 2.28/**3.18** | 2.21/**3.04** | 2.15/**2.94** | 2.10/**2.86** | 2.06/**2.78** | 2.03/**2.72** | 1.98/**2.62** | 1.93/**2.54** | 1.87/**2.43** | 1.82/**2.35** | 1.78/**2.26** | 1.72/**2.17** | 1.69/**2.12** | 1.65/**2.04** | 1.62/**2.00** | 1.59/**1.94** | 1.56/**1.90** | 1.55/**1.87** |
| 38 | 4.10/**7.35** | 3.25/**5.21** | 2.85/**4.34** | 2.62/**3.86** | 2.46/**3.54** | 2.35/**3.32** | 2.26/**3.15** | 2.19/**3.02** | 2.14/**2.91** | 2.09/**2.82** | 2.05/**2.75** | 2.02/**2.69** | 1.96/**2.59** | 1.92/**2.51** | 1.85/**2.40** | 1.80/**2.32** | 1.76/**2.22** | 1.71/**2.14** | 1.67/**2.08** | 1.63/**2.00** | 1.60/**1.97** | 1.57/**1.90** | 1.54/**1.86** | 1.53/**1.84** |
| 40 | 4.08/**7.31** | 3.23/**5.18** | 2.84/**4.31** | 2.61/**3.83** | 2.45/**3.51** | 2.34/**3.29** | 2.25/**3.12** | 2.18/**2.99** | 2.12/**2.88** | 2.07/**2.80** | 2.04/**2.73** | 2.00/**2.66** | 1.95/**2.56** | 1.90/**2.49** | 1.84/**2.37** | 1.79/**2.29** | 1.74/**2.20** | 1.69/**2.11** | 1.66/**2.05** | 1.61/**1.97** | 1.59/**1.94** | 1.55/**1.88** | 1.53/**1.84** | 1.51/**1.81** |
| 42 | 4.07/**7.27** | 3.22/**5.15** | 2.83/**4.29** | 2.59/**3.80** | 2.44/**3.49** | 2.32/**3.26** | 2.24/**3.10** | 2.17/**2.96** | 2.11/**2.86** | 2.06/**2.77** | 2.02/**2.70** | 1.99/**2.64** | 1.94/**2.54** | 1.89/**2.46** | 1.82/**2.35** | 1.78/**2.26** | 1.73/**2.17** | 1.68/**2.08** | 1.64/**2.02** | 1.60/**1.94** | 1.57/**1.91** | 1.54/**1.85** | 1.51/**1.80** | 1.49/**1.78** |
| 44 | 4.06/**7.24** | 3.21/**5.12** | 2.82/**4.26** | 2.58/**3.78** | 2.43/**3.46** | 2.31/**3.24** | 2.23/**3.07** | 2.16/**2.94** | 2.10/**2.84** | 2.05/**2.75** | 2.01/**2.68** | 1.98/**2.62** | 1.92/**2.52** | 1.88/**2.44** | 1.81/**2.32** | 1.76/**2.24** | 1.72/**2.15** | 1.66/**2.06** | 1.63/**2.00** | 1.58/**1.92** | 1.56/**1.88** | 1.52/**1.82** | 1.50/**1.78** | 1.48/**1.75** |
| 46 | 4.05/**7.21** | 3.20/**5.10** | 2.81/**4.24** | 2.57/**3.76** | 2.42/**3.44** | 2.30/**3.22** | 2.22/**3.05** | 2.14/**2.92** | 2.09/**2.82** | 2.04/**2.73** | 2.00/**2.66** | 1.97/**2.60** | 1.91/**2.50** | 1.87/**2.42** | 1.80/**2.40** | 1.75/**2.22** | 1.71/**2.13** | 1.65/**2.04** | 1.62/**1.98** | 1.57/**1.90** | 1.54/**1.86** | 1.51/**1.80** | 1.48/**1.76** | 1.46/**1.72** |
| 48 | 4.04/**7.19** | 3.19/**5.08** | 2.80/**4.22** | 2.56/**3.74** | 2.41/**3.42** | 2.30/**3.20** | 2.21/**3.04** | 2.14/**2.90** | 2.08/**2.80** | 2.03/**2.71** | 1.99/**2.64** | 1.96/**2.58** | 1.90/**2.48** | 1.86/**2.40** | 1.79/**2.28** | 1.74/**2.20** | 1.70/**2.11** | 1.64/**2.02** | 1.61/**1.96** | 1.56/**1.88** | 1.53/**1.84** | 1.50/**1.78** | 1.47/**1.73** | 1.45/**1.70** |

239

TABLE 6—*Continued*

|  |  |  |  |  |  |  |  |  |  |  |  |  |  |  |  |  |  |  |  |  |  |  |  |  |
|---|---|---|---|---|---|---|---|---|---|---|---|---|---|---|---|---|---|---|---|---|---|---|---|---|
| 50 | 4.03/7.17 | 3.18/5.06 | 2.79/4.20 | 2.56/3.72 | 2.40/3.41 | 2.29/3.18 | 2.20/3.02 | 2.13/2.88 | 2.07/2.78 | 2.02/2.70 | 1.98/2.62 | 1.95/2.56 | 1.90/2.46 | 1.85/2.39 | 1.78/2.26 | 1.74/2.18 | 1.69/2.10 | 1.63/2.00 | 1.60/1.94 | 1.55/1.86 | 1.52/1.82 | 1.48/1.76 | 1.46/1.71 | 1.44/1.68 |
| 55 | 4.02/7.12 | 3.17/5.01 | 2.78/4.16 | 2.54/3.68 | 2.38/3.37 | 2.27/3.15 | 2.18/2.98 | 2.11/2.85 | 2.05/2.75 | 2.00/2.66 | 1.97/2.59 | 1.93/2.53 | 1.88/2.43 | 1.83/2.35 | 1.76/2.23 | 1.72/2.15 | 1.67/2.06 | 1.61/1.96 | 1.58/1.90 | 1.52/1.82 | 1.50/1.78 | 1.46/1.71 | 1.43/1.66 | 1.41/1.64 |
| 60 | 4.00/7.08 | 3.15/4.98 | 2.76/4.13 | 2.52/3.65 | 2.37/3.34 | 2.25/3.12 | 2.17/2.95 | 2.10/2.82 | 2.04/2.72 | 1.99/2.63 | 1.95/2.56 | 1.92/2.50 | 1.86/2.40 | 1.81/2.32 | 1.75/2.20 | 1.70/2.12 | 1.65/2.03 | 1.59/1.93 | 1.56/1.87 | 1.50/1.79 | 1.48/1.74 | 1.44/1.68 | 1.41/1.63 | 1.39/1.60 |
| 65 | 3.99/7.04 | 3.14/4.95 | 2.75/4.10 | 2.51/3.62 | 2.36/3.31 | 2.24/3.09 | 2.15/2.93 | 2.08/2.79 | 2.02/2.70 | 1.98/2.61 | 1.94/2.54 | 1.90/2.47 | 1.85/2.37 | 1.80/2.30 | 1.73/2.18 | 1.68/2.09 | 1.63/2.00 | 1.57/1.90 | 1.54/1.84 | 1.49/1.76 | 1.46/1.71 | 1.42/1.64 | 1.39/1.60 | 1.37/1.56 |
| 70 | 3.98/7.01 | 3.13/4.92 | 2.74/4.08 | 2.50/3.60 | 2.35/3.29 | 2.22/3.07 | 2.14/2.91 | 2.07/2.77 | 2.01/2.67 | 1.97/2.59 | 1.93/2.51 | 1.89/2.45 | 1.84/2.35 | 1.79/2.28 | 1.72/2.15 | 1.67/2.07 | 1.62/1.98 | 1.56/1.88 | 1.53/1.82 | 1.47/1.74 | 1.45/1.69 | 1.40/1.63 | 1.37/1.56 | 1.35/1.53 |
| 80 | 3.96/6.96 | 3.11/4.88 | 2.72/4.04 | 2.48/3.56 | 2.33/3.25 | 2.21/3.04 | 2.12/2.87 | 2.05/2.74 | 1.99/2.64 | 1.95/2.55 | 1.91/2.48 | 1.88/2.41 | 1.82/2.32 | 1.77/2.24 | 1.70/2.11 | 1.65/2.03 | 1.60/1.94 | 1.54/1.84 | 1.51/1.78 | 1.45/1.70 | 1.42/1.65 | 1.38/1.57 | 1.35/1.52 | 1.32/1.49 |
| 100 | 3.94/6.90 | 3.09/4.82 | 2.70/3.98 | 2.46/3.51 | 2.30/3.20 | 2.19/2.99 | 2.10/2.82 | 2.03/2.69 | 1.97/2.59 | 1.92/2.51 | 1.88/2.43 | 1.85/2.36 | 1.79/2.26 | 1.75/2.19 | 1.68/2.06 | 1.63/1.98 | 1.57/1.89 | 1.51/1.79 | 1.48/1.73 | 1.42/1.64 | 1.39/1.59 | 1.34/1.51 | 1.30/1.46 | 1.28/1.43 |
| 125 | 3.92/6.84 | 3.07/4.78 | 2.68/3.94 | 2.44/3.47 | 2.29/3.17 | 2.17/2.95 | 2.08/2.79 | 2.01/2.65 | 1.95/2.56 | 1.90/2.47 | 1.86/2.40 | 1.83/2.33 | 1.77/2.23 | 1.72/2.15 | 1.65/2.03 | 1.60/1.94 | 1.55/1.85 | 1.49/1.75 | 1.45/1.68 | 1.39/1.59 | 1.36/1.54 | 1.31/1.46 | 1.27/1.40 | 1.25/1.37 |
| 150 | 3.91/6.81 | 3.06/4.75 | 2.67/3.91 | 2.43/3.44 | 2.27/3.13 | 2.16/2.92 | 2.07/2.76 | 2.00/2.62 | 1.94/2.53 | 1.89/2.44 | 1.85/2.37 | 1.82/2.30 | 1.76/2.20 | 1.71/2.12 | 1.64/2.00 | 1.59/1.91 | 1.54/1.83 | 1.47/1.72 | 1.44/1.66 | 1.37/1.56 | 1.34/1.51 | 1.29/1.43 | 1.25/1.37 | 1.22/1.33 |
| 200 | 3.89/6.76 | 3.04/4.71 | 2.65/3.88 | 2.41/3.41 | 2.26/3.11 | 2.14/2.90 | 2.05/2.73 | 1.98/2.60 | 1.92/2.50 | 1.87/2.41 | 1.83/2.34 | 1.80/2.28 | 1.74/2.17 | 1.69/2.09 | 1.62/1.97 | 1.57/1.88 | 1.52/1.79 | 1.45/1.69 | 1.42/1.62 | 1.35/1.53 | 1.32/1.48 | 1.26/1.39 | 1.22/1.33 | 1.19/1.28 |
| 400 | 3.86/6.70 | 3.02/4.66 | 2.62/3.83 | 2.39/3.36 | 2.23/3.06 | 2.12/2.85 | 2.03/2.69 | 1.96/2.55 | 1.90/2.46 | 1.85/2.37 | 1.81/2.29 | 1.78/2.23 | 1.72/2.12 | 1.67/2.04 | 1.60/1.92 | 1.54/1.84 | 1.49/1.74 | 1.42/1.64 | 1.38/1.57 | 1.32/1.47 | 1.28/1.42 | 1.22/1.32 | 1.16/1.24 | 1.13/1.19 |
| 1000 | 3.85/6.66 | 3.00/4.62 | 2.61/3.80 | 2.38/3.34 | 2.22/3.04 | 2.10/2.82 | 2.02/2.66 | 1.95/2.53 | 1.89/2.43 | 1.84/2.34 | 1.80/2.26 | 1.76/2.20 | 1.70/2.09 | 1.65/2.01 | 1.58/1.89 | 1.53/1.81 | 1.47/1.71 | 1.41/1.61 | 1.36/1.54 | 1.30/1.44 | 1.26/1.38 | 1.19/1.28 | 1.13/1.19 | 1.08/1.11 |
| ∞ | 3.84/6.64 | 2.99/4.60 | 2.60/3.78 | 2.37/3.32 | 2.21/3.02 | 2.09/2.80 | 2.01/2.64 | 1.94/2.51 | 1.88/2.41 | 1.83/2.32 | 1.79/2.24 | 1.75/2.18 | 1.69/2.07 | 1.64/1.99 | 1.57/1.87 | 1.52/1.79 | 1.46/1.69 | 1.40/1.59 | 1.35/1.52 | 1.28/1.41 | 1.24/1.36 | 1.17/1.25 | 1.11/1.15 | 1.00/1.00 |

## TABLE 7
### Negative Exponential Function

| x | $e^{-x}$ | x | $e^{-x}$ | x | $e^{-x}$ |
|---|---|---|---|---|---|
| 0 | 1.000 | 1.5 | .223 | 3.0 | .050 |
| .1 | .905 | 1.6 | .202 | 3.1 | .045 |
| .2 | .819 | 1.7 | .183 | 3.2 | .041 |
| .3 | .741 | 1.8 | .165 | 3.3 | .037 |
| .4 | .670 | 1.9 | .150 | 3.4 | .033 |
| .5 | .607 | 2.0 | .135 | 3.5 | .030 |
| .6 | .549 | 2.1 | .122 | 3.6 | .027 |
| .7 | .497 | 2.2 | .111 | 3.7 | .025 |
| .8 | .449 | 2.3 | .100 | 3.8 | .022 |
| .9 | .407 | 2.4 | .091 | 3.9 | .020 |
| 1.0 | .368 | 2.5 | .082 | 4.0 | .018 |
| 1.1 | .333 | 2.6 | .074 | 4.5 | .011 |
| 1.2 | .301 | 2.7 | .067 | 5.0 | .007 |
| 1.3 | .273 | 2.8 | .061 | 6.0 | .002 |
| 1.4 | .247 | 2.9 | .055 | 7.0 | .001 |

Paul G. Hoel and Raymond J. Jessen, *Basic Statistics for Business and Economics* (New York: John Wiley and Sons, Inc., 1971), p. 420 by permission of the publishers.

# INDEX